William R. Leo

Techniques for Nuclear and Particle Physics Experiments

A How-to Approach

With 254 Figures

Springer-Verlag Berlin Heidelberg New York
London Paris Tokyo

Dr. *William R. Leo*

Ecole Polytechnique Fédérale de Lausanne, Institut de Génie Atomique,
Départment de Physique, PHB-Ecublens, CH-1015 Lausanne, Switzerland

ISBN 3-540-17386-2 Springer-Verlag Berlin Heidelberg New York
ISBN 0-387-17386-2 Springer-Verlag New York Berlin Heidelberg

Library of Congress Cataloging-in-Publication Data. Leo, William R., 1948-. Techniques for nuclear and particle physics experiments. 1. Particles (Nuclear physics)–Technique. 2. Particles (Nuclear physics)–Experiments. 3. Nuclear physics–Technique. 4. Nuclear physics–Experiments. 5. Nuclear counters. I. Title. QC 793.46.L46 1987   539.7'21   87-9473

Typesetting: K+V Fotosatz, 6124 Beerfelden
Offsetprinting: Druckhaus Beltz, 6944 Hemsbach. Bookbinding: J. Schäffer GmbH & Co. KG, 6718 Grünstadt
2153/3150-543210

To my wife *Elisabeth*
for her love and encouragement

Preface

This book is an outgrowth of an advanced laboratory course in experimental nuclear and particle physics the author gave to physics majors at the University of Geneva during the years 1978 – 1983. The course was offered to third and fourth year students, the latter of which had, at this point in their studies, chosen to specialize in experimental nuclear or particle physics. This implied that they would go on to do a "diplome" thesis with one of the high- or intermediate-energy research groups in the physics department.

The format of the course was such that the students were required to concentrate on only one experiment during the trimester, rather than perform a series of experiments as is more typical of a traditional course of this type. Their tasks thus included planning the experiment, learning the relevant techniques, setting up and troubleshooting the measuring apparatus, calibration, data-taking and analysis, as well as responsibility for maintaining their equipment, i.e., tasks resembling those in a real experiment. This more intensive involvement provided the students with a better understanding of the experimental problems encountered in a professional experiment and helped instill a certain independence and confidence which would prepare them for entry into a research group in the department. Teaching assistants were present to help the students during the trimester and a series of weekly lectures was also given on various topics in experimental nuclear and particle physics. This included general information on detectors, nuclear electronics, statistics, the interaction of radiation in matter, etc., and a good deal of practical information for actually doing experiments.

Many of the chapters in this book are essentially based on notes which were prepared for these lectures. The information contained in this book, therefore, will hopefully provide the reader with a practical "guide" to some of the techniques, the equipment, the technical jargon, etc., which make up the world of current experimental nuclear and particle physics but which never seem to appear in the literature. As those in the field already know, the art of experimental physics is learned through a type of "apprenticeship" with a more experienced physicist or physicists, not unlike medieval artisans. It is to these "apprentices" that I address these chapters.

The book is laid out in three parts. The first four chapters treat some of the fundamental background knowledge required of experimental nuclear or particle physicists, such as the passage of radiation through matter, statistics, and radiation protection. Since detailed descriptions of the theory can be found elsewhere, these chapters only summarize the basic ideas and present only the more useful formulae. However, references are provided for the reader desiring more information. In this form, then, these chapters may serve as a reference. A basic understanding of quantum mechanics and fundamental nuclear physics is assumed throughout.

Chapters 5 – 10 are primarily concerned with the functioning and operation of the principal types of detectors used in nuclear and particle physics experiments. In addition to the basic principles, sections dealing with modern detectors such as the time-projection chamber or silicon microstrip detectors have also been included. It might be argued, of course, that some of these detectors are too specialized or still too novel to be included in a textbook of this level. However, for the student going on to more advanced work or the experienced researcher, it is these types of detectors he will most likely encounter. Moreover, it gives the student an idea of the state of the art and the incredible advances that have been made. Hopefully, it will provide food for thought on the advances that can still yet be made!

The final chapters, 11 – 18, are concerned with "nuclear electronics" and the logic which is used in setting up electronics systems for experiments. This has always been a difficult point for many students as most approaches have been from a circuit design point of view requiring analysis of analog circuits, which, of course, is a subject unto itself. With the establishment of standardized systems such as NIM and CAMAC and the availability of commercial modules, however, the experimental physicist can function very well with only a knowledge of electronic logic. These chapters thus treat the characteristics of the pulse signals from detectors and the various operations which can be performed on these signals by commercially available modules. Chapter 18 also presents an introduction to the CAMAC system, which, up to a few years ago, was used only in high-energy physics but which now, with the advent of microcomputers, may also be found on smaller experiments undertaken by students.

Although this book is based on a specific laboratory course, the treatment of the topics outlined above is general and was made without specific reference to any particular experiment, except, perhaps, as an example. As such I hope the book will also be of use to researchers and students in other domains who are called upon to work with detectors and radiation.

I would like to thank the many people who have at some point or another helped realize this book. In particular, my very special thanks are due to Dr. Rene Hausammann, Dr. Catherine Lechanoine-Leluc, Dr. Jacques Ligou, and Dr. Trivan Pal for having read some of the chapters and for their helpful comments and suggestions. I am also grateful to J.-C. Bostedeche and J. Covillot who helped construct, establish and maintain the experiments in the laboratory, to C. Jacquat for the many hours spent on the drawings for this book and to the many authors who have kindly allowed me to use figures from their articles or books. Finally, I would like to thank Elisabeth, who, although not a physicist, was the first to have the idea for this book.

Lausanne, March 1987   *William R. Leo*

Contents

1. Basic Nuclear Processes in Radioactive Sources

Radioactive sources provide a convenient means of testing and calibrating detectors and are essential tools in both the nuclear and high energy physics laboratory. An understanding of the basic nuclear processes in radioactive sources is a necessity, therefore, before beginning to work in the laboratory. We shall begin this book then by briefly reviewing these processes and describing the characteristics of the resulting radiations. A more detailed discussion, of course, is better suited for a nuclear theory course and we refer the reader to any of the standard nuclear physics texts for more information.

Nuclei can undergo a variety of processes resulting in the emission of radiation of some form. We can divide the processes into two categories: radioactivity and nuclear reactions. In a radioactive transformation, the nucleus spontaneously disintegrates to a different species of nuclei or to a lower energy state of the same nucleus with the emission of radiation of some sort. The majority of radiation sources found in the laboratory are of this type. In a nuclear reaction, the nucleus interacts with another particle or nucleus with the subsequent emission of radiation as one of its final products. In many cases, as well, some of the products are nuclei which further undergo radioactive disintegration.

The radiation emitted in both of these processes may be electromagnetic or corpuscular. The electromagnetic radiations consist of x-rays and γ-rays while the corpuscular emissions include α-particles, β-electrons and positrons, internal conversion electrons, Auger electrons, neutrons, protons, and fission fragments, among others. While most of these radiations originate in the nucleus itself, some also arise from the electron cloud surrounding the nucleus. Indeed, the nucleus should not be considered as a system isolated from the rest of the atom. Excitations which arise therein may, in fact, make themselves directly felt in the electron cloud, as in the case of internal conversion, and/or indirectly as in the emission of characteristic x-rays following electron capture. Except for their energy characteristics, these radiations are indistinguishable from those arising in the nucleus.

Table 1.1 summarizes some of the more common types of radiation found in laboratory sources. Each radiation type is characterized by an energy spectrum which is indicative of the nuclear process belying it. Note also that a radioactive source may emit several different types of radiation at the same time. This can arise from the fact that the nuclear isotope in question undergoes several different modes of decay. For example, a ^{137}Cs nucleus can de-excite through either γ-ray emission or internal conversion. The output of a given ^{137}Cs sample, therefore, will consist of both photons and electrons in a proportion equal to the relative probabilities for the two decay modes. A more common occurrence, however, is that the daughter nucleus is also radioactive, so that its radiation is also added to the emitted output. This is the case with many β-sources, where the β-disintegration results in an excited daughter nucleus which then immediately decays by γ-emission.

Table 1.1. Characteristics of nuclear radiations

Type	Origin	Process	Charge	Mass [MeV]	Spectrum (energy)
α-particles	Nucleus	Nuclear decay or reaction	+2	3727.33	Discrete [MeV]
β^--rays	Nucleus	Nuclear decay	−1	0.511	Continuous [keV − MeV)
β^+-rays (positrons)	Nuclear	Nuclear decay	+1	0.511	Continuous [keV − MeV]
γ-rays	Nucleus	Nuclear deexcitation	0	0	Discrete [keV − MeV]
x-rays	Electron cloud	Atomic deexcitation	0	0	Discrete [eV − keV]
Internal conversion electrons	Electron cloud	Nuclear deexcitation	−1	0.511	Discrete [high keV]
Auger electrons	Electron cloud	Atomic deexcitation	−1	0.511	Discrete [eV − keV]
Neutrons	Nucleus	Nuclear reaction	0	939.57	Continuous or discrete [keV − MeV]
Fission fragments	Nucleus	Fission	$\simeq 20$	80 − 160	Continuous 30 − 150 MeV

1.1 Nuclear Level Diagrams

Throughout this text, we will be making use of nuclear energy level diagrams, which provide a compact and convenient way of representing the changes which occur in nuclear transformations. These are usually plotted in the following way. For a given nucleus with atomic number Z and mass A, the energy levels are plotted as horizontal lines on some arbitrary vertical scale. The spin and parity of each of these states may also be indicated. Keeping the same mass number A, the energy levels of neighboring nuclei $(Z-1, A)$, $(Z+1, A)$, ... are now also plotted on this energy scale with Z ordered in the horizontal direction, as illustrated in Fig. 1.1. This reflects the fact that nuclei with different Z but the same A may simply be treated as different states of a system of A *nucleons*. The relation of the energy levels of a nucleus (Z, A) to other nuclei in the same A system is therefore made apparent. With the exception of α-decay, radioactive decay may now be viewed as simply a transition from a higher energy state to a lower energy state within the same system of A nucleons. For example, consider the β-decay process, which will be discussed in the next section. This reaction involves the decay

$$(Z, A) \rightarrow (Z+1, A) + e^- + \bar{\nu},$$

where the final state in the nucleus $(Z+1, A)$ may be the ground state or some excited state. This is shown in Fig. 1.1 by the arrow descending to the right. The atomic number Z increases by one, but A remains constant. The changes that occur can be immedi-

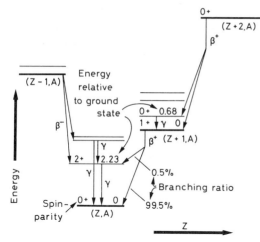
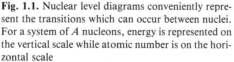

Fig. 1.1. Nuclear level diagrams conveniently represent the transitions which can occur between nuclei. For a system of A nucleons, energy is represented on the vertical scale while atomic number is on the horizontal scale

ately seen: for example, the energy available for the reaction as given by the difference in height between the two levels, the spin and parity changes, etc. If it is possible for the same initial state to make transitions to several different final states, this can also be represented by several arrows emanating from the initial state to the various possible final states. The relative probability of each decay branch (i.e., the branching ratio) may also be indicated next to the corresponding arrow.

In a similar way, transitions which follow the first may also be diagrammed. For example, suppose the final state of the above example is an excited state of the $(Z+1, A)$ nucleus, it may then make a gamma transition to the ground state or to another excited state. This type of transition is indicated by a vertical line since Z remains unchanged (see Fig. 1.1). Other transitions, such as a further β-decay or some other process, may be represented in a similar manner. In this way, the types of radiation emitted by a particular radioactive source and their origins may be easily displayed.

Several tabulations of the radioactive isotopes and their level diagrams as deduced from experiment are available for reference purposes. The most complete of these is the *Table of Isotopes* edited by *Lederer* and *Shirley* [1.1] which is updated periodically. We urge the reader to become familiar with their interpretation.

1.2 Alpha Decay

Alpha particles are ^{4}He nuclei, i.e., a bound system of two protons and two neutrons, and are generally emitted by very heavy nuclei containing too many nucleons to remain stable. The emission of such a nucleon cluster as a whole rather than the emission of single nucleons is energetically more advantageous because of the particularly high binding energy of the α-particle. The parent nucleus (Z, A) in the reaction is thus transformed via

$$(Z, A) \rightarrow (Z-2, A-4) + \alpha \, . \tag{1.1}$$

Theoretically, the process was first explained by Gamow and Condon and by Gurney as the tunneling of the α-particle through the potential barrier of the nucleus. Alpha particles, therefore, show a monoenergetic energy spectrum. As well, since barrier transmission is dependent on energy, all α-sources are generally limited to the range $\simeq 4-6$ MeV with the higher energy sources having the higher transmission probability and thus the shorter half-life. For this reason also, most α-decays are directly to the ground state of the daughter nucleus since this involves the highest energy change. Decays to excited states of the daughter nucleus are nevertheless possible, and in such nuclei, the energy spectrum shows several monoenergetic lines each corresponding to a decay to one of these states. Some of the more commonly used sources are listed below in Table 1.2.

Table 1.2. Characteristics of some alpha emitters

Isotope	Half-life	Energies [MeV]	Branching
^{241}Am	433 days	5.486	85%
		5.443	12.8%
^{210}Po	138 days	5.305	100%
^{242}Cm	163 days	6.113	74%
		6.070	26%

Because of its double charge, $+2e$, alpha particles have a very high rate of energy loss in matter. The range of a 5 MeV α-particle in air is only a few centimeters, for example. For this reason it is necessary to make α sources as thin as possible in order to minimize energy loss and particle absorption. Most α-sources are made, in fact, by depositing the isotope on the surface of a suitable backing material and protecting it with an extremely thin layer of metal foil.

1.3 Beta Decay

Beta particles are fast electrons or positrons which result from the weak-interaction decay of a neutron or proton in nuclei which contain an excess of the respective nucleon. In a neutron-rich nucleus, for example, a neutron can transform itself into a proton via the process

$$n \rightarrow p + e^- + \bar{\nu}, \tag{1.2}$$

where an electron and antineutrino are emitted. (The proton remains bound to the nucleus.) The daughter nucleus now contains one extra proton so that its atomic number is increased by 1.

Similarly, in nuclei with too many protons, the decay

$$p \rightarrow n + e^+ + \nu \tag{1.3}$$

can occur, where a positron and a neutrino are now emitted and the atomic number is decreased by 1. Both are mediated by the same weak interaction.

A basic characteristic of the β-decay process is the continuous energy spectrum of the β-particle. This is because the available energy for the decay (the *Q-value*) is shared between the β-particle and the neutrino (or antineutrino) which usually goes undetected. A typical spectrum is shown in Fig. 1.2. If the small recoil energy of the daughter nucleus is ignored, the maximum energy of this spectrum should correspond to the Q-value for the reaction. For most beta sources, this maximum value ranges from a few tens of keV to a few MeV.

In very many β-sources, the daughter nucleus is left in an excited state which decays immediately with the emission of one or more γ photons (see Sect. 1.5). This is illustrated in the level diagram shown in Fig. 1.3. These sources, therefore, are also emitters of γ radiation. Most β-sources are of this type. *Pure β-emitters* exist but the list is astonishingly short as is seen in Table 1.3.

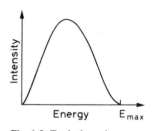

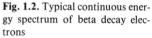

Fig. 1.2. Typical continuous energy spectrum of beta decay electrons

Table 1.3. List of pure β^- emitters

Source	Half-life	E_{max} [MeV]
^{3}H	12.26 yr	0.0186
^{14}C	5730 yr	0.156
^{32}P	14.28 d	1.710
^{33}P	24.4 d	0.248
^{35}S	87.9 d	0.167
^{36}Cl	3.08×10^5 yr	0.714
^{45}Ca	165 d	0.252
^{63}Ni	92 yr	0.067
^{90}Sr/^{90}Y	27.7 yr/64 h	0.546/2.27
^{99}Tc	2.12×10^5 yr	0.292
^{147}Pm	2.62 yr	0.224
^{204}Tl	3.81 yr	0.766

Some β sources may also have more than one decay branch, i.e., they can decay to different excited states of the daughter nucleus. Each branch constitutes a separate β-decay with an end-point energy corresponding to the energy difference between the initial and final states and is in competition with the other branches. The total β-spectrum from such a source is then a superposition of all the branches weighted by their respective decay probabilities.

Since electrons lose their energy relatively easily in matter, it is important that β-sources be thin in order to allow the β's to escape with a minimum of energy loss and absorption. This is particularly important for positron sources since the positron can annihilate with the electrons in the source material or surrounding container. A too

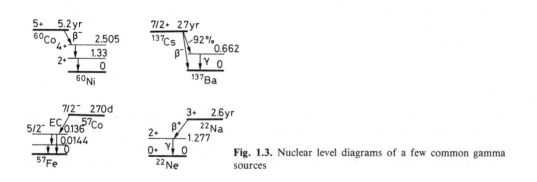

Fig. 1.3. Nuclear level diagrams of a few common gamma sources

thick β^+ source will exhibit a distorted β^+ spectrum and an enormous background of 511 keV annihilation photons.

1.4 Electron Capture (EC)

As an alternative to β^+ emission, proton-rich nuclei may also transform themselves via the capture of an electron from one of the atomic orbitals

$$e^- + p \rightarrow n + \nu. \tag{1.4}$$

This reaction is essentially the same as β^+-decay but with the β particle transposed to the left side. The nuclear level diagram for EC is therefore identical to that for β^+ emission. Since only the neutrino is emitted, electron capture would seem to be a reaction almost impossible to observe, given the well-known difficulty of detecting such a particle! The capture of the electron, however, leaves a hole in the atomic shell which is filled by another atomic electron, giving rise to the emission of a characteristic x-ray or Auger electrons (see Sect. 1.8). These radiations are, of course, much more amenable to detection and can be used to signal the capture reaction. In general, it is the K electron which is most likely captured, although, L-capture is also possible but with a much smaller probability.

1.5 Gamma Emission

Like the electron shell structure of the atom, the nucleus is also characterized by discrete energy levels. Transitions between these levels can be made by the emission (or absorption) of electromagnetic radiation of the correct energy, i.e., with an energy equal to the energy difference between the levels participating in the transition. The energies of these photons, from a few hundred keV to a few MeV, characterize the high binding energy of nuclei. These high-energy photons were historically named γ-rays, and, like atoms, show spectral lines characteristic of the emitting nucleus. Level diagrams illustrating the specific energy structure of some typical γ-ray sources are shown in Fig. 1.3.

Most γ-sources are *"placed"* in their excited states as the result of a β-disintegration, although excited nuclear states are often created in nuclear reactions also. Since electrons and positrons are more easily absorbed in matter, the β-particles in such sources can be *"filtered"* out by enveloping them with sufficient absorbing material, leaving only the more penetrating γ-ray.

1.5.1 Isomeric States

Although most excited states in nuclei make almost immediate transitions to a lower state, some nuclear states may live very much longer. Their de-excitation is usually hindered by a large spin difference between levels (i.e., a *forbidden* transition) resulting in lifetimes ranging from seconds to years. A nuclide which is *"trapped"* in one of these metastable states will thus show radioactive properties different from those in more normal states. Such nuclei are called *isomers* and are denoted by an m next to the mass number in their formulae, e.g. ^{60m}Co or ^{69m}Zn.

1.6 Annihilation Radiation

Another source of high-energy photons is the annihilation of positrons. If a positron source such as ^{22}Na is enclosed or allowed to irradiate an absorbing material, the positrons will annihilate with the absorber electrons to produce two photons, each with an energy equal to the electron mass: 511 keV. In order to conserve momentum, these two photons are always emitted in opposite directions. The γ spectrum from a thick positron source will thus show a peak at 511 keV (corresponding to the detection of one of the annihilation photons) in addition to the peaks characteristic of transitions in the daughter nucleus. Figure 1.4 shows the spectrum observed with a thick ^{22}Na source.

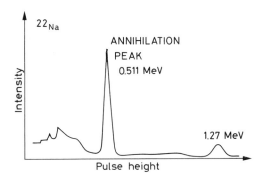

Fig. 1.4. Gamma-ray spectrum of a ^{22}Na source as observed with a NaI detector. Because of positron annihilation in the detector and the source itself, a peak at 511 keV is observed corresponding to the detection of one of the annihilation photons

1.7 Internal Conversion

While the emission of a γ-ray is usually the most common mode of nuclear de-excitation, transitions may also occur through internal conversion. In this process, the nuclear excitation energy is directly transferred to an atomic electron rather than emitted as a photon. The electron is ejected with a kinetic energy equal to the excitation energy minus its atomic binding energy. Unlike β-decay, therefore, internal conversion electrons are monoenergetic having approximately the same energy as the competing γ's, i.e., a few hundred keV to a few MeV.

While the K-shell electrons are the most likely electrons to be ejected, electrons in other orbitals may also receive the excitation energy. Thus, an internal conversion source will exhibit a group of internal conversion lines, their differences in energy being equal to the differences in the binding energies of their respective orbitals.

Internal conversion sources are one of the few nuclear sources of monoenergetic electrons and are thus very useful for calibration purposes. Some internal conversion sources readily found in the laboratory are given in Table 1.4.

Table 1.4. Some internal conversion sources

Source	Energies [keV]
^{207}Bi	480, 967, 1047
^{137}Cs	624
^{113}Sn	365
^{133}Ba	266, 319

1.8 Auger Electrons

As in internal conversion, an excitation which arises in the electron shell can also be transferred to an atomic electron rather than to a characteristic x-ray. Such a process can occur after a reaction such as electron-capture, for example. The electrons emitted are called *Auger* electrons and are monoenergetic. Like internal conversion lines they can occur in groups, however, their energies are more typical of atomic processes being not more than a few keV. They are thus very susceptible to self-absorption and are difficult to detect.

1.9 Neutron Sources

While it is possible to artificially produce isotopes which emit neutrons, natural neutron emitters which can be used practically in the lab do not exist. Laboratory neutron sources, instead, are based on either spontaneous fission or nuclear reactions.

1.9.1 Spontaneous Fission

Spontaneous fission can occur in many transuranium elements with the release of neutrons along with the fission fragments. These fragments, as well, can promptly decay emitting β and γ radiation. If the fission source is enveloped in a sufficiently thick container, however, much of this latter radiation can be absorbed leaving only the more penetrating neutrons.

The most common neutron source of this type is ^{252}Cf which has a half-life of 265 years. The energy spectrum of the neutrons is continuous up to about 10 MeV and exhibits a Maxwellian shape. Figure 1.5 shows this spectrum. The distribution is described very precisely by the form [1.2]

$$\frac{dN}{dE} = \sqrt{E} \, \exp\left(\frac{-E}{T}\right),$$
(1.5)

where $T = 1.3$ MeV for ^{252}Cf.

1.9.2 Nuclear Reactions

A more convenient method of producing neutrons is with the nuclear reactions (α, n) or (γ, n).[1] Reactions of this type occur with many nuclei, however, only those with the highest yield are used. Such sources are generally made by mixing the target material with a suitably strong α or γ emitter. The most common target material is beryllium. Under bombardment by α's, beryllium undergoes a number of reactions which lead to the production of free neutrons:

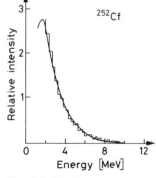

Fig. 1.5. Neutron energy spectrum from ^{252}Cf (from *Lorch* et al. [1.2]). The form of the spectrum can be described by a Maxwellian distribution

---

[1] A common method for denoting nuclear reactions is $A(x, y)B$ where x is the bombarding particle, A the target nucleus, B the resulting nucleus and y the the outgoing particle or particles. Note that the ingoing and outgoing particles are always on the inside of the parentheses. The abbreviated notation (x, y), therefore, indicates any nuclear reaction in which x is the incident particle and y the resulting, outgoing particle.

$$\alpha + {}^9\text{Be} \rightarrow {}^{13}\text{C}^* \rightarrow \begin{cases} {}^{12}\text{C}^{(*)} + \text{n} \\ {}^8\text{Be} + \alpha + \text{n} \\ 3\alpha + \text{n} . \end{cases} \qquad (1.6)$$

Here the excited compound nucleus ${}^{13}\text{C}^*$ is formed which then decays through a variety of modes depending on the excitation energy. In general, the dominant reaction is the decay to ${}^{12}\text{C}$ or to the 4.44 MeV excited state of ${}^{12}\text{C}$. With ${}^{241}\text{Am}$ as an α-source, a neutron yield of about 70 neutrons per 10^6 α's [1.3] is generally obtained. With ${}^{242}\text{Cm}$, which emits α's at a higher energy, the yield is $\simeq 106$ neutrons/10^6-α [1.3]. Other (α, n) neutron sources include ${}^{238}\text{Pu/Be}$, ${}^{226}\text{Ra/Be}$ and ${}^{227}\text{Ac/Be}$. Targets such as B, F, and Li are also used although the neutron yields are somewhat lower. The half-life of these sources, of course, depends on the half-life of the α-emitter.

For incident α's of a fixed energy, the energy spectrum of neutrons emitted in these sources should theoretically show monoenergetic lines corresponding to the different transitions which are made. In mixed sources, however, there is a smearing of the alpha-particle spectrum due to energy loss, so that a large smearing in neutron energy results. There is also considerable Doppler broadening which can amount to as much as 2 MeV. Figure 1.6 shows the energy spectrum of neutrons for several sources of this type.

In the case of the photo-reaction (γ, n), only two target materials are suitable: beryllium and deuterium. The respective reactions are

$$ {}^9\text{Be} + \gamma \rightarrow {}^8\text{Be} + \text{n} , \qquad (1.7)$$

$$ {}^2\text{H} + \gamma \rightarrow {}^1\text{H} + \text{n} . \qquad (1.8)$$

These sources have the advantage of emitting neutrons which are more or less monoenergetic since the γ's are not slowed down as in the case of α's. The neutrons are, of course, not strictly monoenergetic if one works out the kinematics; however, the spread is generally small. The disadvantage of these sources is that the reaction yield per γ is 1 – 2 orders of magnitude lower than that of the α-type sources. As well, the nonreacting gammas are not absorbed as easily as α-particles, so that these sources are also accompanied by a large background of γ radiation.

A more detailed description of these and other neutron sources may be found in the article by *Hanson* [1.4].

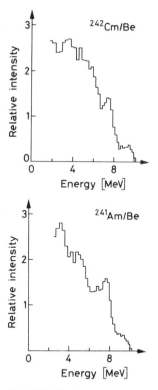

Fig. 1.6. Neutron energy spectrum from ${}^{242}\text{Cm/Be}$ and (*above*) and ${}^{241}\text{Am/Be}$ (*below*) sources (from *Lorch* et al. [1.2])

1.10 Source Activity Units

The *activity* or strength of a radioactive sample is defined as the mean number of *decay processes* it undergoes per unit time. This is an extrinsic quantity which depends on the amount of source material contained in the sample – the larger the sample the greater the total number of decays. Moreover, it should be noted that the activity of a source is not necessarily synonymous with the amount of radiation emitted per unit time by the source, although it is certainly related to it. For example, some nuclear transformations result in an unstable daughter nucleus which also disintegrates. Its radiations would then appear with the radiation from the original decay, but would not be included in the activity. Similarly, some nuclides decay through several competing processes, for example, β^+-emission or electron capture, where only a fraction of the decays appears as a particular emitted radiation. The relation between radiation output and activity, in fact, depends on the specific nuclear decay scheme, and only in the case of a unique radiative transition is the activity identical to the radiation output.

Activity has traditionally been measured in units of *Curies* (Ci). Originally defined as the activity of 1 g of pure radium-226, this unit is equivalent to

$$1 \text{ Curie (Ci)} = 3.7 \times 10^{10} \text{ disintegrations/s (dps)}. \qquad (1.9)$$

This is, in fact, a very *large* unit and one generally works in the laboratory with sources on the order of tens or hundreds of microCuries (μCi).

Because of its rather awkward definition in terms of dps, a new unit, the *Becquerel*, defined as

$$1 \text{ Becquerel (Bq)} = 1 \text{ disintegration/s} \qquad (1.10)$$

has been designated by the General Conference on Weights and Measures to replace the Curie.

For the beginning student, it is important to distinguish units of *activity* from those of *dose* such as the *rad* or the *rem*. These latter units essentially measure the effects of radiation *received* by an object or person, whereas the Curie or Becquerel are concerned with the disintegrations in the source itself. Units of dose will be treated in Chap. 3.

1.11 The Radioactive Decay Law

The *radioactive decay law* was first established experimentally near the beginning of the century by Rutherford and Soddy and states that the activity of a radioactive sample decays exponentially in time. In terms of modern quantum mechanics, this can easily be derived by considering the fact that a nuclear decay process is governed by a transition probability per unit time, λ, characteristic of the nuclear species. If a nuclide has more than one mode of decay, then λ is the sum of the separate constants for each mode

$$\lambda = \lambda_1 + \lambda_2 + \ldots . \qquad (1.11)$$

In a sample of N such nuclei, the *mean* number of nuclei decaying in a time dt would then be

$$dN = -\lambda N \, dt , \qquad (1.12)$$

where N is the number of nuclei and λ is the decay constant. We have assumed here that N is large so that it may be considered as continuous. Equation (1.12) may be considered as the differential form of the radioactive decay law. Integrating (1.12) then results in the exponential,

$$N(t) = N(0) \exp(-\lambda t) , \qquad (1.13)$$

where $N(0)$ is the number of the nuclei at $t = 0$. The exponential decrease in activity of a radioactive sample is thus governed by the constant λ. In practice, it is more habitual to use the inverse of λ,

$$\tau_m = 1/\lambda \, , \tag{1.14}$$

which is known as the *mean lifetime*. This is just the time it takes for the sample to decay to $1/e$ of its initial activity. Equally in use is the *half-life*, $T_{1/2}$, which is defined as the time it takes for the sample to decay to one-half of its original activity. Thus,

$$\tfrac{1}{2} = \exp(-\lambda T_{1/2}) \, , \tag{1.15}$$

which implies

$$T_{1/2} = \frac{1}{\lambda} \ln 2 = \tau_m \ln 2 \, . \tag{1.16}$$

1.11.1 Fluctuations in Radioactive Decay

Consider now the number of decays undergone by a radioactive source in a period of time Δt which is short compared to the half-life of the source. The activity of the source may then be considered as constant. If repeated measurements of the number of decays, n, in the interval Δt are now made, fluctuations will be observed from measurement to measurement. This is due to the statistical nature of the decay process; indeed, from quantum mechanics we know that the exact number of decays at any given time can never be predicted, only the probability of such an event. From the radioactive decay law, it can be shown, in fact, (see *Segre* [1.5], for example) that the probability of observing n counts in a period Δt is given by a Poisson distribution,

$$P(n, \Delta t) = \frac{m^n}{n!} \exp(-m) \, , \tag{1.17}$$

where m is the average number of counts in the period Δt. The standard deviation of the distribution is then

$$\sigma = \sqrt{m} \tag{1.18}$$

as is characteristic of Poisson statistics.

Example 1.1   A source is observed for a period of 5 s during which 900 counts are accumulated by the detector. What is the count rate per second and error from this measurement?

Take the measurement as a single trial for the determination of the mean count rate in 5 s, i.e., $m = 900$ for $\Delta t = 5$ s. The standard deviation is then

$$\sigma = \sqrt{900} = 30.$$

The count rate per second is then

$$\text{rate/s} = (900 \pm 30)/5 = (180 \pm 6) \text{ cts/s} \, .$$

Example 1.2   A weak radioactive source is found to have a mean count rate of 1 counts/s. What is the probability of observing no counts at all in a period of 4 s? One count in 4 s?

For a period $\Delta t = 4$ s, the mean count rate is obviously $m = 4$. Using the Poisson distribution, we find

$$P = (4)^0 \frac{\exp(-4)}{0!} = 0.0183 \ .$$

Similarly, the probability of observing 1 count in 4 s is

$$P = 4^1 \frac{\exp(-4)}{1!} = 0.0733 \ .$$

1.11.2 Radioactive Decay Chains

A very often encountered situation is a radioactive decay chain in which a nuclide decays to a daughter nucleus which itself disintegrates to another unstable nucleus and so on. In the simple case of a three-nucleus chain, i.e.,

$$A \rightarrow B \rightarrow C \ ,$$

where C is stable, application of the radioactive decay law gives the equations

$$\frac{dN_a}{dt} = -\lambda_a N_a \ ,$$

$$\frac{dN_b}{dt} = \lambda_a N_a - \lambda_b N_b \ , \qquad\qquad (1.19)$$

$$\frac{dN_c}{dt} = \lambda_b N_b \ ,$$

where λ_a and λ_b are the corresponding decay constants. For longer chains, the equations for the additional nuclides are derived in the same manner. If initially $N_b(0) = N_c(0) = 0$, solution of (1.19) results in

$$N_a(t) = N_a(0) \exp(-\lambda_a t) \ ,$$

$$N_b(t) = N_a(0) \frac{\lambda_a}{\lambda_b - \lambda_a} [\exp(-\lambda_a t) - \exp(-\lambda_b t)] \ , \qquad\qquad (1.20)$$

$$N_c(t) = N_a(0) \left\{ 1 + \frac{1}{\lambda_b - \lambda_a} [\lambda_a \exp(-\lambda_b t) - \lambda_b \exp(-\lambda_a t)] \right\} \ .$$

The behavior in time of the three nuclear species is graphed in Fig. 1.7. Note that the activity of B here is *not* given by dN_b/dt but $\lambda_b N_b$. This is because dN_b/dt now also includes the rate of B created by A. We might note also that N_b goes through a maximum. By setting the derivative to zero, we find

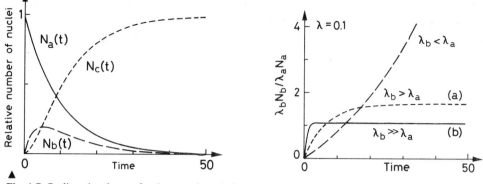

Fig. 1.7. Radioactive decay of a three nucleus chain

Fig. 1.8. Ratio of daughter to parent radionuclide activity. Curve (a) shows the condition known as *transient equilibrium*, while curve (b) illustrates *secular equilibrium*

$$t_{max} = \frac{\ln \frac{\lambda_b}{\lambda_a}}{\lambda_b - \lambda_a} . \tag{1.21}$$

At this point, the activity of B is a maximum equal to

$$\lambda_b N_b(t_{max}) = \lambda_a N_a(t_{max}) \tag{1.22}$$

as seen from from (1.20). This is known as *ideal equilibrium*. At any other time, the ratio of the activity of B to A (or the ratio of any daughter to its immediate parent in longer chains) is given by

$$\frac{\lambda_b N_b}{\lambda_a N_a} = \frac{\lambda_b}{\lambda_b - \lambda_a} \{1 - \exp[-(\lambda_b - \lambda_a)t]\} . \tag{1.23}$$

Three cases may be distinguished:

1) If $\lambda_a > \lambda_b$ then the ratio increases with time.
2) If $\lambda_b > \lambda_a$, then (1.23) becomes almost constant > 1 at large t. This is known as *transient* equilibrium.
3) If $\lambda_b \gg \lambda_a$, then the ratio rapidly levels off to $\simeq 1$ and reaches a state of *secular* equilibrium.

These three cases are illustrated in Fig. 1.8. In secular equilibrium, note that the number of daughter nuclei B stays constant relative to A. This means that the rate of disintegration of B is the same as its rate of creation. An example is the β-decay of ^{90}Sr:

$$^{90}\text{Sr} \xrightarrow[28\,\text{y}]{\beta^-} {}^{90}\text{Y} \xrightarrow[64.8\,\text{h}]{\beta^-} {}^{90}\text{Zr}$$

where the end-point energies for the two β's are 0.546 MeV and 2.27 MeV respectively. Since the number of ^{90}Y nuclei is kept constant by regeneration from ^{90}Sr, we essentially have a ^{90}Y source with a half-life of 28 yrs rather than 65 hours!

1.11.3 Radioisotope Production by Irradiation

A useful application of our example above is in the production of radioactive isotopes by irradiation of a stable element. In such a case, we have the nuclear reaction

$$A(x, y)B \to C,$$

where the isotope B is produced and decays to C with some constant λ_b. If $\sigma(A \to B)$ is the reaction cross section, F, the flux of irradiating particles x and N_a, the number of nuclei A, then the radioactive decay equations result in

$$\frac{dN_a}{dt} = -F\sigma(A \to B)N_a = -\lambda_a N_a$$

$$\frac{dN_b}{dt} = -\lambda_b N_b + \lambda_a N_a,$$

(1.24)

which is in strict analogy to our example above. The maximum yield of isotope B, therefore, is obtained at a time t_{max} given by (1.21).

Example 1.3   Ordinary copper consists of $\simeq 69\%$ ^{63}Cu and 31% ^{65}Cu. When irradiated by slow neutrons from a reactor, for example, radioactive ^{64}Cu and ^{66}Cu are formed. The half-lives of these two isotopes are 12.7 h and 5.1 min respectively. What is the activity of each of these isotopes if 1 g of ordinary copper is irradiated in a thermal neutron flux of 10^9 neutrons/cm^2-s for 15 min?

From the isotope tables, we find the thermal capture cross sections

$$\sigma(^{63}\text{Cu} + \text{n} \to {}^{64}\text{Cu}) = 4.4 \text{ barns}$$

$$\sigma(^{65}\text{Cu} + \text{n} \to {}^{66}\text{Cu}) = 2.2 \text{ barns}.$$

The rate at which ^{64}Cu and ^{66}Cu are formed is then

$$\lambda_a = F\sigma = \begin{cases} 10^9 \times (4.4 \times 10^{-24}) = 4.4 \times 10^{-15} \text{ s}^{-1} & {}^{64}\text{Cu} \\ 10^9 \times (2.2 \times 10^{-24}) = 2.2 \times 10^{-15} \text{ s}^{-1} & {}^{66}\text{Cu} \end{cases}$$

and the rate at which they decay is

$$\lambda_b = \frac{1}{\tau_m} = \begin{cases} \dfrac{\ln 2}{12.7} = 0.054 \text{ h}^{-1} & {}^{64}\text{Cu} \\ \dfrac{\ln 2}{5.1} = 0.136 \text{ min}^{-1} & {}^{66}\text{Cu}. \end{cases}$$

The activity of each isotope after a time t is then $\lambda_b N_b(t)$ where $N_b(t)$ is given by (1.20). Since $\lambda_a \ll \lambda_b$ and $\lambda_a t \ll 1$, we can approximate (1.20) to obtain

$$\lambda_b N_b(t) \simeq N_a(0)\lambda_a[1 - \exp(-\lambda_b t)] .$$

Since $N_a(0) = (6.02 \times 10^{23}/A) \times (\text{abundance}) \times (1 \text{ g})$, we have

$$\lambda_b N_b(15 \text{ min}) = \begin{cases} 3.86 \times 10^5 \text{ dps} = 10.43 \ \mu\text{Ci} & \text{for} \quad ^{64}\text{Cu} \\ 5.62 \times 10^6 \text{ dps} = 152 \ \mu\text{Ci} & \text{for} \quad ^{66}\text{Cu} . \end{cases}$$

It is interesting also to calculate the optimum time t_{max}. From (1.21) then

$$t_{max} = \begin{cases} 16.8 \text{ days} & \text{for} \quad ^{64}\text{Cu} \\ 3.4 \text{ h} & \text{for} \quad ^{66}\text{Cu}. \end{cases}$$

The corresponding activities at this time are then 778 μCi and 175 μCi respectively.

2. Passage of Radiation Through Matter

This chapter concerns the basic reactions which occur when radiation encounters matter and the effects produced by these processes. For the experimental nuclear or particle physicist, knowledge of these interactions is of paramount importance. Indeed, as will be seen in the following chapters, these processes are the basis of all current particle detection devices and thus determine the sensitivity and efficiency of a detector. At the same time, these same reactions may also interfere with a measurement by disturbing the physical state of the radiation: for example, by causing energy information to be lost, or deflecting the particle from its original path, or absorbing the particle before it can be observed. A knowledge of these reactions and their magnitudes is thus necessary for experimental design and corrections to data. Finally, these are also the processes which occur when living matter is exposed to radiation.

Penetrating radiation, of course, sees matter in terms of its basic constituents, i.e., as an aggregate of electrons and nuclei (and their constituents as well!). Depending on the type of radiation, its energy and the type of material, reactions with the atoms or nuclei as a whole, or with their individual constituents may occur through whatever channels are allowed. An alpha particle entering a gold foil, for example, may scatter elastically from a nucleus via the Coulomb force, or collide electromagnetically with an atomic electron, or be absorbed in a nuclear reaction to produce other types of radiation, among other processes. These occur with a certain probability governed by the laws of quantum mechanics and the relative strengths of the basic interactions involved. For charged particles and photons, the most common processes are by far the electromagnetic interactions, in particular, inelastic collisions with the atomic electrons. This is not too surprising considering the strength and long range of the Coulomb force relative to the other interactions. For the the neutron, however, processes involving the strong interaction will preferentially occur, although it is also subject to electromagnetic (through its magnetic moment!) and weak processes as well. The type of processes allowed to each type of radiation explain, among other things, their penetrability through matter, their difficulty or ease of detection, their danger to biologial organisms, etc.

The theory behind the principal electromagnetic and neutron processes is well developed and is covered in many texts on experimental nuclear and particle physics. In this chapter, therefore, we will only briefly survey the relevant ideas and concentrate instead on those results useful for nuclear and particle physics. As well, we restrict ourselves only to the energy range of nuclear and particle physics, i.e., a few keV and higher.

2.1 Preliminary Notions and Definitions

To open our discussion of radiation in matter, we first review a few basic notions concerning the interaction of particles.

2.1.1 The Cross Section

The collision or interaction of two particles is generally described in terms of the *cross section*. This quantity essentially gives a measure of the probability for a reaction to occur and may be calculated if the form of the basic interaction between the particles is known. Formally, the cross-section is defined in the following manner. Consider a beam of particles *1* incident upon a target particle *2* as shown in Fig. 2.1. Assume that the beam is much broader than the target and that the particles in the beam are uniformly distributed in space and time. We can then speak of a *flux* of *F* incident particles per unit area per unit time. Now look at the number of particles scattered[1] into the solid angle $d\Omega$ per unit time. Because of the randomness of the impact parameters, this number will fluctuate over different finite periods of measuring time. However, if we average many finite measuring periods, this number will tend towards a fixed $dN_s/d\Omega$, where N_s is the average number scattered per unit time. The *differential* cross section is then defined as the ratio

$$\frac{d\sigma}{d\Omega}(E, \Omega) = \frac{1}{F}\frac{dN_s}{d\Omega}, \tag{2.1}$$

that is, $d\sigma/d\Omega$ is the average fraction of the particles scattered into $d\Omega$ per unit time per unit flux *F*. In terms of a single quantum mechanical particle, this may be reformulated as the scattered probability current in the angle $d\Omega$ divided by the total incident probability passing through a unit area in front of the target.

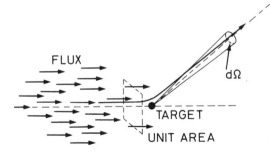

Fig. 2.1. Definition of the scattering cross section

Note that because of the dimensions of *F*, $d\sigma$ has dimensions of area, which leads to the heuristic interpretation of $d\sigma$ as the geometric cross sectional area of the target intercepting the beam. That fraction of the flux incident on this area will then obviously interact while all those missing $d\sigma$ will not. This is only a visual aid, however, and should in *no* way be taken as a real measure of the physical dimensions of the target.

In general, the value of $d\sigma/d\Omega$ will vary with the energy of the reaction and the angle at which the particle is scattered. We can calculate a *total* cross section for any scattering whatsoever at an energy *E* defined as the integral of $d\sigma/d\Omega$ over all solid angles,

$$\sigma(E) = \int d\Omega \frac{d\sigma}{d\Omega}. \tag{2.2}$$

---

[1] By *scattering* here, we mean *any* reaction in which an outgoing particle is emitted into Ω. The incident particle need not retain its identity.

While the above example is easily visualized, it is not a practical case. In real situations, of course, the target is usually a slab of material containing many scattering centers and it is desired to know how many interactions occur on the average. Assuming that the target centers are uniformly distributed and the slab is not too thick so that the likelihood of one center sitting in front of another is low, the number of centers per unit perpendicular area which will be seen by the beam is then $N \delta x$ where N is the density of centers and δx is the thickness of the material along the direction of the beam. If the beam is broader than the target and A is the total perpendicular area of the target, the number of incident particles which are eligible for an interaction is then FA. The average number scattered into $d\Omega$ per unit time is then

$$N_s(\Omega) = FA\,N\,\delta x\,\frac{d\sigma}{d\Omega}\,. \tag{2.3}$$

The total number scattered into all angles is similarly

$$N_{\text{tot}} = FA\,N\,\delta x\,\sigma\,. \tag{2.4}$$

If the beam is smaller than the target, then we need only set A equal to the area covered by the beam. Then $FA \rightarrow n_{\text{inc}}$, the total number of incident particles per unit time. In both cases, now, if we divide (2.4) by FA, we have the probability for the scattering of a *single* particle in a thickness δx,

$$\text{Prob. of interaction in } \delta x = N\sigma\,\delta x\,. \tag{2.5}$$

This is an important quantity and we will come back to this probability later.

2.1.2 Interaction Probability in a Distance x. Mean Free Path

In the previous section, we discussed the probability for the interaction of a particle traveling through a thin slab of matter containing many interaction centers. Let us consider the more general case of *any* thickness x. To do this, we ask the opposite question: what is the probability for a particle *not* to suffer an interaction in a distance x? This is known as the *survival* probability and may be calculated in the following way. Let

$P(x)$: probability of *not* having an interaction after a distance x,
$w\,dx$: probability of having an interaction between x and $x + dx$.

The probability of *not* having an interaction between x and $x + dx$ is then

$$P(x + dx) = P(x)(1 - w\,dx)\,,$$
$$P(x) + \frac{dP}{dx}\,dx = P - Pw\,dx\,,$$
$$dP = -wP\,dx\,, \tag{2.6}$$
$$P = C\exp(-w)\,,$$

where C is a constant. Requiring that $P(0) = 1$, we find $C = 1$. The probability of the particle surviving a distance x is thus exponential in distance. From this, of course, we see immediately that the probability of suffering an interaction *anywhere* in the distance x is just

$$P_{\text{int}}(x) = 1 - \exp(-wx), \tag{2.7}$$

while the probability of the particle suffering a collision between x and $x + dx$ after surviving the distance x is

$$F(x)\,dx = \exp(-wx)\,w\,dx. \tag{2.8}$$

Now let us calculate the mean distance, λ, traveled by the particle without suffering a collision. This is known as the *mean free path*. Thus,

$$\lambda = \frac{\int x P(x)\,dx}{\int P(x)\,dx} = \frac{1}{w}. \tag{2.9}$$

Intuitively, λ must be related to the density of interaction centers and the cross-section, for as we have seen, this governs the probability of interaction. To find this relation, let us return to our slab of material. For a small thickness δx, the interaction probability (2.7) can then be approximated as

$$P_{\text{int}} = 1 - \left(1 - \frac{\delta x}{\lambda} + \dots\right) \simeq \frac{\delta x}{\lambda}, \tag{2.10}$$

where we have expanded the exponential. Comparing with (2.5), we thus find,

$$\lambda = 1/N\sigma, \tag{2.11}$$

so that our survival probability becomes

$$P(x) = \exp\left(\frac{-x}{\lambda}\right) = \exp(-N\sigma x), \tag{2.12}$$

and the interaction probabilities

$$P_{\text{int}}(x) = 1 - \exp\left(\frac{-x}{\lambda}\right) = 1 - \exp(-N\sigma x), \tag{2.13}$$

$$F(x)\,dx = \exp\left(\frac{-x}{\lambda}\right)\frac{dx}{\lambda} = \exp(-N\sigma x)\,N\sigma\,dx. \tag{2.14}$$

2.1.3 Surface Density Units

A unit very often used for expressing thicknesses of absorbers is the *surface density* or *mass thickness*. This is given by the mass density of the material times its thickness in normal units of length, i.e.,

$$\text{mass thickness} \triangleq \rho \cdot t \tag{2.15}$$

with ρ: mass density, t: thickness, which, of course, yields dimensions of mass per area, e.g. g/cm^2.

For discussing the interaction of radiation in matter, mass thickness units are more convenient than normal length units because they are more closely related to the density of interaction centers. They thus have the effect of normalizing materials of differing mass densities. As will be seen later, equal mass thicknesses of different materials will have roughly the same effect on the same radiation.

2.2 Energy Loss of Heavy Charged Particles by Atomic Collisions

In general, two principal features characterize the passage of charged particles through matter: (1) a loss of energy by the particle and (2) a deflection of the particle from its incident direction. These effects are primarily the result of two processes:

1) inelastic collisions with the atomic electrons of the material
2) elastic scattering from nuclei.

These reactions occur many times per unit path length in matter and it is their cumulative result which accounts for the two principal effects observed. These, however, are by no means the only reactions which can occur. Other processes include

3) emission of Cherenkov radiation
4) nuclear reactions
5) bremsstrahlung.

In comparison to the atomic collision processes, they are extremely rare, however, and with the exception of Cherenkov radiation, will be ignored in this treatment.

For reasons which will become clearer in the following sections, it is necessary to separate charged particles into two classes: (1) electrons and positrons, and (2) heavy particles, i.e., particles heavier than the electron. This latter group includes the muons, pions, protons, α-particles and other light nuclei. Particles heavier than this, i.e., the heavy ions, although technically part of this latter group, are excluded in this discussion because of additional effects which arise.

Of the two electromagnetic processes, the inelastic collisions are almost solely responsible for the energy loss of heavy particles in matter. In these collisions ($\sigma \simeq 10^{-17} - 10^{-16}\,\text{cm}^2$!), energy is transferred from the particle to the atom causing an ionization or excitation of the latter. The amount transferred in each collision is generally a very small fraction of the particle's total kinetic energy; however, in normally dense matter, the number of collisions per unit path length is so large, that a substantial cumulative enery loss is observed even in relatively thin layers of material. A 10 MeV proton, for example, already loses *all* of its energy in only 0.25 mm of copper! These atomic collisions are customarily divided into two groups: *soft* collisions in which only an excitation results, and *hard* collisions in which the energy transferred is sufficient to cause ionization. In some of the *hard* reactions, enough energy is, in fact, transferred such that the electron itself causes substantial secondary ionization. These high-energy recoil electrons are sometimes referred to as *δ-rays* or *knock-on* electrons.

Elastic scattering from nuclei also occurs frequently although not as often as electron collisions. In general very little energy is transferred in these collisions since the masses of the nuclei of most materials are usually large compared to the incident particle. In cases where this is not true, for example, an α-particle in hydrogen, some energy is also lost through this mechanism. Nevertheless, the major part of the energy loss is still due to atomic electron collisions.

The inelastic collisions are, of course, statistical in nature, occurring with a certain quantum mechanical probability. However, because their number per macroscopic pathlength is generally large, the fluctuations in the total energy loss are small and one can meaningfully work with the average energy loss per unit path length. This quantity, often called the *stopping power* or simply dE/dx, was first calculated by Bohr using classical arguments and later by Bethe, Bloch and others using quantum mechanics. Bohr's calculation is, nevertheless, very instructive and we will briefly present a simplified version due to *Jackson* [2.1] here.

2.2.1 Bohr's Calculation – The Classical Case

Consider a heavy particle with a charge ze, mass M and velocity v passing through some material medium and suppose that there is an atomic electron at some distance b from the particle trajectory (see Fig. 2.2). We assume that the electron is free and initially at rest, and furthermore, that it only moves very slightly during the interaction with the heavy particle so that the electric field acting on the electron may be taken at its initial position. Moreover, after the collision, we assume the incident particle to be essentially undeviated from its original path because of its much larger mass ($M \gg m_e$). This is one reason for separating electrons from heavy particles!

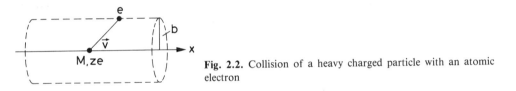

Fig. 2.2. Collision of a heavy charged particle with an atomic electron

Let us now try to calculate the energy gained by the electron by finding the momentum impulse it receives from colliding with the heavy particle. Thus

$$I = \int F\,dt = e\int E_\perp\,dt = e\int E_\perp \frac{dt}{dx}\,dx = e\int E_\perp \frac{dx}{v}\,, \tag{2.16}$$

where only the component of the electric field $E_\perp$ perpendicular to the particle trajectory enters because of symmetry. To calculate the integral $\int E_\perp\,dx$, we use Gauss' Law over an infinitely long cylinder centered on the particle trajectory and passing through the position of the electron. Then

$$\int E_\perp\,2\pi b\,dx = 4\pi ze\,, \qquad \int E_\perp\,dx = \frac{2ze}{b}\,, \tag{2.17}$$

so that

$$I = \frac{2ze^2}{bv} \tag{2.18}$$

and the energy gained by the electron is

$$\Delta E(b) = \frac{I^2}{2m_e} = \frac{2z^2e^4}{m_e v^2 b^2}\,. \tag{2.19}$$

If we let N_e be the density of electrons, then the energy lost to all the electrons located at a distance between b and $b + db$ in a thickness dx is

$$- dE(b) = \Delta E(b) N_e \, dV = \frac{4 \pi z^2 e^4}{m_e v^2} N_e \frac{db}{b} \, dx \, , \qquad (2.20)$$

where the volume element $dV = 2 \pi b \, db \, dx$. Continuing in a straight forward manner, one would at this point be tempted to integrate (2.20) from $b = 0$ to ∞ to get the total energy loss; however, this is contrary to our original assumptions. For example, collisions at very large b would not take place over a short period of time, so that our impulse calculation would not be valid. As well, for $b = 0$, we see that (2.19) gives an infinite energy transfer, so that (2.19) is not valid at small b either. Our integration, therefore, must be made over some limits b_{min} and b_{max} between which (2.19) holds. Thus,

$$- \frac{dE}{dx} = \frac{4 \pi z^2 e^4}{m_e v^2} N_e \ln \frac{b_{max}}{b_{min}} \, . \qquad (2.21)$$

To estimate values for b_{min} and b_{max}, we must make some physical arguments. Classically, the maximum energy transferable is in a head-on collision where the electron obtains an energy of $\frac{1}{2} m_e (2 v)^2$. If we take relativity into account, this becomes $2 \gamma^2 m_e v^2$, where $\gamma = (1 - \beta^2)^{-1/2}$ and $\beta = v/c$. Using (2.19) then, we find

$$\frac{2 z^2 e^4}{m_e v^2 b_{min}^2} = 2 \gamma^2 m v^2, \qquad b_{min} = \frac{z e^2}{\gamma m_e v^2} \, . \qquad (2.22)$$

For b_{max}, we must recall now that the electrons are not free but bound to atoms with some orbital frequency v. In order for the electron to absorb energy, then, the perturbation caused by the passing particle must take place in a time short compared to the period $\tau = 1/v$ of the bound electron, otherwise, the perturbation is adiabatic and no energy is transferred. This is the principle of *adiabatic invariance*. For our collisions the typical interaction time is $t \simeq b/v$, which relativistically becomes $t \Rightarrow t/\gamma = b/(\gamma v)$, so that

$$\frac{b}{\gamma v} \leq \tau = \frac{1}{v} \, . \qquad (2.23)$$

Since there are several bound electron states with different frequencies v, we have used here a mean frequency, $\bar{v}$, averaged over all bound states. An upper limit for b, then, is

$$b_{max} = \frac{\gamma v}{\bar{v}} \, . \qquad (2.24)$$

Substituting into (2.21), we find

$$- \frac{dE}{dx} = \frac{4 \pi z^2 e^4}{m_e v^2} N_e \ln \frac{\gamma^2 m v^3}{z e^2 \bar{v}} \, . \qquad (2.25)$$

This is essentially Bohr's classical formula. It gives a reasonable description of the energy loss for very heavy particles such as the α-particle or heavier nuclei. However, for lighter particles, e.g. the proton, the formula breaks down because of quantum effects. It nevertheless contains all the essential features of electronic collision loss by charged particles.

2.2.2 The Bethe-Bloch Formula

The correct quantum-mechanical calculation was first performed by Bethe, Bloch and other authors. In the calculation the energy transfer is parametrized in terms of momentum transfer rather than the impact parameter. This, of course, is more realistic since the momentum transfer is a measurable quantity whereas the impact parameter is not. The formula obtained is then

$$-\frac{dE}{dx} = 2\pi N_a r_e^2 m_e c^2 \rho \frac{Z}{A} \frac{z^2}{\beta^2} \left[\ln \left(\frac{2 m_e \gamma^2 v^2 W_{max}}{I^2} \right) - 2\beta^2 \right] . \qquad (2.26)$$

Equation (2.26) is commonly known as the *Bethe-Bloch formula* and is the basic expression used for energy loss calculations. In practice, however, two corrections are normally added: the *density effect* correction δ, and the *shell* correction C, so that

$$-\frac{dE}{dx} = 2\pi N_a r_e^2 m_e c^2 \rho \frac{Z}{A} \frac{z^2}{\beta^2} \left[\ln \left(\frac{2 m_e \gamma^2 v^2 W_{max}}{I^2} \right) - 2\beta^2 - \delta - 2\frac{C}{Z} \right] , \qquad (2.27)$$

with

$$2\pi N_a r_e^2 m_e c^2 = 0.1535 \; \text{MeV cm}^2/\text{g}$$

r_e: classical electron
        radius $= 2.817 \times 10^{-13}$ cm
m_e: electron mass
N_a: Avogadro's
        number $= 6.022 \times 10^{23}$ mol^{-1}
I: mean excitation potential
Z: atomic number of absorbing
        material
A: atomic weight of absorbing material

ρ: density of absorbing material
z: charge of incident particle in
             units of e
$\beta =$ v/c of the incident particle
$\gamma =$ $1/\sqrt{1-\beta^2}$
δ: density correction
C: shell correction
W_{max}:   maximum energy transfer in a
             single collision.

The maximum energy transfer is that produced by a head-on or *knock-on* collision. For an incident particle of mass M, kinematics gives

$$W_{max} = \frac{2 m_e c^2 \eta^2}{1 + 2s\sqrt{1+\eta^2} + s^2} , \qquad (2.28)$$

where $s = m_e/M$ and $\eta = \beta\gamma$. Moreover, if $M \gg m_e$, then

$$W_{max} \simeq 2 m_e c^2 \eta^2 .$$

The Mean Excitation Potential. The mean excitation potential, I, is the main parameter of the Bethe-Bloch formula and is essentially the average orbital frequency $\bar{v}$ from Bohr's formula times Planck's constant, $h\bar{v}$. It is theoretically a logarithmic average of

v weighted by the so-called oscillator strengths of the atomic levels. In practice, this is a very difficult quantity to calculate since the oscillator strengths are unknown for most materials. Instead, values of I for several materials have been deduced from actual measurements of dE/dx and a semi-empirical formula for I vs Z fitted to the points. One such formula is

$$\frac{I}{Z} = 12 + \frac{7}{Z}\,\text{eV} \qquad\qquad Z < 13$$

$$\frac{I}{Z} = 9.76 + 58.8\,Z^{-1.19}\,\text{eV} \quad Z \geq 13\,. \tag{2.29}$$

It has been shown, however, that I actually varies with Z in a more complicated manner [2.2]. In particular, there are local irregularities or *wiggles* due to the closing of certain atomic shells. Improved values of I are given in Table 2.1 for several materials. A more extensive list may be found in the articles by *Sternheimer* et al. [2.2 – 3].

The Shell and Density Corrections. The quantities δ and C are corrections to the Bethe-Bloch formula which are important at high and low energies respectively.

The *density effect* arises from the fact that the electric field of the particle also tends to polarize the atoms along its path. Because of this polarization, electrons far from the path of the particle will be shielded from the full electric field intensity. Collisions with these outer lying electrons will therefore contribute less to the total energy loss than predicted by the Bethe-Bloch formula. This effect becomes more important as the particle energy increases, as can be seen from the expression for b_{max} in (2.24). Clearly as the velocity increases, the radius of the cylinder over which our integration is performed also increases, so that distant collisions contribute more and more to the total energy loss. Moreover, it is clear that this effect depends on the density of the material (hence the term *"density"* effect), since the induced polarization will be greater in condensed materials than in lighter substances such as gases. A comparison of the Bethe-Bloch formula with and without corrections is shown in Fig. 2.3.

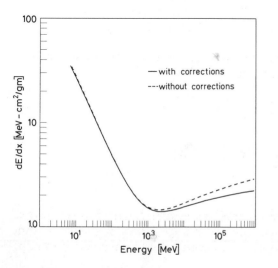

Fig. 2.3. Comparison of the Bethe-Bloch formula with and without the shell and density corrections. The calculation shown here is for copper

Values for δ are given by a formula due to Sternheimer:

$$\delta = \begin{cases} 0 & X < X_0 \\ 4.6052X + C + a(X_1 - X)^m & X_0 < X < X_1 \\ 4.6052X + C & X > X_1, \end{cases} \tag{2.30}$$

where $X = \log_{10}(\beta\gamma)$.

The quantities X_0, X_1, C, a and m depend on the absorbing material. The parameter C is defined as

$$C = -\left(2\ln\frac{I}{h\nu_p} + 1\right), \tag{2.31}$$

where $h\nu_p$ is the so-called plasma frequency of the material, i.e.,

$$\nu_p = \sqrt{\frac{N_e e^2}{\pi m_e}}, \tag{2.32}$$

where N_e (density of electrons) $= N_a \rho Z/A$. The remaining constants are determined by fitting (2.30) to experimental data. Values for several materials are presented in Table 2.1. A more complete listing may be found in *Sternheimer* et al. [2.3].

The *shell* correction accounts for effects which arise when the velocity of the incident particle is comparable or smaller than the orbital velocity of the bound electrons. At such energies, the assumption that the electron is stationary with respect to the incident particle is no longer valid and the Bethe-Bloch formula breaks down. The correction is generally small as can be seen in Fig. 2.3. We give there an empirical formula [2.4] for this correction, valid for $\eta \geq 0.1$:

$$C(I, \eta) = (0.422377\,\eta^{-2} + 0.0304043\,\eta^{-4} - 0.00038106\,\eta^{-6}) \times 10^{-6} I^2$$
$$+ (3.850190\,\eta^{-2} - 0.1667989\,\eta^{-4} + 0.00157955\,\pi^{-6}) \times 10^{-9} I^3, \tag{2.33}$$

where $\eta = \beta\gamma$ and I is the mean excitation potential in eV.

Table 2.1. Constants for the density effect correction

Material	I [eV]	$-C$	a	m	X_1	X_0
Graphite density = 2	78	2.99	0.2024	3.00	2.486	-0.0351
Mg	156	4.53	0.0816	3.62	3.07	0.1499
Cu	322	4.42	0.1434	2.90	3.28	-0.0254
Al	166	4.24	0.0802	3.63	3.01	0.1708
Fe	286	4.29	0.1468	2.96	3.15	-0.0012
Au	790	5.57	0.0976	3.11	3.70	0.2021
Pb	823	6.20	0.0936	3.16	3.81	0.3776
Si	173	4.44	0.1492	3.25	2.87	0.2014
NaI	452	6.06	0.1252	3.04	3.59	0.1203
N_2	82	10.5	0.1534	3.21	4.13	1.738
O_2	95	10.7	0.1178	3.29	4.32	1.754
H_2O	75	3.50	0.0911	3.48	2.80	0.2400
Iucite	74	3.30	0.1143	3.38	2.67	0.1824
Air	85.7	10.6	0.1091	3.40	4.28	1.742
BGO	534	5.74	0.0957	3.08	3.78	0.0456
Plastic Scint.	64.7	3.20	0.1610	3.24	2.49	0.1464

Other Corrections. In addition to the shell and density effects, the validity and accuracy of the Bethe-Bloch formula may be extended by including a number of other corrections pertaining to radiation effects at ultrarelativistic velocities, kinematic effects due to the assumption of an infinite mass for the projectile, higher-order QED processes, higher-order terms in the scattering cross-section, corrections for the internal structure of the particle, spin effects and electron capture at very slow velocities. With the exception of electron-capture effects with heavy ions, these are usually negligible to within $\simeq 1\%$. An outline of these additional factors may be found in the articles by *Ahlen* [2.5 – 6]. For *"elementary"* particles, the Bethe-Bloch formula with the shell and density corrections is more than sufficient however.

2.2.3 Energy Dependence

An example of the energy dependence of *dE/dx* is shown in Fig. 2.4 which plots the Bethe-Bloch formula as a function of kinetic energy for several different particles. At non-relativistic energies, *dE/dx* is dominated by the overall $1/\beta^2$ factor and decreases with increasing velocity until about $v \simeq 0.96\,c$, where a minimum is reached. Particles at this point are known as *minimum ionizing*. Note that the minimum value of *dE/dx* is almost the same for all particles of the same charge. As the energy increases beyond this point, the term $1/\beta^2$ becomes almost constant and *dE/dx* rises again due to the logarithmic dependence of (2.27). This *relativistic rise* is cancelled, however, by the density correction as seen in Fig. 2.3.

For energies below the minimum ionizing value, each particle exhibits a *dE/dx* curve which, in most cases, is distinct from the other particle types. This characteristic is often exploited in particle physics as a means for identifying particles in this energy range.

Not shown in Fig. 2.4, is the very low energy region, where the Bethe-Bloch formula breaks down. At low velocities comparable to the velocity of the orbital electrons of the material, *dE/dx*, in fact, reaches a maximum and then drops sharply again. Here, a number of complicated effects come into play. The most important of these is the tendency of the particle to pick up electrons for part of the time. This lowers the effective charge of the particle and thus its stopping power. Calculating this effective charge can be a difficult problem especially for heavy ions.

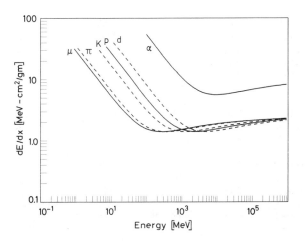

Fig. 2.4. The stopping power *dE/dx* as function of energy for different particles

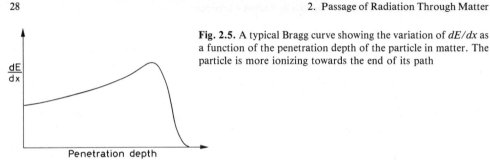

Fig. 2.5. A typical Bragg curve showing the variation of dE/dx as a function of the penetration depth of the particle in matter. The particle is more ionizing towards the end of its path

From Fig. 2.4, it is clear that as a heavy particle slows down in matter, its rate of energy loss will change as its kinetic energy changes. And indeed, more energy per unit length will be deposited towards the end of its path rather than at its beginning. This effect is seen in Fig. 2.5 which shows the amount of ionization created by a heavy particle as a function of its position along it slowing-down path. This is known as a *Bragg* curve, and, as can be seen, most of the energy is deposited near the end of the trajectory. At the very end, however, it begins to pick up electrons and the dE/dx drops. This behavior is particularly used in medical applications of radiation where it is desired to deliver a high dose of radiation to deeply embedded malignant growths with a minimum of destruction to the overlaying tissue.

2.2.4 Scaling Laws for dE/dx

For particles in the same material medium, the Bethe-Bloch formula can be seen to be of the form

$$-\frac{dE}{dx} = z^2 f(\beta) \,, \tag{2.34}$$

where $f(\beta)$ is a function of the particle velocity only. Thus, the energy loss in any given material is dependent only on the charge and velocity of the particle. Since the kinetic energy $T = (\gamma - 1)Mc^2$, the velocity is a function of T/M, so that $\beta = g(T/M)$. We can therefore transform (2.34) to

$$-\frac{dE}{dx} = z^2 f'\left(\frac{T}{M}\right) . \tag{2.35}$$

This immediately suggests a scaling law: if we know the dE/dx for a particle of mass M_1 and charge z_1, then the energy loss of a particle of mass M_2, charge z_2 and energy T_2 in the same material may be found from the values of particle 1 by scaling the energy of particle 2 to $T = T_2(M_1/M_2)$ and multiplying by the charge ratio $(z_2/z_1)^2$, i.e.,

$$-\frac{dE_2}{dx}(T_2) = -\frac{z_2^2}{z_1^2}\frac{dE_1}{dx}\left(T_2\frac{M_1}{M_2}\right) . \tag{2.36}$$

2.2.5 Mass Stopping Power

When dE/dx is expressed in units of mass thickness, it is found to vary little over a wide range of materials. Indeed, if we make the dependence on material type more evident in the Bethe-Bloch formula, we find

$$-\frac{dE}{d\varepsilon} = -\frac{1}{\rho}\frac{dE}{dx} = z^2\frac{Z}{A}f(\beta, I) , \qquad (2.37)$$

where $d\varepsilon = \rho\, dx$. For not too different Z, the ratio (Z/A), in fact, varies little. This is also true of the dependence on $I(Z)$ since it appears in a logarithm. $dE/d\varepsilon$, therefore, is almost independent of material type. A 10 MeV proton, for example, will lose about the same amount of energy in 1 g/cm^2 of copper as it will in 1 g/cm^2 of aluminium or iron, etc. As will also be seen, these units are also more convenient when dE/dx's are combined for mixed materials.

2.2.6 dE/dx for Mixtures and Compounds

The dE/dx formula which we have given so far applies to pure elements. What about dE/dx for compounds and mixtures? Here, if accurate values are desired, one must usually resort to direct measurements; however, a good approximate value can be found in most cases by averaging dE/dx over each element in the compound weighted by the fraction of electrons belonging to each element (*Bragg's Rule*). Thus

$$\frac{1}{\rho}\frac{dE}{dx} = \frac{w_1}{\rho_1}\left(\frac{dE}{dx}\right)_1 + \frac{w_2}{\rho_2}\left(\frac{dE}{dx}\right)_2 + \ldots , \qquad (2.38)$$

where w_1, w_2, etc. are the fractions by *weight* of elements 1, 2, ... in the compound. More explicitly, if a_i is the number of atoms of the ith element in the molecule M, then

$$w_i = \frac{a_i A_i}{A_m} , \qquad (2.39)$$

where A_i is the atomic weight of ith element, $A_m = \sum a_i A_i$.

By expanding (2.38) explicitly and regrouping terms, we can define effective values for Z, A, I, etc. which may be used directly in (2.27),

$$Z_{\mathrm{eff}} = \sum a_i Z_i , \qquad (2.40)$$

$$A_{\mathrm{eff}} = \sum a_i A_i , \qquad (2.41)$$

$$\ln I_{\mathrm{eff}} = \sum \frac{a_i Z_i \ln I_i}{Z_{\mathrm{eff}}} , \qquad (2.42)$$

$$\delta_{\mathrm{eff}} = \sum \frac{a_i Z_i \delta_i}{Z_{\mathrm{eff}}} , \qquad (2.43)$$

$$C_{\mathrm{eff}} = \sum a_i C_i . \qquad (2.44)$$

Note here the convenience of working with the mass stopping power, $1/\rho(dE/dx)$, rather than the linear stopping power dE/dx.

2.2.7 Limitations of the Bethe-Bloch Formula and Other Effects

The Bethe-Bloch formula as given in (2.27) with the shell and density effect corrections is the usual expression employed in most dE/dx calculations. For *elementary* particles

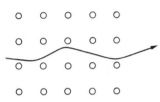

Fig. 2.6. Schematic diagram of channeling in crystalline materials. The particle suffers a series of correlated scatterings which guides it down an open channel of the lattice

and nuclei up to the α-particle, this formula generally gives results accurate to within a few percent for velocities ranging from the relativistic region down to $\beta \simeq 0.1$. This accuracy may be increased and extended to higher-Z nuclei up to $Z \simeq 26$ by including the charge-dependent corrections mentioned earlier [2.5 − 6].

For $\beta \leq 0.05$, many of the assumptions inherent in the Bethe-Bloch formula are no longer valid even with the corrections. Between $0.01 < \beta < 0.05$, in fact, there is still no satisfactory theory for protons. For heavier nuclei, this is even more the case because of electron capture effects. Some empirical formulae for this energy range may be found in [2.7]. Below $\beta \simeq 0.01$, however, a successful explanation of energy loss is given by the theory of *Lindhard* [2.8].

2.2.8 Channeling

An important exception to the applicability of the Bethe-Bloch formula is in the case of *channeling* in materials having a spatially symmetric atomic structure, i.e., crystals. This is an effect which occurs only when the particle is incident at angles less than some critical angle with respect to a symmetry axis of the crystal. As it passes through the crystal planes, the particle, in fact, suffers a series of correlated small-angle scatterings which guide it down an open crystal channel. Figure 2.6 illustrates this schematically. As can be seen, the correlated scatterings cause the particle to follow a slowly oscillating trajectory which keeps it within the open channel over relatively long distances. The wavelength of the trajectory is generally many lattice lengths long. The net effect of this, of course, is that the particle encounters less electrons than it normally would in an amorphous material (which is assumed by the Bethe-Bloch calculation). When the particle undergoes channeling, therefore, its rate of energy loss is greatly reduced. When working with crystalline materials, it is important therefore to be aware of the crystal orientation with respect to the incident particles so as to avoid (or achieve, if that is the case) channeling effects.

In general, the critical angle necessary for channeling is small ($\simeq 1°$ for $\beta \simeq 0.1$) and decreases with energy. It can be estimated by the formula [2.5]

$$\phi_c \simeq \frac{\sqrt{zZa_0Ad}}{1670\,\beta\sqrt{\gamma}}, \tag{2.45}$$

where a_0 is the Bohr radius, and d the interatomic spacing. For $\phi > \phi_c$, channeling does not occur and the material may be treated as amorphous. A more detailed discussion of channeling and the stopping power under such conditions can be found in the review by *Gemmell* [2.8].

2.2.9 Range

Knowing that charged particles lose their energy in matter, a natural question to ask is: How far will the particles penetrate before they lose all of their energy? Moreover, if we assume that the energy loss is continuous, this distance must be a well defined number, the same for all identical particles with the same initial energy in the same type of material. This quantity is called the *range* of the particle, and depends on the type of material, the particle type and its energy.

Experimentally, the range can be determined by passing a beam of particles at the desired energy through different thicknesses of the material in question and measuring

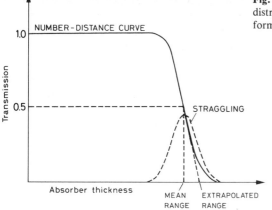

Fig. 2.7. Typical range number-distance curve. The distribution of ranges is approximately Gaussian in form

the ratio of transmitted to incident particles. A typical curve of this ratio versus absorber thickness, known as a *range number-distance* curve, is shown in Fig. 2.7. As can be seen, for small thicknesses, all (or practically all) the particles manage to pass through. As the range is approached this ratio drops. The surprising thing, however, is that the ratio does not drop immediately to the background level, as expected of a well defined quantity. Instead the curve slopes down over a certain spread of thicknesses. This result is due to the fact that the energy loss is *not* in fact continuous, but statistical in nature. Indeed, two identical particles with the same initial energy will *not* in general suffer the same number of collisions and hence the same energy loss. A measurement with an ensemble of identical particles, therefore, will show a statistical distribution of ranges centered about some mean value. This phenomenon is known as *range straggling*. In a first approximation, this distribution is gaussian in form. The mean value of the distribution is known as the *mean range* and corresponds to the midpoint on the descending slope of Fig. 2.7. This is the thickness at which roughly half the particles are absorbed. More commonly, however, what is desired is the thickness at which all the particles are absorbed, in which case the point at which the curve drops to the background level should be taken. This point is usually found by taking the tangent to the curve at the midpoint and extrapolating to the zero-level. This value is known as the *extrapolated* or *practical range* (see Fig. 2.7).

From a theoretical point of view, we might be tempted to calculate the mean range of a particle of a given energy, T_0, by integrating the dE/dx formula,

$$S(T_0) = \int_0^{T_0} \left(\frac{dE}{dx}\right)^{-1} dE. \tag{2.46}$$

This yields the approximate pathlength travelled. Equation (2.46) ignores the effect of multiple Coulomb scattering, however, which causes the particle to follow a zigzag path through the absorber (see Fig. 2.14). Thus, the range, defined as the straight-line thickness, will generally be smaller than the total zigzag pathlength.

As it turns out, however, the effect of multiple scattering is generally small for heavy charged particles, so that the total path length is, in fact, a relatively good approximation to the straight-line range. In practice, a semi-empirical formula must be used,

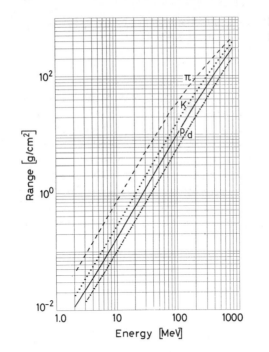

Fig. 2.8. Calculated range curves of different heavy particles in aluminium

$$R(T_0) = R_0(T_{min}) + \int_{T_{min}}^{T_0} \left(\frac{dE}{dx}\right)^{-1} dE \, , \tag{2.47}$$

where T_{min} is the minimum energy at which the dE/dx formula is valid, and $R_0(T_{min})$ is an empirically determined constant which accounts for the remaining low energy behavior of the energy loss. Results accurate to within a few percent can be obtained in this manner.[2] Figure 2.8 shows some typical range-energy curves for different particles calculated by a numerical integration of the Bethe-Bloch formula. From its almost linear form on the log-log scale, one might expect a relation of the type

$$R \propto E^b \, . \tag{2.48}$$

This can also be seen from the stopping power formula, which at not too high energies, is dominated by the β^{-2} term,

$$-dE/dx \propto \beta^{-2} \propto T^{-1} \, , \tag{2.49}$$

where T is the kinetic energy. Integrating, we thus find

$$R \propto T^2 \, , \tag{2.50}$$

---

[2] We might emphasize here that the range as calculated by (2.47) only takes into account energy losses due to atomic collisions and is valid only as long as atomic collisions remain the principal means of energy loss. At very high energies, where the range becomes larger than the mean free path for a nuclear interaction or for bremsstrahlung emission, this is no longer true and one must take into account these latter interactions as well.

which is consistent with our rough guess. A more accurate fit in this energy range, in fact, gives

$$R \propto T^{1.75} ,\tag{2.51}$$

which is not too far from our simple calculation. This is only one of many theoretical and semi-empirical formulas which cover many energy ranges and materials. A discussion of some of these relations is given in the article by *Bethe* and *Ashkin* [2.10].

Range-energy relations of this type are extremely useful as they provide an accurate means of measuring the energy of the particles. This was one of the earliest uses of range measurements. As we will see later, they are also necessary for deciding the sizes of detectors to be used in an experiment or in determining the thickness of radiation shielding, among other things.

Because of the scaling of dE/dx, a scaling law for ranges may also be derived. Using (2.36), it is easy then to see

$$R_2(T_2) = \frac{M_2}{M_1} \frac{z_1^2}{z_2^2} R_1 \left(T_2 \frac{M_1}{M_2} \right)\tag{2.52}$$

for different particles in the same medium.

For the same particle in different materials, a rough relation known as the *Bragg-Kleeman* rule also exists

$$\frac{R_1}{R_2} = \frac{\rho_2}{\rho_1} \frac{\sqrt{A_1}}{\sqrt{A_2}} ,\tag{2.53}$$

where ρ and A are the densities and atomic numbers of the materials. For compounds, a rough approximation for the range can also be found from the formula

$$R_{comp} = \frac{A_{comp}}{\sum \frac{a_i A_i}{R_i}} ,\tag{2.54}$$

where A_{comp} is the molecular weight of the compound, A_i and R_i are the atomic weight and range of the ith constituent element, respectively, and a_i is the number of atoms of the ith element in the compound molecule.

2.3 Cherenkov Radiation

Cherenkov radiation arises when a charged particle in a material medium moves faster than the speed of light in that same medium. This speed is given by

$$\beta c = v = c/n ,\tag{2.55}$$

where n is the index of refraction and c is the speed of light in a vacuum. A particle emitting Cherenkov radiation must therefore have a velocity

$$v_{part} > c/n .\tag{2.56}$$

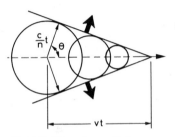

Fig. 2.9. Cherenkov radiation: an electromagnetic shock wave is formed when the particle travels faster than the speed of light in the same medium

In such a case, an electromagnetic *shock wave* is created, just as a faster-than-sound aircraft creates a sonic shock wave. This is illustrated in Fig. 2.9. The coherent wave-front formed is conical in shape and is emitted at an angle

$$\cos \theta = 1/\beta n \tag{2.57}$$

with respect to the trajectory of the particle. In general, a continuous spectrum of frequencies is radiated with the photons being linearly polarized.

The energy carried off by Cherenkov radiation was first calculated by Tamm and Frank to be

$$-\frac{dE}{dx} = \frac{4\pi e^2}{c^2} \int \omega \, d\omega \left(1 - \frac{1}{\beta^2 n^2}\right), \tag{2.58}$$

where the integration is only over those frequencies for which $\beta n(\omega) > 1$. This energy loss is *already included* in the Bethe-Bloch formula (2.27) and is greatest at relativistic velocities. Even at these energies, however, its contribution is small compared to collision loss. Indeed, for condensed materials,

$$-\left(\frac{dE}{dx}\right)_c \simeq 10^{-3} \, \text{MeVcm}^2\text{g}^{-1}, \tag{2.59}$$

which is negligible with respect to the collisional loss. For gases such as H_2,

$$-\frac{dE}{dx} \simeq 0.1 \, \text{MeVcm}^2\text{g}^{-1} \tag{2.60}$$

while for He and higher-Z gases

$$-\frac{dE}{dx} \simeq 0.01 \, \text{MeVcm}^2\text{g}^{-1}. \tag{2.61}$$

The dependence of the emission angle of Cherenkov radiation on particle velocity has been particularly exploited by particle physicists in the form of *Cherenkov counters*. Such devices provide the most accurate measurement of particle velocities and are reviewed in [2.11].

2.4 Energy Loss of Electrons and Positrons

Like heavy charged particles, electrons and positrons also suffer a collisional energy loss when passing through matter. However, because of their small mass an additional energy loss mechanism comes into play: the emission of electromagnetic radiation arising from scattering in the electric field of a nucleus (*bremsstrahlung*). Classically, this may be understood as radiation arising from the acceleration of the electron (or positron) as it is deviated from its straight-line course by the electrical attraction of the nucleus. At energies of a few MeV or less, this process is still a relatively small factor. However, as the energy is increased, the probability of bremsstrahlung quickly shoots up so that at a few 10's of MeV, loss of energy by radiation is comparable to or greater than the collision-ionization loss. At energies above this *critical energy,* bremsstrahlung dominates completely.

The total energy loss of electrons and positrons, therefore, is composed of two parts:

$$\left(\frac{dE}{dx}\right)_{\text{tot}} = \left(\frac{dE}{dx}\right)_{\text{rad}} + \left(\frac{dE}{dx}\right)_{\text{coll}} .$$ (2.62)

2.4.1 Collision Loss

While the basic mechanism of collision loss outlined for heavy charged particles is also valid for electrons and positrons, the Bethe-Bloch formula must be modified somewhat for two reasons. One, as we have already mentioned, is their small mass. The assumption that the incident particle remains undeflected during the collision process is therefore invalid. The second is that for electrons the collisions are between identical particles, so that the calculation must take into account their indistinguishability. These considerations change a number of terms in the formula, in particular, the maximum allowable energy transfer becomes $W_{\text{max}} = T_{\text{e}}/2$ where T_{e} is the kinetic energy of the incident electron or positron. If one redoes the calculation, the Bethe-Bloch formula then becomes

$$-\frac{dE}{dx} = 2\pi N_a r_e^2 m_e c^2 \rho \; \frac{Z}{A} \; \frac{1}{\beta^2} \left[\ln \frac{\tau^2(\tau+2)}{2(I/m_e c^2)^2} + F(\tau) - \delta - 2\frac{C}{Z} \right],$$ (2.63)

where τ is the kinetic energy of particle in units of $m_e c^2$,

$$F(\tau) = 1 - \beta^2 + \frac{\dfrac{\tau^2}{8} - (2r+1)\ln 2}{(\tau+1)^2} \qquad \text{for } e^-$$

$$F(\tau) = 2\ln 2 - \frac{\beta^2}{12}\left(23 + \frac{14}{\tau+2} + \frac{10}{(\tau+2)^2} + \frac{4}{(\tau+2)^3}\right) \quad \text{for } e^+ \qquad .$$

The remaining quantities are as described previously in (2.27 – 33).

2.4.2 Energy Loss by Radiation: Bremsstrahlung

As a practical matter, electrons and positrons are the only particles in which radiation contributes substantially to the energy loss of the particle — at least at the current energies of particle physics today. This can easily be seen from the bremsstrahlung cross-sections which we will present in the following section. The emission probability, in fact, varies as the inverse square of the particle mass, i.e, $\sigma \propto r_e^2 = (e^2/mc^2)^2$. Radiation loss by muons ($m = 106$ MeV), the next lightest particle, for example, is thus some 40000 times smaller than that for electrons!

Since bremsstrahlung emission depends on the strength of the electric field felt by the electron, the amount of screening from the atomic electrons surrounding the nucleus plays an important role. The cross section is thus dependent not only on the incident electron energy but also on its impact parameter and the atomic number, Z, of the material.

The effect of screening can be parametrized by the quantity

$$\xi = \frac{100 \, m_e c^2 h v}{E_0 E Z^{1/3}} \tag{2.64}$$

with E_0: initial *total* energy of electron (or positron); E: final *total* energy of electron; hv: energy of photon emitted, $(E_0 - E)$. This parameter is related to the radius of the Thomas-Fermi atom and is small, $\xi \simeq 0$, for complete screening and large, $\xi \gg 1$, for no screening.

For relativistic energies greater than a few MeV, the bremsstrahlung cross section is given [2.12] by the formula

$$d\sigma = 4 Z^2 r_e^2 \alpha \frac{dv}{v} \left\{ (1 + \varepsilon^2) \left[\frac{\phi_1(\xi)}{4} - \frac{1}{3} \ln Z - f(Z) \right] \right.$$
$$\left. - \frac{2}{3} \varepsilon \left[\frac{\phi_2(\xi)}{4} - \frac{1}{3} \ln Z - f(Z) \right] \right\}, \tag{2.65}$$

with ε: E/E_0, α: $1/137$, $f(Z)$: Coulomb correction, $\phi_1(\xi)$, $\phi_2(\xi)$ are screening functions depending on ξ. This expression is the result of a Born approximation calculation and is not valid at low energies.

For heavy elements ($Z \gtrsim 5$), the screening functions ϕ_1 and ϕ_2 are usually calculated using a Thomas-Fermi model of the atom and the values given numerically. A useful approximation accurate to $\simeq 0.5\%$ is given [2.13] by the empirical formulae

$$\phi_1(\xi) = 20.863 - 2 \ln [1 + (0.55846 \, \xi)^2] - 4 [1 - 0.6 \exp(-0.9 \, \xi) - 0.4 \exp(-1.5 \, \xi)]$$
$$\phi_2(\xi) = \phi_1(\xi) - \tfrac{2}{3} (1 + 6.5 \, \xi + 6 \, \xi^2)^{-1}, \tag{2.66}$$

where

$$\phi_1(0) = \phi_2(0) + \tfrac{2}{3} = 4 \ln 183 \qquad \text{as } \xi \to 0$$
$$\phi_1(\infty) = \phi_2(\infty) \to 19.19 - 4 \ln \xi \qquad \text{as } \xi \to \infty.$$

The function $f(Z)$ is a small correction to the Born approximation which takes into account the Coulomb interaction of the emitting electron in the electric field of the nucleus. *Davies* et al. [2.14] give the formula

$$f(Z) \simeq a^2 [(1 + a^2)^{-1} + 0.20206 - 0.0369 \, a^2 + 0.0083 \, a^4 - 0.002 \, a^6], \tag{2.67}$$

where $a = Z/137$.

In the limiting cases of no screening and complete screening, (2.65) can be expressed in simpler analytic forms. For $\xi \gg 1$ (no screening), (2.65) becomes

$$d\sigma = 4 Z^2 r_e^2 \alpha \frac{dv}{v} \left(1 + \varepsilon^2 - \frac{2\varepsilon}{3} \right) \left[\ln \frac{2 E_0 E}{m_e c^2 h v} - \frac{1}{2} - f(Z) \right]. \tag{2.68}$$

For $\xi \simeq 0$, (complete screening),

$$d\sigma = 4 Z^2 r_e^2 \alpha \frac{dv}{v} \left\{ \left(1 + \varepsilon^2 - \frac{2\varepsilon}{3}\right) [\ln(183 Z^{-1/3}) - f(Z)] + \frac{\varepsilon}{9} \right\}. \tag{2.69}$$

The energy loss due to radiation can now be calculated by integrating the cross-section times the photon energy over the allowable energy range, i.e.,

$$-\left(\frac{dE}{dx}\right)_{\text{rad}} = N \int_0^{v_0} h v \frac{d\sigma}{dv}(E_0, v)\, dv \tag{2.70}$$

with N: number of atoms/cm^3, $N = \rho N_a / A$; $v_0 = E_0 / h$.
We can rewrite this as

$$-\left(\frac{dE}{dx}\right)_{\text{rad}} = N E_0 \Phi_{\text{rad}}, \quad \text{where}$$

$$\Phi_{\text{rad}} = \frac{1}{E_0} \int h v \frac{d\sigma}{dv}(E_0, v)\, dv. \tag{2.71}$$

The motivation behind this is that $d\sigma/dv$ is approximately proportional to v^{-1}; the integral Φ_{rad} is therefore practically independent of v and is a function of the material only.
For $m_e c^2 \ll E_0 \ll 137 m_e c^2 Z^{1/3}$, $\xi \gg 1$, we have no screening, so that integration yields

$$\Phi_{\text{rad}} = 4 Z^2 r_e^2 \alpha \left(\ln \frac{2 E_0}{m_e c^2} - \frac{1}{3} - f(Z) \right). \tag{2.72}$$

For $E_0 \gg 137 m_e c^2 Z^{1/3}$, $\xi \simeq 0$ (complete screening)

$$\Phi_{\text{rad}} = 4 Z^2 r_e^2 \alpha \left[\ln(183 Z^{-1/3}) + \frac{1}{18} - f(Z) \right]. \tag{2.73}$$

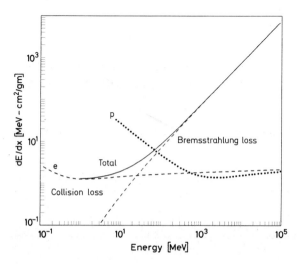

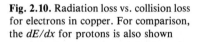

Fig. 2.10. Radiation loss vs. collision loss for electrons in copper. For comparison, the dE/dx for protons is also shown

At intermediate values of ξ, (2.70) must be integrated numerically.

It is interesting to compare (2.71) to the ionization loss formula in (2.63) (see Fig. 2.10). Whereas the ionization loss varies logarithmically with energy and linearly with Z, the radiation loss increases almost linearly with E and quadratically with Z. This dependence explains the rapid rise of radiation loss.

Another difference is that unlike the ionization loss which is quasicontinuous along the path of the electron or positron, almost all the radiation energy can be emitted in one or two photons. There are thus large fluctuations observed for a beam of mono-energetic electrons or positrons.

2.4.3 Electron-Electron Bremsstrahlung

The above formulae represent the mean energy loss from radiation in the field of the nucleus. There is, however, also a contribution from bremsstrahlung which arises in the field of the atomic electrons. Formulas for *electron-electron* bremsstrahlung have been worked out by several authors and it can be shown that the cross sections are essentially given by those above except that Z^2 is replaced by Z. This contribution can thus be approximately taken into account by simply replacing Z^2 by $Z(Z+1)$ in all of the above cross-section formulae.

2.4.4 Critical Energy

As we have seen the energy loss by radiation depends strongly on the absorbing material. For each material, we can define a *critical* energy, E_c, at which the radiation loss equals the collision loss. Thus,

$$\left(\frac{dE}{dx}\right)_{\text{rad}} = \left(\frac{dE}{dx}\right)_{\text{coll}} \quad \text{for} \quad E = E_c. \tag{2.74}$$

Above this energy, radiation loss will dominate over collision-ionization losses and vice-versa below E_c. An approximate formula for E_c due to *Bethe* and *Heitler* [2.10] is,

$$E_c \simeq \frac{1600\, m_e c^2}{Z}, \tag{2.75}$$

Table 2.2. Critical energies of some materials

Material	Critical energy [MeV]
Pb	9.51
Al	51.0
Fe	27.4
Cu	24.8
Air (STP)	102
Lucite	100
Polystyrene	109
NaI	17.4
Anthracene	105
H_2O	92

where $m_e c^2$ is the mass of the electron. Table 2.2 gives a short list of critical energies for various materials so as to give some feeling for the order of magnitudes.

2.4.5 Radiation Length

A similar quantity known as the *radiation length* of the material is even more frequently used. This parameter is defined as the distance over which the electron energy is reduced by a factor $1/e$ due to radiation loss only. Indeed, if we rearrange (2.71), we get the differential equation

$$-dE/E = N\,\Phi_{\text{rad}}\,dx\,. \tag{2.76}$$

Considering the high energy limit where collision loss can be ignored relative to radiation loss, Φ_{rad} in (2.73) is independent of E, so that

$$E = E_0 \exp\left(\frac{-x}{L_{\text{rad}}}\right). \tag{2.77}$$

where x is the distance travelled and $L_{\text{rad}} = 1/N\,\Phi_{\text{rad}}$ is the radiation length. Using (2.73), we thus find the formula

$$\frac{1}{L_{\text{rad}}} \simeq \left[4Z(Z+1)\,\frac{\rho N_a}{A}\right] r_e^2\,\alpha\,[\ln(183\,Z^{-1/3}) - f(Z)]\,, \tag{2.78}$$

where we have included the contribution from electron-electron bremsstrahlung and ignored the small constant term. Some values of L_{rad} are given in Table 2.3 for several materials.

The usefulness of the radiation length becomes evident when material thicknesses are measured in these units. Clearly, if x is expressed in units of L_{rad}, then (2.71) becomes

$$-(dE/dt) \simeq E_0\,, \tag{2.79}$$

where t is the distance in radiation lengths. Thus, the radiation energy loss when expressed in terms of radiation length is roughly independent of the material type.

Table 2.3. Radiation lengths for various absorbers

Material	[gm/cm^2]	[cm]
Air	36.20	30050
H_2O	36.08	36.1
NaI	9.49	2.59
Polystyrene	43.80	42.9
Pb	6.37	0.56
Cu	12.86	1.43
Al	24.01	8.9
Fe	13.84	1.76
BGO	7.98	1.12
BaF$_2$	9.91	2.05
Scint.	43.8	42.4

For compounds and mixtures, the radiation lengths may be computed by applying Bragg's rule. Expressing L_{rad} in mass thickness units, we then have

$$\frac{1}{L_{rad}} = w_1 \left(\frac{1}{L_{rad}}\right)_1 + w_2 \left(\frac{1}{L_{rad}}\right)_2 + \dots , \qquad (2.80)$$

where $w_1, w_2, \dots$ are the fractions by weight of each element in the mixture as defined in (2.39).

2.4.6 Range of Electrons

Because of the electron's greater susceptibility to multiple scattering by nuclei, the range of electrons is generally very different from the calculated path length obtained from an integration of the dE/dx formula. Differences ranging from $20 - 400\%$ depending on the energy and material are often found. In addition, the energy loss by electrons fluctuates much more than for heavy particles. This is due to the much greater energy transfer per collision allowed for electrons and to the emission of bremsstrahlung. In both cases, it is possible for a few single collisions (or photons) to absorb the major part of the electron's energy. This, of course, results in greater range straggling as illustrated by Fig. 2.11 which shows some measured range curves.

As for heavy particles, a number of empirical range-energy relations have been formulated. Figure 2.12 presents some typical range-energy curves for electrons in various materials as calculated assuming a continuous slowing-down process. A tabulation of ranges for different materials is also given by *Pages* et al. [2.16].

2.4.7 The Absorption of β Electrons

Because of their continuous spectrum of energies, the absorption of β-decay electrons exhibits a behavior which is very well approximated by an exponential form. This is illustrated in Fig. 2.13 which shows the number-distance curves for different absorbers plotted on a semi-logarithmic scale. As can be seen, the curves are almost linear and are easily fit by

$$I = I_0 \exp(-\mu x) . \qquad (2.81)$$

The constant μ is known as the *β-absorption coefficient* and is found to be directly related to the endpoint energy of the β-decay. One of the earliest uses of this behavior was, in fact, to measure β endpoint energies and the thicknesses of thin foils. It is important to note, however, that exponential absorption is *not* a general characteristic of β-decay. Indeed, this behavior only holds in the case of simple allowed decays. In more complicated forbidden decays where the shape of the β-spectrum is different, deviations become apparent.

2.5 Multiple Coulomb Scattering

In addition to inelastic collisions with the atomic electrons, charged particles passing through matter also suffer repeated elastic Coulomb scatterings from nuclei although

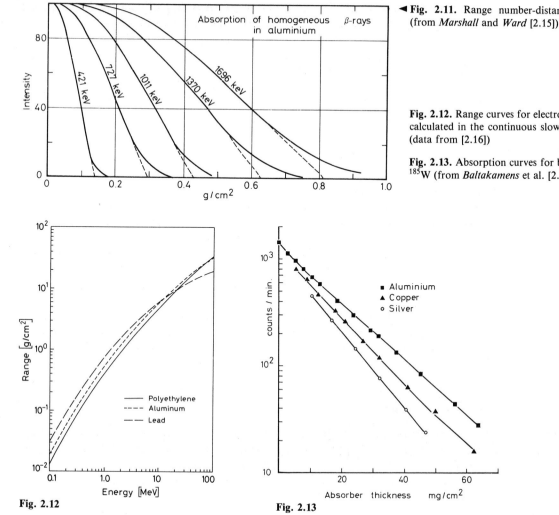

◀ **Fig. 2.11.** Range number-distance curves for electrons (from *Marshall* and *Ward* [2.15])

Fig. 2.12. Range curves for electrons in several materials as calculated in the continuous slowing down approximation (data from [2.16])

Fig. 2.13. Absorption curves for beta decay electrons from ^{185}W (from *Baltakamens* et al. [2.17])

Fig. 2.12

Fig. 2.13

with a somewhat smaller probability. Ignoring spin effects and screening, these collisions are individually governed by the well-known Rutherford formula

$$\frac{d\sigma}{d\Omega} = z_2^2 z_1^2 r_e^2 \frac{mc/\beta p}{4 \sin^4(\theta/2)} . \tag{2.82}$$

Because of its $1/\sin^4(\theta/2)$ dependence, the vast majority of these collisions result, therefore, in a small angular deflection of the particle. We assume here that the nuclei are much more massive than the incident particles so that the small energy transfer to the nucleus is negligible. The particle thus follows a random zigzag path as it traverses the material. The cumulative effect of these small angle scatterings is, however, a net deflection from the original particle direction, as shown in Fig. 2.14.

In general, the treatment of Coulomb scattering in matter is divided into three regions:

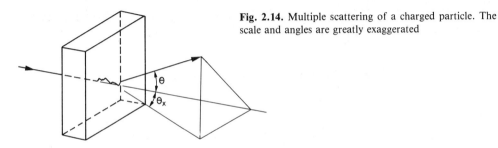

Fig. 2.14. Multiple scattering of a charged particle. The scale and angles are greatly exaggerated

1) *Single Scattering.* If the absorber is very thin such that the probability of more than one Coulomb scattering is small, then the angular distribution will be given by the simple Rutherford formula in (2.82).
2) *Plural Scattering.* If the average number of scatterings $N<20$, then we have *plural scattering.* This is the most difficult case to treat as neither the simple Rutherford formula nor statistical methods can be simply applied. Some work in this region has been done by *Keil* et al. [2.18] and the reader is referred there for further information.
3) *Multiple Scattering.* If the average number of independent scatterings is $N>20$, and energy loss is small or negligible, the problem can be treated statistically to obtain a probability distribution for the net angle of deflection as a function of the thickness of material traversed. This is the most common case encountered and we devote the remainder of this section to this topic.

In general, rigorous calculations of multiple scattering are extremely complicated and several formulations and formulae with different levels of sophistication exist. These are reviewed by *Scott* [2.19] and by *Hemmer* and *Farquahr* [2.20]. Among the most used are the small-angle approximations by Moliere and by Snyder and Scott. Their formulations are essentially equivalent and have been demonstrated to be generally valid for all particles up to angles of $\theta \simeq 30°$ with the exception of slow electrons ($\beta<0.05$) and electrons in very heavy elements.

Moliere expresses the polar angle distribution as a series

$$P(\theta)d\Omega = \eta \, d\eta \left(2 \exp(-\eta^2) + \frac{F_1(\eta)}{B} + \frac{F_2(\eta)}{B^2} + \dots \right), \qquad (2.83)$$

where $\eta = \theta/(\theta_1 \sqrt{B})$ and $\theta_1 = 0.3965 \, (zQ/p\beta) \sqrt{(\rho \delta x/A)}$.

The parameter B is defined by the equation:

$$g(B) = \ln B - B + \ln \gamma - 0.154 = 0, \quad \text{where}$$

$$\gamma = 8.831 \times 10^3 \frac{qz^2 \rho \, \delta x}{\beta^2 A \Delta} \quad \text{and} \quad \Delta = 1.13 + 3.76 \cdot \left(\frac{Zz}{137\beta} \right)^2.$$

For a given γ, B may be found numerically by using, for example, Newton's Method for finding the zeros of $g(B)$. The functions $F_k(\eta)$ are defined by the integral

$$F_k(\eta) = \frac{1}{k!} \int J_0(\eta y) \exp\left(\frac{-y^2}{4} \right) \left[\frac{y^2}{4} \ln\left(\frac{y^2}{4} \right) \right]^k y \, dy,$$

where J_0 = Bessel function. For convenience, some values of F_1 and F_2 for various values of η are tabulated in Table 2.4.

Table 2.4. Values of F_1 and F_2 for the Moliere distribution (from [2.21])

η	$F_1(\eta)$	$F_2(\eta)$	η	$F_1(\eta)$	$F_2(\eta)$
0.0	0.8456	2.49	2.2	0.106	0.02
0.2	0.700	2.07	2.4	0.101	−0.046
0.4	0.343	1.05	2.6	0.082	−0.064
0.6	−0.073	−0.003	2.8	0.062	−0.055
0.8	−0.396	−0.606	3.0	0.045	−0.036
1.0	−0.528	−0.636	3.2	0.033	−0.019
1.2	−0.477	−0.305	3.5	0.0206	0.0052
1.4	−0.318	0.052	4.0	0.0105	0.0011
1.6	−0.147	0.243	5.0	0.00382	0.000836
1.8	0.000	0.238	6.0	0.00174	0.000345
2.0	0.080	0.131	7.0	0.00091	0.000157

The remaining variables are:

Z: atomic weight of material
A: atomic weight of material
δx: thickness of scatterer [cm]
ρ: density of scatterer [g/cm^3]
p: momentum of incident particle [MeV/c]
$\beta = v/c$ of incident particle
z: charge of particle in units of e

$$Q = \begin{cases} \sqrt{Z(Z+1)} & \text{for electron and positrons} \\ Z & \text{for other particles} \end{cases}$$

$$q = \begin{cases} (Z+1)Z^{1/3} & \text{for electrons and positrons} \\ Z^{4/3} & \text{for other particles.} \end{cases}$$

For most calculations, it is usually not necessary to go beyond the first three terms.

Figure 2.15 shows an example of this distribution of 15 MeV electrons passing through a thin gold foil. At small angles, this space angle distribution (with respect to solid angle!) is close to that of a Gaussian, but as angle increases, corrective terms come into play to form a long broad tail. The deflections at larger angles are generally due to one single, large angle Coulomb scattering in the material rather than to the cumulative effect of many small angle scatterings. The broad tail, therefore, should roughly follow that of the Rutherford $1/\sin^4(\theta/2)$ form for single scattering rather than that of a Gaussian. The transition between the small and larger angle regions is governed by plural scattering. This is given by Moliere as a correction to the small angle distribution.

2.5.1 Multiple Scattering in the Gaussian Approximation

If we ignore the small probability of large-angle single scattering, a good idea of the effect of multiple scattering in a given material can be obtained by considering the

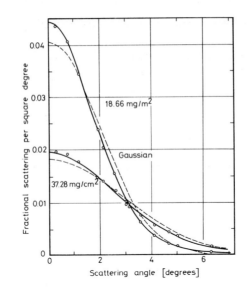

Fig. 2.15. Angular distribution of 15.7 MeV electrons scattered from a thin Au foil (from *Hanson* et al. [2.22]). The experimental values are compared with the Gaussian approximation to multiple scattering

distribution resulting from the small angle ($<10°$) single scatterings *only*. In such a case, as we have seen, the probability distribution is approximately Gaussian in form,

$$P(\theta)\,d\Omega \simeq \frac{2\theta}{\langle \theta^2 \rangle} \exp\left(\frac{-\theta^2}{\langle \theta^2 \rangle}\right) d\theta, \tag{2.84}$$

where we have used the small-angle approximation $d\Omega \simeq 2\pi\theta\,d\theta$. The parameter $\langle \theta^2 \rangle$ represents the mean squared scattering angle, as can be shown by integrating $\int \theta^2 P(\theta)\,d\Omega$ from $\theta = 0$ to ∞. The square root $\sqrt{\langle \theta^2 \rangle}$ is known as the RMS scattering angle and should be equal to the RMS scattering angle of the full multiple scattering angle distribution. By comparing (2.84) to the first term of Moliere's expansion in (2.83), this angle should be approximately given by $\sqrt{\langle \theta^2 \rangle} \simeq \theta_1 \sqrt{B}$. However, because the Moliere distribution has a long tail, the true value is somewhat larger.

A better estimate is obtained by using an empirical formula proposed by *Highland* [2.23] which is valid to within 5% for $Z>20$ and for target thicknesses $10^{-3}L_{\rm rad}<x<10\,L_{\rm rad}$,

$$\sqrt{\langle \theta^2 \rangle} = z\,\frac{20\,[\text{MeV}/c]}{p\beta} \sqrt{\frac{x}{L_{\rm rad}}} \left(1 + \frac{1}{9}\log_{10}\frac{x}{L_{\rm rad}}\right) \text{rad} \tag{2.85}$$

with $L_{\rm rad}$: radiation length of material; x: thickness of material; p: momentum of particle; z: charge of particle.

For low velocities and heavy elements somewhat larger errors ($10-20\%$) are obtained. It should be noted that the appearance of $L_{\rm rad}$ is totally fortuitous; it, of course, has nothing to do with multiple scattering and is only used as a simplification.

A sometimes useful quantity is the angular deflection projected onto a perpendicular plane containing the incident trajectory (see Fig. 2.14). Here the projected distribution is also approximately Gaussian,

$$P(\theta_x)\,d\theta_x = (2\pi \langle \sigma_x^2 \rangle)^{-1/2} \exp\left(\frac{-\theta_x^2}{2\langle \theta_x^2 \rangle}\right) d\theta_x. \tag{2.86}$$

The mean squared projected scattering angle $\langle \theta_x^2 \rangle$ is related to the space scattering angle by $\langle \theta_x^2 \rangle = \langle \theta^2 \rangle / 2$.

From Fig. 2.14, there is also a lateral displacement of the particle. This is usually very small, however, if one calculates this distribution, one finds

$$P(r)\,dr = 6r(\langle \theta^2 \rangle t^2)^{-1} \exp\left(\frac{-3r^2}{\langle \theta^2 \rangle t^2}\right) dr \tag{2.87}$$

with r: displacement; $t = x/L_{rad}$: thickness in radiation lengths. By comparison, the mean squared displacement is then

$$\langle r^2 \rangle = \langle \theta^2 \rangle t^2 / 3 \ . \tag{2.88}$$

2.5.2 Backscattering of Low-Energy Electrons

Because of its small mass, electrons are particularly susceptible to large angle deflections by scattering from nuclei. This probability is so high, in fact, that multiply scattered electrons may be turned around in direction altogether, so that they are backscattered out of the absorber. This is illustrated schematically in Fig. 2.16. The effect is particularly strong for low energy electrons, and increases with the atomic number Z of the material. Backscattering also depends on the angle of incidence. Obviously electrons entering at obliques angles to the surface of the absorber have a greater probability of being scattered out than those incident along the perpendicular.

The ratio of the number of backscattered electrons to incident electrons is known as the *backscattering coefficient* or *albedo*. Figure 2.17 shows some measured coefficients for various materials and electron energies. Backscattering is an important consideration for electron detectors where depending on the geometry and energy, a large fraction of electrons may be scattered out before being able to produce a usable signal. For non-collimated electrons on a high-Z material such as NaI, for example, as much as 80% may be reflected back.

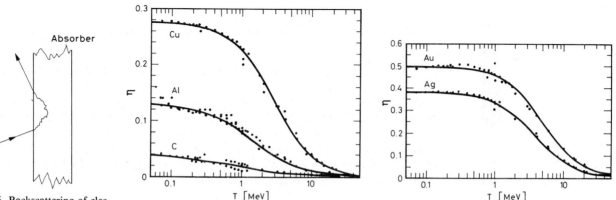

Fig. 2.16. Backscattering of electrons due to large angle multiple scatterings

Fig. 2.17. Some measured electron backscattering coefficients for various materials. The electrons are perpendicularly incident on the surface of the sample (from *Tabata* et al. [2.24])

2.6 Energy Straggling: The Energy Loss Distribution

Our discussion of energy loss up until now has been concerned mainly with the *mean* energy loss suffered by charged particles when passing through a thickness of matter. For any given particle, however, the amount of energy lost will *not*, in general, be equal to this mean value because of the statistical fluctuations which occur in the number of collisions suffered and in the the energy transferred in each collision. An initially monoenergetic beam, after passing through a fixed thickness of material, will therefore show a distribution in energy rather than a delta-function peak shifted down by the mean energy loss as given by the dE/dx formula. We have already seen these fluctuations in the form of range straggling. This, in fact, is the same problem viewed from a different angle: instead of observing the fluctuations in energy loss for a fixed thickness of absorber, we observe the fluctuations in thickness of pathlength for a fixed loss in energy.

Theoretically, calculating the distribution of energy losses for a given thickness of absorber is a difficult mathematical problem and is generally divided into two cases: thick absorbers and thin absorbers.

2.6.1 Thick Absorbers: The Gaussian Limit

For relatively thick absorbers such that the number of collisions is large, the energy loss distribution can easily be shown to be Gaussian in form. This follows directly from the *Central Limit Theorem* in statistics which states that the sum of N random variables, all following the same statistical distribution, approaches that of Gaussian-distributed variable in the limit $N \to \infty$. If we take our random variable to be δE, the energy lost in a single atomic collision, and assume that the energy lost in each collision is such that the velocity of the particle is negligibly altered (so that the velocity-dependent collision cross-section stays constant), then the total energy lost is the sum of many independent δE, all commonly distributed. Assuming there are a sufficient number of collisions N, then the total will approach the Gaussian form,

$$f(x, \Delta) \propto \exp\left(\frac{-(\Delta - \bar{\Delta})^2}{2\sigma^2}\right) \tag{2.89}$$

with x: thickness of absorber; Δ: energy loss in absorber; $\bar{\Delta}$: mean energy loss; σ: standard deviation.

For nonrelativistic heavy particles the spread σ_0 of this Gaussian was calculated by Bohr to be

$$\sigma_0^2 = 4\pi N_a r_e^2 (m_e c^2)^2 \rho \frac{Z}{A} x = 0.1569 \rho \frac{Z}{A} x \, [\text{MeV}^2] \,, \tag{2.90}$$

where N_a is Avogadro's number, r_e and m_e the classical electron radius and mass, and ρ, Z, A are the density, atomic number and atomic weight of the material respectively. This formula can easily be extended to relativistic particles

$$\sigma^2 = \frac{(1 - \frac{1}{2}\beta^2)}{1 - \beta^2} \,. \tag{2.91}$$

2.6.2 Very Thick Absorbers

A critical assumption in the above analysis was that the energy loss was small compared to the initial energy so that the velocity change of the particle could be ignored. For very thick absorbers where a substantial amount of energy is lost, this assumption, of course, breaks down. This case has been treated in depth by *Tschalar* [2.25, 26] and the reader is referred to his articles and to the resume by *Bichsel* [2.27] for details of this distribution.

2.6.3 Thin Absorbers: The Landau and Vavilov Theories

In contrast to the thick absorber case, the distribution for thin absorbers or gases where the number of collisions N is too small for the Central Limit Theorem to hold is extremely complicated to calculate. This is because of the possibility of large energy transfers in a single collision. For heavy particles, this W_{max} is kinematically limited to the expression given in (2.28), while for electrons, as much as one-half the initial energy can be transferred. In this latter case, there is also the additional possibility of a large *"one-shot"* energy loss from bremsstrahlung as well. While these events are rare, their possibility adds a long tail to the high energy side of the energy-loss probability distribution thus giving it a skewed, asymmetric form. Figure 2.18 illustrates this general shape. Note that the mean energy loss no longer corresponds to the peak but is displaced because of the high energy tail. In contrast, the position of the peak now defines the *most probable* energy loss. These two quantities may be used to parametrize the distribution.

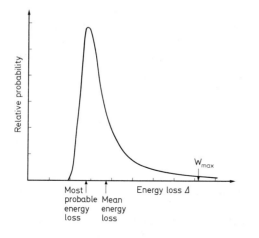

Fig. 2.18. Typical distribution of energy loss in a thin absorber. Note that it is asymmetric with a long high energy tail

Theoretically, basic calculations of this distribution have been carried out by Landau, Symon and Vavilov; each of these, however, has a somewhat different region of applicability. The distinguishing parameter in all these theories is the ratio

$$\kappa = \bar{\Delta}/W_{max}, \tag{2.92}$$

that is the the ratio between the mean energy loss and the maximum energy transfer allowable in a single collision. The mean energy loss may be calculated from the Bethe-Bloch formula, however, for most purposes it is usually approximated by taking the first multiplicative term only and ignoring the logarithmic term, i.e.,

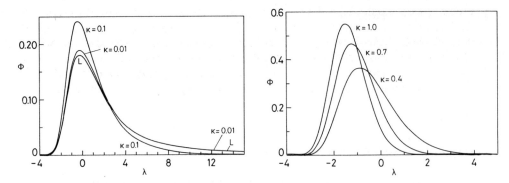

Fig. 2.19. Vavilov distributions for various κ. For comparison, Landau's distribution (denoted by the L) for $\kappa = 0$ is also shown (from *Seltzer* and *Berger* [2.29])

$$\bar{\Delta} \simeq \xi = 2\pi N_a r_e^2 m_e c^2 \rho \frac{Z}{A} \left(\frac{z}{\beta}\right)^2 x \,. \tag{2.93}$$

Following the literature, we denote this quantity by ξ. The thin absorber region is generally taken to be $\kappa < 10$, although for $\kappa > 1$, the distribution already begins to approach the the Gaussian limit (see Fig. 2.19). By $\kappa > 10$, there, of course, is only a very negligible difference.

Landau's Theory: $\kappa \leq 0.01$. *Landau* [2.28] was the first to calculate the energy loss distribution for the case of very thin absorbers, that is, $\kappa \leq 0.01$. In this theory, Landau makes the assumptions that:

1) the maximum energy transfer permitted it infinite, $W_{max} \to \infty$, in essence taking $\kappa \to 0$,
2) the individual energy transfers are sufficiently large such that the electrons may be treated as free. Small energy transfers from so-called *distant* collisions are ignored,
3) the decrease in velocity of the particle is negligible, i.e., the particle maintains a constant velocity.

The distribution is then expressed as

$$f(x, \Delta) = \phi(\lambda)/\xi, \quad \text{where} \tag{2.94}$$

$$\phi(\lambda) = \frac{1}{\pi} \int_0^\infty \exp(-u \ln u - u\lambda) \sin \pi u \, du$$

$$\lambda = \frac{1}{\xi}[\Delta - \xi(\ln \xi - \ln \varepsilon + 1 - C)]$$

$C = $ Euler's Const $= 0.577 \ldots$ and

$$\ln \varepsilon = \ln \frac{(1 - \beta^2) I^2}{2mc^2\beta^2} + \beta^2 \,.$$

The quantity ε essentially represents the minimum energy transfer allowed by assumption 2. The function $\phi(\lambda)$ is a universal function depending only on the parameter λ and must be evaluated numerically. A tabulation for various λ may be found in *Seltzer* and *Berger* [2.29] or *Borsh-Supan* [2.30], for example. A computer program for the calculation of the Landau distribution has also been developed by *Schorr* [2.31].

From an evaluation of $\phi(\lambda)$, the most probable energy loss is found to be

$$\Delta_{mp} = \xi \left[\ln(\xi/\varepsilon) + 0.198 - \delta \right], \tag{2.95}$$

where we have also added on the density effect for completeness.

Symon's Theory and Vavilov's Theory: Intermediate κ. The region between small κ covered by Landau and the Gaussian limit is treated by Symon and by Vavilov. Using the limiting distribution derived by Landau, Symon was able to make a number of ingenious approximations in deriving the energy-loss distributions. His results, unfortunately, are expressed in graphic form, which in today's world of computers, make them inconvenient to use. These graphs and the procedure for employing them may be found in the book by *Rossi* [2.32].

Vavilov's theory, in contrast, is along the line of Landau's formulation and in fact, generalizes the latter's calculation by taking into account the correct expressions for maximum allowable energy transfer. The latter two assumptions made by Landau are kept however. His results are somewhat more complicated, but reduce to the Landau distribution in the limit $\kappa \to 0$ and to a Gaussian form in the limit $\kappa \to \infty$. Rather than give these formulas here, the reader is referred to *Seltzer* and *Berger* [2.29] or to Vavilov's original paper [2.33]. A computer program for its evaluation is also given by *Schorr* [2.31].

To give an idea of Vavilov's results, we show Vavilov's distributions for various values of κ in Fig. 2.19. These should be compared to the Landau distribution (denoted

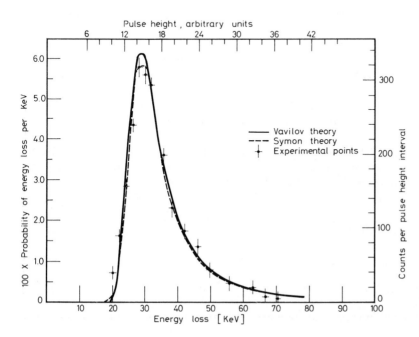

Fig. 2.20. Comparison of Vavilov's and Symon's theories with experiment (from *Seltzer* and *Berger* [2.29])

by L) at $\kappa = 0$, also shown in Fig. 2.19. Note also how the distribution already resembles a Gaussian form for $\kappa = 1$. In the Gaussian limit, Vavilov gives the variance as

$$\sigma^2 = \frac{\xi^2}{\kappa} \frac{1 - \beta^2}{2} , \qquad (2.96)$$

which agrees with Bohr's formula for heavy particles in (2.91).

To see how theory compares with experiment, some measured results are also shown in Fig. 2.20.

Corrections to the Landau and Vavilov Distributions. Supplementing the calculations by Landau and Vavilov, are also a number of limited modifications made by various authors. *Blunck* and *Leisegang* [2.34], in particular, have modified Landau's theory to include binding effects of the atomic electrons (assumption 2). Needless to say, the result is complicated, however, a suitable form for calculation may be found in *Matthews* et al. [2.35]. For the Vavilov distribution, a similar modification has been made by *Shulek* et al. [2.36]. Details may be found in their original article.

2.7 The Interaction of Photons

The behavior of photons in matter (in our case, x-rays and γ-rays) is dramatically different from that of charged particles. In particular, the photon's lack of an electric charge makes impossible the many inelastic collisions with atomic electrons so characteristic of charged particles. Instead, the main interactions of x-rays and γ-rays in matter are:

1) Photoelectric Effect
2) Compton Scattering (including Thomson and Rayleigh Scattering)
3) Pair Production.

Also possible, but much less common, are nuclear dissociation reactions, for example, (γ, n), which we will neglect in our discussion.

These reactions explain the two principal qualitative features of x-rays and γ-rays: (1) x-rays and γ-rays are many times more penetrating in matter than charged particles, and (2) a beam of photons is *not* degraded in energy as it passes through a thickness of matter, only attenuated in intensity. The first feature is, of course, due to the much smaller cross section of the three processes relative to the inelastic electron collision cross section. The second characteristic, however, is due to the fact the three processes above remove the photon from the beam entirely, either by absorption or scattering. The photons which pass straight through, therefore, are those which have not suffered any interactions at all. They therefore retain their original energy. The total number of photons is, however, reduced by the number which have interacted. The attenuation suffered by a photon beam can be shown, in fact, to be exponential with respect to the thickness, i.e.,

$$I(x) = I_0 \exp(-\mu x) \qquad (2.97)$$

with I_0: incident beam intensity; x: thickness of absorber; μ: absorption coefficient.

The absorption coefficient is a quantity which is characteristic of the absorbing material and is directly related to the total interaction cross-section. This is a quantity often referred to when discussing γ-ray detectors. However, let us first discuss the three processes individually before turning to the calculation of the absorption coefficient.

2.7.1 Photoelectric Effect

The photoelectric effect involves the absorption of a photon by an atomic electron with the subsequent ejection of the electron from the atom. The energy of the outgoing electron is then

$$E = h\nu - \text{B.E.}, \tag{2.98}$$

where B.E. is the binding energy of the electron.

Since a free electron cannot absorb a photon and also conserve momentum, the photoelectric effect always occurs on bound electrons with the nucleus absorbing the recoil momentum. Figure 2.21 shows a typical photoelectric cross section as a function of incident photon energy. As can be seen, at energies above the highest electron binding energy of the atom (the K shell), the cross section is relatively small but increases rapidly as the K-shell energy is approached. Just after this point, the cross section drops drastically since the K-electrons are no longer available for the photoelectric effect. This drop is known as the K absorption edge. Below this energy, the cross section rises once again and dips as the L, M, levels, etc. are passed. These are known respectively as the L-absorption edges, M-absorption edge, etc.

Theoretically, the photoelectric effect is difficult to treat rigorously because of the complexity of the Dirac wavefunctions for the atomic electrons. For photon energies above the K-shell, however, it is almost always the K electrons which are involved. If this is assumed and the energy is nonrelativistic, i.e., $h\nu \ll m_e c^2$, the cross-section can then be calculated using a Born approximation. In such a case, one obtains

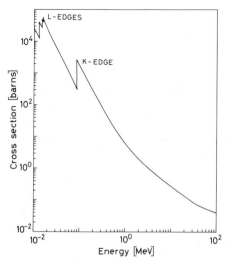

Fig. 2.21. Calculated photoelectric cross section for lead

$$\Phi_{\text{photo}} = 4\alpha^4 \sqrt{2}\, Z^5 \phi_0 (m_e c^2/h\nu)^{7/2} \text{ per atom}, \tag{2.99}$$

with $\phi_0 = 8\pi r_e^2/3 = 6.651 \times 10^{-25} \text{ cm}^2$; $\alpha = 1/137$.

For energies closer to the K-edge, (2.99) must be multiplied by a correction factor to give

$$\Phi_{\text{photo}} = \phi_0 \frac{2^7 \pi (137)^3}{Z^2} \left[\frac{\nu_k}{\nu} \right]^4 \frac{\exp(-4\xi \cot^{-1}\xi)}{1 - \exp(-2\pi\xi)} \text{ per atom}, \tag{2.100}$$

where $h\nu_k = (Z - 0.03)^2 m_e c^2 \alpha^2/2$ and $\xi = \sqrt{\nu_k/(\nu - \nu_k)}$. For ν very close to ν_k, $\xi^{-1} \gg 1$, so that (2.100) can be simplified to

$$\Phi_{\text{photo}} = \frac{6.3 \times 10^{-18}}{Z^2} \left(\frac{\nu_k}{\nu} \right)^{8/3}. \tag{2.101}$$

Formulas for the L and M shells have also been calculated, but these are even more complicated than those above. The reader is referred to *Davisson* [2.37] for these results.

It is interesting to note the dependence of the cross section on the atomic number Z. This varies somewhat depending on the energy of the photon, however, at MeV energies, this dependence goes as Z to the 4th or 5th power. Clearly, then, the higher Z materials are the most favored for photoelectric absorption, and, as will be seen in later chapters, are an important consideration when choosing γ-ray detectors.

2.7.2 Compton Scattering

Compton scattering is probably one of the best understood processes in photon interactions. As will be recalled, this is the scattering of photons on free electrons. In matter, of course, the electrons are bound; however, if the photon energy is high with respect to the binding energy, this latter energy can be ignored and the electrons can be considered as essentially free.

Figure 2.22 illustrates this scattering process. Applying energy and momentum conservation, the following relations can be obtained.

$$h\nu' = \frac{h\nu}{1 + \gamma(1 - \cos\theta)},$$

$$T = h\nu - h\nu' = h\nu \frac{\gamma(1 - \cos\theta)}{1 + \gamma(1 - \cos\theta)},$$

$$\cos\theta = 1 - \frac{2}{(1+\gamma)^2 \tan^2\varphi + 1}, \tag{2.102}$$

$$\cot\varphi = (1 + \gamma) \tan\frac{\theta}{2},$$

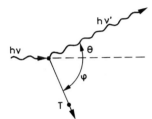

Fig. 2.22. Kinematics of Compton scattering

where $\gamma = h\nu/m_e c^2$. Other relations between the various variables may be found by substitution in the above formulae.

The cross section for Compton scattering was one of the first to be calculated using quantum electrodynamics and is known as the *Klein-Nishina* formula:

$$\frac{d\sigma}{d\Omega} = \frac{r_e^2}{2} \frac{1}{[1+\gamma(1-\cos\theta)]^2} \left(1 + \cos^2\theta + \frac{\gamma^2(1-\cos\theta)^2}{1+\gamma(1-\cos\theta)} \right), \qquad (2.103)$$

where r_e is the classical electron radius. Integration of this formula over $d\Omega$, then, gives the total probability per electron for a Compton scattering to occur.

$$\sigma_c = 2\pi r_e^2 \left\{ \frac{1+\gamma}{\gamma^2} \left[\frac{2(1+\gamma)}{1+2\gamma} - \frac{1}{\gamma}\ln(1+2\gamma) \right] + \frac{1}{2\gamma}\ln(1+2\gamma) - \frac{1+3\gamma}{(1+2\gamma)^2} \right\}. \qquad (2.104)$$

Figure 2.23 plots this total cross section as a function of energy.

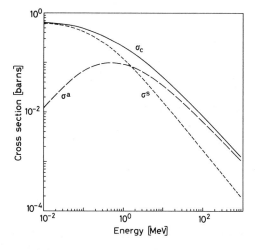

Fig. 2.23. Total Compton scattering cross sections

Two useful quantities which can be calculated from the Klein-Nishina formula are the Compton *scattered* and Compton *absorption* cross sections. The Compton scattered cross section, σ^s, is defined as the average fraction of the total energy contained in the scattered photon, while the absorption cross section, σ^a, is the average energy transferred to the recoil electron. Since the electron is stopped by the material, this is the average energy-fraction absorbed by the material in Compton scattering. Obviously, the sum must be equal to σ_c

$$\sigma_c = \sigma^s + \sigma^a . \qquad (2.105)$$

To calculate σ^s, we form

$$\frac{d\sigma^s}{d\Omega} = \frac{h\nu'}{h\nu} \frac{d\sigma}{d\Omega}, \qquad (2.106)$$

which after integration yields

$$\sigma^s = \pi r_e^2 \left[\frac{1}{\gamma^3}\ln(1+2\gamma) + \frac{2(1+\gamma)(2\gamma^2-2\gamma-1)}{\gamma^2(1+2\gamma)^2} + \frac{8\gamma^2}{3(1+2\gamma)^3} \right]. \qquad (2.107)$$

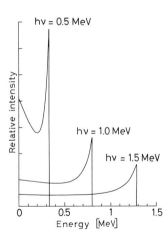

Fig. 2.24. Energy distribution of Compton recoil electrons. The sharp drop at the maximum recoil energy is known as the *Compton edge*

The absorption cross section can then be simply calculated by

$$\sigma^a = \sigma_c - \sigma^s .$$

(2.108)

Another formula which we will make use of very often when discussing detectors is the energy distribution of the Compton recoil electrons. By substituting into the Klein-Nishina formula, one obtains

$$\frac{d\sigma}{dT} = \frac{\pi r_e^2}{m_e c^2 \gamma^2} \left[2 + \frac{s^2}{\gamma^2 (1-s)^2} + \frac{s}{1-s} \left(s - \frac{2}{\gamma} \right) \right],$$

(2.109)

where $s = T/h\nu$. Figure 2.24 shows this distribution for several incident photon energies. The maximum recoil energy allowed by kinematics is given by

$$T_{max} = h\nu \left(\frac{2\gamma}{1+2\gamma} \right)$$

(2.110)

[see (2.102)] and is known as the Compton *edge*.

Thomson and Rayleigh Scattering. Related to Compton scattering are the classical processes of *Thomson* and *Rayleigh* scattering. Thomson scattering is the scattering of photons by free electrons in the classical limit. At low energies with respect to the electron mass, the Klein-Nishina formula, in fact, reduces to the Thomson cross-section,

$$\sigma = \frac{8\pi}{3} r_e^2 .$$

(2.111)

Rayleigh scattering, on the other hand, is the scattering of photons by atoms as a whole. In this process, all the electrons in the atom participate in a coherent manner. For this reason it is also called *coherent* scattering.

In both processes, the scattering is characterized by the fact that *no* energy is transferred to the medium. The atoms are neither excited nor ionized and only the direction of the photon is changed. At the relatively high energies of x-rays and γ-rays, Thomson and Rayleigh scattering are very small and for most purposes can be neglected.

2.7.3 Pair Production

The process of pair production involves the transformation of a photon into an electron-positron pair. In order to conserve momentum, this can only occur in the presence of a third body, usually a nucleus. Moreover, to create the pair, the photon must have at least an energy of 1.022 MeV.

Theoretically, pair production is related to bremsstrahlung by a simple substitution rule, so that once the calculations for one process are made, results for the other immediately follow. As for bremsstrahlung, the screening by the atomic electrons surrounding the nucleus plays an important role in pair production. The cross sections are thus dependent on the parameter ξ [see (2.64)], which is now defined by

$$\xi = \frac{100 m_e c^2 h\nu}{E_+ E_- Z^{1/3}}$$

(2.112)

with E_+: total energy of outgoing positron; E_-: total energy of outgoing electron.

At extreme relativistic energies and arbitrary screening, a Born approximation calculation gives the formula

$$d\tau = 4Z^2 r_e^2 \alpha \frac{dE_+}{(h\nu)^3} \left\{ (E_+^2 + E_-^2) \left[\frac{\phi_1(\xi)}{4} - \frac{1}{3} \ln Z - f(Z) \right] \right.$$

$$\left. + \frac{2}{3} E_+ E_- \left[\frac{\phi_2(\xi)}{4} - \frac{1}{3} \ln Z - f(Z) \right] \right\}$$

(2.113)

where ϕ_1 and ϕ_2 are the screening functions used in (2.66) and the other variables are as defined in (2.65).

As before, this formula simplifies in the limiting cases of no screening and complete screening. Thus for no screening ($\xi \gg 1$), we obtain

$$d\tau = 4Z^2 \alpha r_e^2 \, dE_+ \frac{E_+^2 + E_-^2 + 2E_+ E_- /3}{(h\nu)^3} \left[\ln \frac{2E_+ \acute{E}_-}{h\nu m_e c^2} - \frac{1}{2} - f(Z) \right],$$

(2.114)

while for complete screening, $\xi \to 0$,

$$d\tau = 4Z^2 \alpha r_e^2 \frac{dE_+}{(h\nu)^3} \left\{ \left(E_+^2 + E_-^2 + \frac{2E_+ E_-}{3} \right) [\ln (183 Z^{1/3}) - f(Z)] - \frac{E_+ E_-}{9} \right\}.$$

(2.115)

Because of the Born approximation, these formulae are not very accurate for high Z or low energy. A more complicated formula valid for low energies and no screening has been derived by Bethe and Heitler and is given in the article by *Bethe* and *Ashkin* [2.10] along with a somewhat simpler formula from Hough.

To obtain the total pair production cross section, a numerical integration of the above expressions must generally be performed. In the case of no screening with $m_e c^2 \ll h\nu \ll 137 m_e c^2 Z^{1/3}$, an analytic integration is possible yielding

$$\tau_{\text{pair}} = 4Z^2 \alpha r_e^2 \left[\frac{7}{9} \left(\ln \frac{2h\nu}{m_e c^2} - f(Z) \right) - \frac{109}{54} \right].$$

(2.116)

Similarly for complete screening, $h\nu \gg 137 m_e c^2 Z^{-1/3}$,

$$\tau_{\text{pair}} = 4Z^2 \alpha r_e^2 \{ \tfrac{7}{9} \, [\ln (183 Z^{1/3}) - f(Z)] - 1/54 \}.$$

(2.117)

For all other cases, a numerical integration of (2.113) must be performed. Figure 2.25 illustrates the energy dependence of the total pair cross section.

As for bremsstrahlung, pair production may also occur in the field of an atomic electron. Not surprisingly, a similar result is obtained for the cross section, but smaller by about a factor Z. To approximately account for this interaction, then, one need only replace Z^2 by $Z(Z+1)$ in the above formulae.

From the total cross section, it is interesting to calculate the mean free path, λ_{pair}, of a γ-ray for pair production. Thus, using (2.117)

$$1/\lambda_{\text{pair}} = N \tau_{\text{pair}} \simeq \tfrac{7}{9} 4Z(Z+1) N r_e^2 \alpha [\ln (183 Z^{-1/3}) - f(Z)],$$

(2.118)

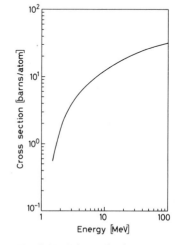

Fig. 2.25. Pair production cross section in lead

where N is the density of atoms and we have ignored the small constant term. This may be recognized as being very similar to the radiation length, and, in fact, comparison with (2.78) shows

$$\lambda_{\text{pair}} \simeq \tfrac{9}{7} L_{\text{rad}} .$$

(2.119)

2.7.4 Electron-Photon Showers

One of the most impressive results of the combined effect of pair production and bremsstrahlung emission by high energy photons and electrons is the formation of *electron-photon showers*. A high energy photon in matter converts into an electron and positron pair which then emit energetic bremsstrahlung photons. These, in turn, will convert into further $e^+ e^-$ pairs, and so on. The result is a cascade or *shower* of photons, electrons and positrons. This continues until the energy of the pair-produced electrons and positrons drops below the critical energy. At this point, the $e^+ e^-$ pairs will preferentially lose their energy via atomic collisions rather than bremsstrahlung emission, thus halting the cascade.

The development of the cascade is, of course, a statistical process. Using the notion of radiation length, however, we may construct a simple model to describe the mean number of particles produced and their mean energies as a function of penetration depth in the converting material. Suppose we begin with an energetic photon of energy E_0. On the average, then, the photon will convert into an $e^+ e^-$ pair after one radiation length. The energy of each member of the pair is then $E_0/2$. After two radiation lengths, the electron and positron will then each emit a bremsstrahlung photon with approximately half the energy of the charged particle. At this point there are 4 particles present: two photons and an electron-positron pair, each with energy $E_0/4$. At the end of three radiation lengths, the bremsstrahlung photons will have converted into two more $e^+ e^-$ pairs, while the original pair will have emitted another set of bremsstrahlung photons. The number of particles present is thus 8 and their energy $E_0/8$. Continuing in this manner, it is easy to see that at the end of t radiation lengths, the total number of particles (i.e., photons, electrons and positrons) present will be

$$N \simeq 2^t$$

(2.120)

each with an average energy of

$$E(t) \simeq \frac{E_0}{2^t} .$$

(2.121)

The same result result would also be obtained had we started with an electron rather than a photon.

Now what is the maximum penetration depth of the cascade? If we assume that the shower stops abruptly at the critical energy E_c, then, we have

$$E(t_{\text{max}}) = \frac{E_0}{2^{t_{\text{max}}}} = E_c$$

(2.122)

which, solving for t_{max}, yields,

$$t_{\text{max}} = \frac{\ln \dfrac{E_0}{E_c}}{\ln 2} .$$

(2.123)

The maximum number of particles produced is thus,

$$N_{max} \simeq \frac{E_0}{E_c} \,. \tag{2.124}$$

This simple model, of course, only gives a rough qualitative picture of the shower. In reality, the number of particles in an electron-photon cascade rises exponentially to a broad maximum after which it declines gradually. In general, a precise calculation cannot be made analytically and recourse must be made to such techniques as Monte Carlo simulations.

For $E_0 \geq 100$ MeV, a fit to the results of such a calculation yields the formula [2.38],

$$N = N_0 t^a \exp(-bt), \quad \text{where} \tag{2.125}$$

$$N_0 = 5.51 E_0 \,[\text{GeV}] \, \sqrt{Z} \, b^{a+1} / \Gamma(a+1)$$

$$b = 0.634 - 0.0021 Z$$

$$a = \begin{cases} 1.77 - 0.52 \ln E_0 & Z = 13 \\ 2.0 - Z/340 + (0.664 - Z/340)\ln E_0 & Z \geq 26 \end{cases}$$

for the number of particles as a function of distance t.

2.7.5 The Total Absorption Coefficient and Photon Attenuation

The total probability for a photon interaction in matter is the sum of the individual cross sections outlined above. If we calculate the cross-section per atom, this yields

$$\sigma = \Phi_{photo} + Z\sigma_c + \tau_{pair}, \tag{2.126}$$

where we have multiplied the Compton cross-section by Z to take into account the Z electrons per atom. This is shown in Fig. 2.26 for the case of lead. If we now multiply σ by the density of atoms, N, we then obtain the probability per unit length for an interaction,

$$\mu = N\sigma = \sigma(N_a \rho / A) \tag{2.127}$$

with N_a: Avogadro's Number; ρ: density of the material; A: molecular weight.

This is more commonly known as the *total absorption coefficient* and is just the inverse of the mean free path of the photon. From (2.12), then, it follows that the fraction of photons surviving a distance x is then

$$I/I_0 = \exp(-\mu x), \tag{2.128}$$

where I_0 is the incident intensity.

For compounds and mixtures, the total absorption coefficient may be calculated using Bragg's rule (2.38),

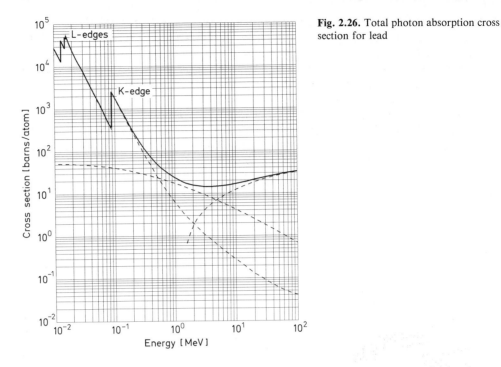

Fig. 2.26. Total photon absorption cross section for lead

$$\frac{\mu}{\rho} = w_1 \frac{\mu_1}{\rho_1} + w_2 \frac{\mu_2}{\rho_2} + \dots, \tag{2.129}$$

where w_i is the weight fraction of each element in the compound.

2.8 The Interaction of Neutrons

Like the photon, the neutron lacks an electric charge, so that it is not subject to Coulomb interactions with the electrons and nuclei in matter. Instead, its principal means of interaction is through the strong force with nuclei. These reactions are, of course, much rarer in comparison because of the short range of this force. Neutrons must come within $\simeq 10^{-13}$ cm of the nucleus before anything can happen, and since normal matter is mainly empty space, it is not surprising that the neutron is observed to be a very penetrating particle.

When the neutron does interact, however, it may undergo a variety of nuclear processes depending on its energy. Among these are:

1) Elastic scattering from nuclei, i.e., $A(n, n)A$. This is the principal mechanism of energy loss for neutrons in the MeV region.
2) Inelastic scattering, e.g., $A(n, n')A^*$, $A(n, 2n')B$, etc. In this reaction, the nucleus is left in an excited state which may later decay by gamma-ray or some other form of radiative emission. In order for the inelastic reaction to occur, the neutron must, of course, have sufficient energy to excite the nucleus, usually on the order of 1 MeV or more. Below this energy threshold, only elastic scattering may occur.

3) Radiative neutron capture, i.e., $n + (Z, A) \rightarrow \gamma + (Z, A+1)$. In general, the cross-section for neutron capture goes approximately as $\simeq 1/v$ where v is the velocity of the neutron. Absorption is most likely, therefore, at low energies. Depending on the element, there may also be resonance peaks superimposed upon this $1/v$ dependence. At these energies, of course, the probability of neutron capture is greatly enhanced.

4) Other nuclear reactions, such as (n, p), (n, d), (n, α), (n, t), (n, αp), etc. in which the neutron is captured and charged particles are emitted. These generally occur in the eV to keV region. Like the radiative capture reaction, the cross section generally falls as $1/v$. Resonances may also occur depending on the element.

5) Fission, i.e., (n, f). Again this is most likely at thermal energies.

6) High energy hadron shower production. This occurs only for very high energy neutrons with $E > 100$ MeV.

Because of the strong energy dependence of neutron interactions, it has become customary to classify neutrons according to their energy, although no specific boundaries are prescribed between classes. In general, high energy neutrons are considered to be those with energies above $\simeq 100$ MeV or so, whereas those between a few ten's of MeV and a few hundred keV are known as *fast* neutrons. Between $\simeq 100$ keV and $\simeq 0.1$ eV, where nuclear resonance reactions occur, neutrons are referred to as *epithermal*. At lower energies comparable to the thermal agitation energy at room temperature, (i.e., $E \simeq kT \simeq 1/40$ eV), neutrons are known as *thermal* or *slow*. Going even lower to energies of milli- or micro-eV, neutrons come under the appellation of *cold* or *ultra-cold*.

The total probability for a neutron to interact in matter is given by the sum of the individual cross sections, i.e.,

$$\sigma_{\text{tot}} = \sigma_{\text{elastic}} + \sigma_{\text{inelastic}} + \sigma_{\text{capture}} + \dots . \tag{2.130}$$

Figure 2.27 gives an example of the total reaction cross-section for neutrons on a few materials versus neutron energy. Here the energy dependence is quite smooth. A compilation of cross sections for other materials may be found in the bibliography.

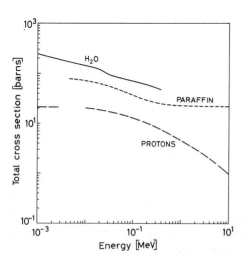

Fig. 2.27. Total reaction cross-sections for neutrons in water, paraffin and protons (data from [2.29])

If we multiply (2.130) by the density of atoms we can obtain the mean free path length

$$\frac{1}{\lambda} = N\sigma_{\text{tot}} = \frac{N_a \rho}{A}\,\sigma_{\text{tot}}\,.$$

(2.131)

In analogy to photons, then, a beam of neutrons passing through matter will be exponentially attenuated

$$N = N_0 \exp(-x/\lambda)\,,$$

(2.132)

where x is the thickness of the material. Equation (2.132), of course, is useful only for a collimated beam of neutrons. For the more general case of a noncollimated source, a sophisticated transport equation is usually necessary.

2.8.1 Slowing Down of Neutrons. Moderation

The slowing down of fast neutrons is known as *moderation* and is an important process in nuclear physics and engineering. In most cases, a fast neutron entering into matter will scatter back and forth on the nuclei, both elastically and inelastically, losing energy until it comes into thermal equilibrium with the surrounding atoms. At this point, it will diffuse through matter until it is finally captured by a nucleus or enters into some other type of nuclear reaction, e.g. fission. The neutron, of course, may undergo a nuclear reaction or be captured before attaining thermal energies, especially if resonances are present. Barring such reactions, however, the v^{-1} dependence of the cross-section favors the survival of the neutron down to thermal velocities.

Elastic scattering is the principal mechanism of energy loss for fast neutrons. At energies of several MeV, the problem may be treated nonrelativistically and very simply with conservation laws. Consider, therefore, a single collision in the lab frame of reference between a neutron with velocity v_0 and a nucleus at rest with a mass M, as shown in Fig. 2.28. In these calculations, it is customary to work in units of neutron mass, i.e. $m_n = 1$. The mass of the nucleus is then just the atomic mass number A. If we transform to the center-of-mass system, the velocity of the neutron becomes

$$v_{\text{cm}} = \frac{A}{A+1}\,v_0$$

(2.133)

while the nucleus takes on a velocity

$$V = \frac{1}{A+1}\,v_0\,.$$

(2.134)

After the collision, the neutron takes on a new direction but retains its speed in the cm system (see Fig. 2.28). Using the law of cosines, the corresponding velocity of the neutron in the lab system is then

$$(v_{\text{lab}})^2 = (v_{\text{cm}})^2 + V^2 - 2v_{\text{cm}}V\cos(\pi - \theta_{\text{cm}})\,,$$

(2.135)

where θ_{cm} is the center-of-mass scattering angle. If we now substitute (2.133) and (2.134) into (2.135), we find

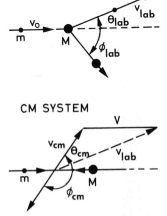

LAB SYSTEM

CM SYSTEM

Fig. 2.28. Elastic scattering of a neutron on a nucleus of mass M

$$(v_{lab})^2 = \left(\frac{A}{A+1}\right)^2 v_0^2 + \left(\frac{1}{A+1}\right)^2 v_0^2 - 2\frac{A}{(A+1)^2}v_0^2\cos(\pi - \theta_{cm}).\tag{2.136}$$

Since the kinetic energy is $E = \frac{1}{2}mv^2$, we thus have

$$\frac{E}{E_0} = \left(\frac{v_{lab}}{v_0}\right)^2 = \frac{A^2 + 1 + 2A\cos\theta_{cm}}{(A+1)^2}.\tag{2.137}$$

Using the cosine law in a similar manner, we can also find the laboratory scattering angle θ_{lab},

$$(v_{cm})^2 = (v_{lab})^2 + V^2 - 2v_{lab}V\cos\theta_{lab}\tag{2.138}$$

which using (2.136) yields

$$\cos\theta_{lab} = \frac{A\cos\theta_{cm} + 1}{\sqrt{A^2 + 1 + 2A\cos\theta_{cm}}}.\tag{2.139}$$

Continuing, we also calculate the scattering parameters for the recoil nucleus

$$E_A = E_0\frac{4A}{(A+1)^2}\cos^2\phi_{lab} = E_0\frac{2A}{(A+1)^2}(1+\cos\phi_{cm})\tag{2.140}$$

$$\cos\phi_{lab} = \sqrt{\frac{1+\cos\phi_{cm}}{2}}.\tag{2.141}$$

From (2.137), it is easy to see now that the energy of the scattered neutron is limited to the range between

$$\left(\frac{A-1}{A+1}\right)^2 E_0 < E < E_0,\tag{2.142}$$

where the limits correspond to scattering at $\cos\theta_{cm} = \pm 1$. In the particular case of scattering on protons, $A = 1$,

$$0 < E < E_0.$$

This is not surprising, as intuitively, the lighter the nucleus the more recoil energy it absorbs from the neutron. This implies of course that the slowing down of neutrons is most efficient when protons or light nuclei are used. This explains the use of hydrogenous materials such as water or paraffin (CH_2) in connection with neutron moderators and shielding.

Let us now calculate the energy distribution of the scattered neutrons. At not too high energies (≤ 15 MeV), neutron scattering is usually restricted to s-wave scattering which is isotropic. Thus the probability of scattering into a solid angle $d\Omega$ is simply

$$dw = \frac{d\Omega}{4\pi} = 2\pi\sin\theta_{cm}\frac{d\theta_{cm}}{4\pi} = \frac{1}{2}\sin\theta_{cm}\,d\theta_{cm}.\tag{2.143}$$

From (2.137), however,

$$\frac{dE}{E_0} = 2 \frac{A}{(A+1)^2} \sin\theta_{cm}\, d\theta_{cm}, \tag{2.144}$$

which after substitution yields

$$\frac{dw_1}{dE} = \frac{(A+1)^2}{4A}\frac{1}{E_0} = \frac{1}{E_0(1-\alpha)}, \tag{2.145}$$

where $\alpha = [(A-1)/(A+1)]^2$. After one scattering, therefore, the energy distribution of an originally monoenergetic neutron is constant over the energy range given by (2.142).

We can use this result now to find the distribution after two scatterings

$$\frac{dw_2}{dE} = \begin{cases} \displaystyle\int_E^{E_0} d\varepsilon \frac{dw_1}{d\varepsilon}\frac{1}{\varepsilon(1-\alpha)} = \frac{1}{E_0(1-\alpha)^2}\ln\frac{E_0}{E} & \alpha E_0 < E < E_0 \\[4mm] \displaystyle\int_{\alpha E_0}^{E/\alpha} d\varepsilon \frac{dw_1}{d\varepsilon}\frac{1}{\varepsilon(1-\alpha)} = -\frac{1}{E_0(1-\alpha)^2}\left[\ln\frac{E_0}{E}+2\ln\alpha\right] & \alpha^2 E_0 < E < \alpha E_0. \end{cases} \tag{2.146}$$

For three collisions, a similar calculation gives the expression

$$\frac{dw_3}{dE} = \begin{cases} \displaystyle\frac{1}{2E_0(1-\alpha)^3}\left(\ln\frac{E_0}{E}\right)^2 & \alpha E_0 < E < E_0 \\[4mm] \displaystyle-\frac{1}{2E_0(1-\alpha)^3}\left[2\left(\ln\frac{E_0}{E}\right)^2 + 6\ln\alpha\ln\frac{E_0}{E} + 3(\ln\alpha)^3\right] & \alpha^2 E_0 < E < \alpha E_0 \\[4mm] \displaystyle\frac{1}{2E_0(1-\alpha)^3}\left(\ln\frac{E_0}{E}+3\ln\alpha\right)^3 & \alpha^3 E_0 < E < \alpha^2 E_0. \end{cases} \tag{2.147}$$

For more collisions the distributions can be worked out by continuing the process although the algebra becomes more and more complicated. Figure 2.29 compares the distributions obtained after several collisions. *Condon* and *Breit* [2.40] have worked out the general problem and for the case of n scatterings on hydrogen give the general formula

$$\frac{dw_n}{dE} = \frac{1}{E_0(n-1)!}\left(\ln\frac{E_0}{E}\right)^{n-1}. \tag{2.148}$$

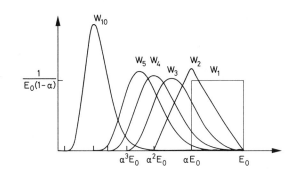

Fig. 2.29. Energy distribution of neutrons after several elastic scatterings

An obvious question which arises now is how many collisions are needed to reduce the average energy of a neutron to some given level? This is most easily found by calculating the logarithmic change in energy:

$$u = \ln E_0 - \ln E = \ln \frac{E_0}{E} , \tag{2.149}$$

where E_0 is the initial energy and the E the final energy. This is also known as the *lethargy* change. From (2.137) we see, in fact, that after a single scattering at angle θ, u is given by

$$u(\theta) = \ln \frac{(A+1)^2}{A^2+1+2A \cos \theta} . \tag{2.150}$$

If we integrate (2.150) over all directions and divide by 4π, we can then find the average $u(\theta)$ for a single scattering,

$$\xi = \langle u(\theta) \rangle = \int u(\theta) \frac{d\Omega}{4\pi} = \frac{1}{2} \int \ln \frac{(A+1)^2}{A^2+1+2A \cos \theta} d(\cos \theta)$$

$$= 1 + \frac{(A-1)^2}{2A} \ln \frac{A-1}{A+1} . \tag{2.151}$$

This leaves us with the interesting result that the average lethargy change, ξ, after one scattering is a constant, independent of the initial energy. Now, for a neutron to slow down from an energy E_0 to an energy E', a total lethargy change of $\ln(E_0/E')$ is required. Since the average lethargy change per collision is ξ, the average number of collisions n required for this total change would then be

$$n = \frac{u}{\xi} = \frac{1}{\xi} \ln \frac{E_0}{E'} . \tag{2.152}$$

With carbon-12 as a moderator, for example, we would have $\xi = 0.158$, so that a 1 MeV neutron slowing down to thermal energies (1/40 eV) would require (1/0.158) $\ln(40 \times 10^6) \simeq 111$ collisions. For hydrogen, $\xi = 1$, so that this number would only be $n \simeq 17.5$. More rigorous approaches are given in [2.41, 42].

3. Radiation Protection.
Biological Effects of Radiation

That radiation can be hazardous to living organisms is well-known to any informed person today. However, except for this simple fact, further knowledge of just how and why this is so appears to be rare even among those who work with radiation professionally. Indeed, it behooves anyone handling radioactive material or working in a radiation environment to have at least a few elementary ideas concerning the effects of exposure to radiation, the permissible limits and the safety precautions to be taken. The nuclear physicist is, of course, no exception. In this chapter, therefore, we will briefly survey the dosimetric units used for discussing the effects of irradiation, the resultant biological effects, the recommended maximum permissible dose and some simple safety precautions to be followed in the nuclear physics laboratory.

3.1 Dosimetric Units

The quantity of radiation received by an object is measured by several different units. Since radiation interacts with matter by ionizing or exciting the atoms and molecules making up the material, these units are either a measure of the quantity of ionization produced or the amount of energy deposited in the material.

3.1.1 The Roentgen

The oldest unit is the *Roentgen*, which is a measure of *exposure* and is defined as

> 1 Roentgen = the quantity of x-rays producing an ionization of 1 esu/cm^3
> (= 2.58 Coul/kg) in air at STP.

Note that the definition refers specifically to x-rays and γ-rays in air. As such, it is an easy quantity to measure with ionization chambers, however, it becomes inconvenient when the irradiated object is living tissue or some other material.

In air, ionization is produced primarily by the slowing down of the recoil electrons resulting from the Compton scattering of the gamma-rays and x-rays. The amount of ionization produced, therefore, depends on both the absorption coefficient for γ-rays and the specific ionization of electrons. If isotropic radiation from a point is assumed and attenuation from air is ignored, the ionization per unit time or *exposure rate* due to a given source may be found from the formula,

$$\text{Exposure rate} = \frac{\Gamma \cdot A}{d^2}, \tag{3.1}$$

Table 3.1. Short list of exposure rate constants
[3.1]

Source	Γ[R-cm^2/hr-mCi]
^{137}Cs	3.3
^{57}Co	13.2
^{22}Na	12.0
^{60}Co	13.2
^{222}Ra	8.25

where A is the activity, d the distance to the source and Γ is an *exposure rate constant* dependent on the decay scheme of the particular source, the energy of the γ's, the absorption coefficient in air and the specific ionization of electrons. This constant has been calculated for a number of common γ-sources and a short list is given in Table 3.1. A more complete list is given in the Radiological Health Handbook [3.1] issued by the U.S. Dept. of Health, Education and Welfare.

3.1.2 Absorbed Dose

A more relevant quantity for discussing the effects of irradiation is the *absorbed dose*. This is a quantity which measures the total *energy* absorbed per unit mass. There are two units:

   1 rad = 100 erg/g   energy absorbed

   1 Gray (Gy) = 1 Joule/kg = 100 rad.

The Gray unit is the more recent of the two and has been proposed by the General Conference on Weights and Measures to replace the rad. It should also be noted that the absorbed dose gives no indication of the rate at which the irradiation occurred nor the specific type of radiation.

Example 3.1   Calculate the absorbed dose in air for 1 Roentgen of γ-rays. Assume that for electrons, the average energy needed to create an ion-electron pair in air is 32 eV.

$$1\ \text{R} = 1\ \text{esu/cm}^3 = \frac{1}{4.8 \times 10^{-10}\ \text{esu/elect}} = 2.08 \times 10^9\ \text{ion-pairs/cm}^3 .$$

The energy expended is thus

$$32\ \text{eV/ion-pr} \times (2.08 \times 10^9\ \text{ion-pr/cm}^3) = 6.66 \times 10^4\ \text{MeV/cm}^3 .$$

Taking the density $\rho_{\text{air}} = 1.2\ \text{mg/cm}^3$ and 1 MeV $= 1.6 \times 10^{-6}$ erg, we find

$$\text{absorbed dose} = (6.66 \times 10^4\ \text{MeV/cm}^3) \times \frac{1.6 \times 10^{-6}}{1.2 \times 10^{-3}} = 88.8\ \text{erg/g} = 0.89\ \text{rad} .$$

Example 3.2   Assuming soft living tissue absorbs $\simeq 93$ erg/g for 1 R of γ radiation, what is the dose rate received from working at an average distance of 50 cm from a 100 μCi ^{22}Na source?

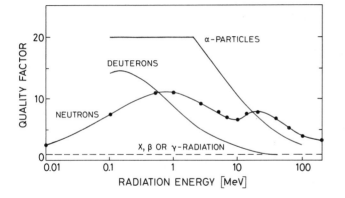

Fig. 3.1. Quality factor for different radiations as a function of energy (from *Myers* [3.2])

From Table 3.1, the exposure rate is

$$\text{Exp. Rate} = \frac{12.0 \times 0.1}{50^2} = 0.48 \text{ mR/hr}$$

$$\text{Dose Rate} = 93 \times 0.48 \times 10^{-3} = 0.447 \text{ erg/g-hr} = 4.47 \text{ mrad/hr} .$$

3.1.3 Relative Biological Effectiveness (RBE)

For discussing biological effects, the nonspecificity of the absorbed dose proves to be inconvenient. Indeed, studies show that the biological damage caused by radiation is a strong function of the specific radiation type. A dose of α-particles, for example, produces more damage than an equal dose of protons and, this, more damage than a similar dose of electrons or γ-rays. The difference lies in the *linear energy transfer* (LET) of the different particles, i.e., the energy locally deposited per unit path length.[1] Thus, the more ionizing the particle the greater is the concentration of ionized and excited molecules along its path and the greater the local biological damage. To account for this effect, a *quality factor,* measuring the *relative biological effectiveness* (RBE) is assigned to each radiation type. Figure 3.1 plots this quantity as function of energy for several different types of particles. When the energy is not known or the particles have a spectrum, it is more customary to use the values shown in Table 3.2. In general, therefore, α-particles may be considered as 2 times more damaging than protons, and these 10 times more than electrons or gamma-rays. The RBE, however, may vary depending on the specific biological effect.

Table 3.2. Quality factors for various radiations

	γ	β	protons	α	fast n	thermal n
Q-factor	1	1	10	20	10	3

---

[1] For most purposes, this is the same as dE/dx. The only difference is the emission of bremsstrahlung, which generally escapes from the region of the particle path. This energy loss is included in dE/dx, but not in the LET.

3.1.4 Dose Equivalent

By multiplying the absorbed dose (the rad or Gray) by the RBE quality factor, a normalized measure of the biological effect can be obtained. This is the *dose equivalent* and its units of measurement are the *rem* (*roentgen-man-equivalent*) or the *Sievert* (Sv) where

rem = Quality Factor × rad

Sievert (Sv) = Quality Factor × Gray   (1 Sv = 100 rem) .

One rem of α-particles therefore produces approximately the same effect as one rem of γ's, etc. The dose equivalent is the most commonly used quantity, however, it should be noted that it is not directly measurable whereas the absorbed dose is.

3.2 Typical Doses from Common Sources in the Environment

As is well-known, we are constantly bathed in radiation coming from a variety of natural and artificial sources. These include cosmic rays, radioactive isotopes found naturally in the environment, (e.g., the ground, building materials, etc.) nuclear fall-out, consumer products, and radioactive sources used in industry. To give an idea of the magnitude of our dose units, Table 3.3 lists the typical doses received per year from some of these natural and artificial sources.

Table 3.3. Typical doses from several common sources (from Upton [3.3])

Natural sources	
Cosmic rays	28 mrem/yr
Natural background (U, Th, Ra)	26
Internal radioactive sources in the body (^{40}K, ^{14}C)	26
Environmental sources	
Technologically enhanced	4 mrem/yr
Global fallout	4
Nuclear power	0.3
Medical	
Diagnostic	78 mrem/yr
1 x-ray	100 – 200 mrem
Pharmaceuticals	14
Occupational	1
Consumer products (TV, etc.)	5

These values may vary by as much as a factor 2 or 3 depending on the region in which the individual lives. At an altitude of 2000 meters, for example, the cosmic ray dose is practically double that at sea level. Similarly, the natural background dose may also be larger or smaller depending on the mineral and geological structure of the region. Nevertheless, the largest sources of radiation are from the natural background and medical diagnosis.

3.3 Biological Effects

Radiation is harmful to living tissue because of its ionizing power in matter. This ionization can damage living cells directly, by breaking the chemical bonds of important biological molecules (particularly DNA), or indirectly, by creating chemical radicals from water molecules in the cells, which can then attack the biological molecules chemically. To a certain extent, these molecules are repaired by natural biological processes, however, the effectiveness of this repair depends on the extent of the damage. Obviously, if the repair is successful then no effect is observed, however, if the repair is faulty or not made at all, the cell may then suffer three possible fates:

1) Death (of the cell)
2) An impairment in the natural functioning of the cell leading to somatic effects (i.e. physical effects suffered by the irradiated individual only) such as cancer
3) A permanent alteration of the cell which is transmitted to later generations, i.e., a genetic effect.

The sequence of events and the time scales involved are briefly outlined below in Fig. 3.2.

Let us now consider, the specific biological consequences which may result in humans. Depending on the dose, these consequences may be immediate or delayed by many years.

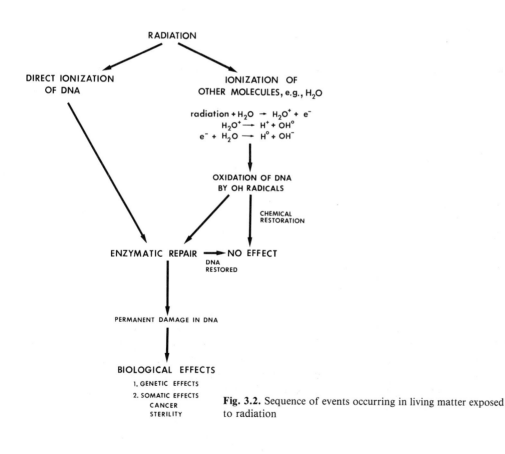

Fig. 3.2. Sequence of events occurring in living matter exposed to radiation

3.3.1 High Doses Received in a Short Time

The effects of high doses of radiation (≥ 100 rad) received in a short time period ($\leq$ a few hours) are generally well-known. The immediate effect is a disruption of the reproductive process in mitotic cells leading to their depletion. The most important of these are the white blood cells, the bone marrow and the cells lining the intestine. The first consequences of a high dose of radiation will thus be noticed in the blood of an individual. If the dose is greater than $200 - 300$ rem, death may occur either due to the radiation itself or to complications arising from the depletion of the mitotic cells, e.g., infections. An outline of the possible sequence of events which might occur after exposure to a dose of $400 - 600$ rem is given in Table 3.4.

Table 3.4. Symptoms after receiving $400 - 600$ rem in a short time

$0 - 48$ h	Loss of appetite, nausea, vomiting, fatigue and prostration
2 days $- 6 - 8$ wks	Above symptoms disappear, patient feels better
$2 - 3$ wks to $6 - 8$ wks	Purpura and hemorrhage, diarrhea, loss of hair, fever, lethargy, death
$6 - 8$ wks	Recovery stage

If the patient survives, a number of other effects may develop at a later time, for example, reddening of the skin, sterility, cataracts, and birth defects. These effects, including death, all exhibit a threshold characteristic, i.e., there exists a safe minimum dose below which these effects do not appear. Above this threshold, there is a certain chance of developing one or more of the effects with the probability increasing with increasing dose. This threshold characteristic appears reasonable as, in general, a minimum number of cells must be damaged before the impairment of a organ is affected. As well, this also explains the dependence on the dose rate. A summary of some of the effects and their threshold doses is also given in Table 3.5. It is perhaps important to note, here, the relative sensitivity of the fetus to radiation!

3.3.2 Low-Level Doses

Low-level doses are taken to be doses of 20 rad or less, or higher doses received at the maximum permissible rates prescribed in the next section. Here, the two principal effects are cancer and genetic effects. In contrast to the high dose situation, however,

Table 3.5. Threshold doses for several effects (from Myers [3.2])

Stage of development	Effect	Threshold dose [rem]
Embryo	Small head circumference	4
Fetus	Diminished body growth Increased infant mortality	20
Child	Hypothyroidism	500
Adult	Opacity of the eye lens	250
Adult	Death	$200 - 300$
Adult	Aging	300
Adult	Erythema (reddening of the skin)	$300 - 1000$
Male adult	Temporary sterility	$50 - 100$
	Permanent stability	> 500
Female adult	Permanent sterility	$300 - 400$

very little is known about the relation of radiation to the occurrence of these two diseases. For cancer, this is due in part to the long delay between irradiation and the appearance of the effect and, in part, to the difficulty of isolating radiation from other possible causes such as drugs, cigarettes, chemicals, etc. In the case of genetic effects, no radiation-induced genetic defect in humans (including the Hiroshima-Nagasaki survivors) has ever been significantly demonstrated, although laboratory experiments on mice and other animals have shown these injuries. Present knowledge of the genetic effects of radiation in man, in fact, is based entirely upon extrapolation from these experiments.

Nevertheless, it is generally accepted that these effects

1) do *not* exhibit a threshold, that is, there is *no safe* level of radiation below which these effects are not observed, and
2) that they do *not* depend on the dose rate, but rather on the total accumulated dose.

Indeed, for a given total dose, one has a certain nonzero probability of developing one or the other of these effects. In general, a linear relation between the total dose and the risk of developing a cancer or a genetic defect is assumed, although there may be deviations from this model at higher doses. Current estimates of this risk as supplied by the International Commission on Radiological Protection (ICRP) are given in Table 3.6. These estimates vary somewhat depending on the source and should be taken as order of magnitude estimates. These risks should be put into perspective by comparing them with the risks taken in some common, everyday occupations. This is shown in Table 3.7 where the risk has been transformed into of an average loss in life expectancy. Normal life expectancy is taken as 73 years.

Table 3.6. Risk of developing cancer or genetic defects (ICRP publication 26 [3.4])

	Risk/Sv $- 10^4$ persons
Fatal cancers	125
Curable cancers	125
Total genetic diseases including subsequent descendents	80

Table 3.7. Comparison of risk from radiation with risk from other occupations (from Myers [3.2])

Occupation	Average loss of life expectancy [months]
20 rem (typical dose of radiation worker in research lab after 47 yrs i.e, age 18 – 65)	0.4
50 rem (typical dose of worker in nuclear power plant after 47 years)	1
235 rem	5
Trade	1
Service industries	1.2
Transportation and public utilities	5
Off-the-job accidents	7.5
Construction	10
Mining and quarrying	11

3.4 Maximum Permissible Dose (MPD)

We now turn to a question important for anyone handling radioactive materials: what is the maximum dose an individual can be permitted to receive in addition to the natural background dose? This is a difficult question to answer. Indeed, as we have seen, no safe level of radiation exists and, moreover, the effects are cumulative. Nevertheless, certain benefits are derived from radiation, e.g. medical diagnosis or cancer therapy, so that abandoning the use of radiation altogether would also result in net loss to society. The setting of a maximum permissible dose thus implies establishing a balance between the benefits to be gained versus the risks incurred. This is obviously a subjective question and indeed the equilibrium point may be different for different people, different localities, etc.

The only internationally recognized body for setting these limits is the International Commission on Radiological Protection (ICRP). Because of the possible differences mentioned above, the ICRP presents its limits as recommendations only. Each country is then free to accept, reject or modify these values as it feels fit. To avoid any implication of a threshold, these limits are technically termed *maximum permissible doses* (MPD) and are defined as "... *that dose accumulated over a long period of time or resulting from a single exposure, which in the light of current knowledge, carries a negligible probability of severe somatic or genetic injuries* ...".

Two sets of MPD's are defined: one for individuals exposed occupationally and one for the general public. Within each set, maximum doses for different parts of the body are given, since some organs are more sensitive than others. It should be stressed that these are allowable doses in *addition* to the natural background dose. Table 3.8 summarizes some of the current MPD's to the various organs. For general exposure to the entire body, the MPD is given by the formula

$$(N-18) \times 5 \text{ rem} = \text{MPD} \tag{3.2}$$

where N is the age of the individual. However, the person must not receive more than 3 rem in 13 weeks or 12 rem in 12 months. Thus, a 38 year old radiation worker, for example, is allowed a total accumulated dose of 100 rem, but not all at one time!

Table 3.8. Summary of MPD's to various organs

	Radiation worker		General public
	rem/yr	rem/13 wks	rem/yr
Gonads, bone marrow	5	3	0.5
Skin, bone, thyroid	30	15	3
Hands, forearms, feet	75	38	7.5
Other	15	8	1.5

3.5 Shielding

To ensure total safety, all radioactive materials in the laboratory or place of work should be surrounded by sufficiently thick shielding material such that the radiation in neighboring work areas is kept at minimum permissible levels. This quantity of shielding is determined by the material chosen, the distance of the work area from the source and the maximum time it is inhabited.

The choice of shielding materials and the design of the shield depends on the type of radiation and its intensity. Gamma rays, for example, are best attenuated by materials with a high atomic number, as we have seen in Chap. 2. Materials such as Pb or iron, therefore, would be more suitable than, say, plastic or water. Similarly, for stopping charged particles, dense materials would be preferred because of their higher dE/dx. For neutron shielding, on the other hand, hydrogenous materials should be chosen in order to facilitate moderation. In these choices, the possibility of secondary radiation from interactions in the shield should also be considered. For example, positrons are easily stopped by a very thin layer of Pb, however, once at rest they annihilate with electrons resulting in the emission of even more penetrating annihilation radiation. The shield, then, must not only be designed for stopping positrons but also for absorbing 511 keV photons! A summary of the recommended shielding schemes for various radiations found in the nuclear physics laboratory is given in Table 3.9.

Table 3.9. Shielding materials for various radiations

Radiation	Shielding
Gamma-rays	High-Z material, e.g. Pb or steel
Electrons	Low-Z materials, e.g., polystyrene or lucite. High-Z materials should be avoided because of bremsstrahlung production. For intense electron sources, a double layer shield consisting of an inner layer of low-Z material followed by a layer of Pb (or some other high-Z material) to absorb bremsstrahlung should be used. The inner layer should, of course, be sufficiently thick to stop the electrons while the outer layer should provide sufficient attenuation of the bremsstrahlung.
Positrons	High-Z material. Since the stopping of positrons is always accompanied by annihilation radiation, the shield should be designed for absorbing annihilation radiation. A double layer design, here, is usually not necessary.
Charged particles	High density materials in order to maximize dE/dx
Neutrons	Hydrogenous materials such as water or paraffin. As for electrons, this shielding should also be followed by a layer of Pb or other high-Z material in order to absorb γ's from neutron capture reactions.

While certain materials are better suited than others for a given type of radiation, cost usually limits the choice of shielding to a few readily available materials. The most used are lead, iron and steel, water, paraffin and concrete. Lead is often used because of its high atomic number and density. As well, it is soft and malleable and easily cast into various forms. When large amounts of Pb are required, it is usually cheaper to use scrap iron or steel. For very large volumes, concrete blocks are generally the most advantageous as far as cost is concerned. In accelerator laboratories, concrete is, in fact, the standard shielding material.

3.6 Radiation Safety in the Nuclear Physics Laboratory

Since our text is concerned with experimental nuclear physics, it behooves us to say a few words concerning safety in the nuclear physics lab. In general, the risks of working in a student nuclear physics laboratory are very small. The radioactive sources are of relatively low intensity and are all normally sealed against any *"rubbing off"* of radioactive material. Nevertheless, needless exposure should be avoided and to ensure that this risk be kept at a minimum, a few simple safety precautions should be followed:

1) Do not eat or smoke in the laboratory. The most dangerous situation, even with a low intensity source, is when radioactive material is ingested. Vital organs, which are otherwise protected by clothing, skin and muscle, would then be exposed.
2) For the same reason as above, always wash hands after handling radioactive material.
3) Wear your dosimeter!

4. Statistics and the Treatment of Experimental Data

Statistics plays an essential part in all the sciences as it is the tool which allows the scientist to treat the uncertainties inherent in all measured data and to eventually draw conclusions from the results. For the experimentalist, it is also a design and planning tool. Indeed, before performing any measurement, one must consider the tolerances required of the apparatus, the measuring times involved, etc., as a function of the desired precision on the result. Such an analysis is essential in order to determine its feasibility in material, time and cost.

Statistics, of course, is a subject unto itself and it is neither fitting nor possible to cover all the principles and techniques in a book of this type. We have therefore limited ourselves to those topics most relevant for experimental nuclear and particle physics. Nevertheless, given the (often underestimated) importance of statistics we shall try to give some view of the general underlying principles along with examples, rather than simple "*recipes*" or "*rules of thumb*". This hopefully will be more useful to the physicist in the long run, if only because it stimulates him to look further. We assume here an elementary knowledge of probability and combinatorial theory.

4.1 Characteristics of Probability Distributions

Statistics deals with random processes. The outcomes of such processes, for example, the throwing of a die or the number of disintegrations in a particular radioactive source in a period of time T, fluctuate from trial to trial such that it is impossible to predict with certainty what the result will be for any given trial. Random processes are described, instead, by a *probability density* function which gives the expected frequency of occurrence for each possible outcome. More formally, the outcome of a random process is represented by a *random variable x*, which ranges over all admissible values in the process. If the process is the throwing of a single die, for instance, then x may take on the integer values 1 to 6. Assuming the die is true, the probability of an outcome x is then given by the density function $P(x) = 1/6$, which in this case happens to be the same for all x. The random variable x is then said to be *distributed* as $P(x)$.

Depending on the process, a random variable may be continuous or discrete. In the first case, it may take on a continuous range of values, while in the second only a finite or denumerably infinite number of values is allowed. If x is discrete, $P(x_i)$ then gives the frequency at each point x_i. If x is continuous, however, this interpretation is not possible and only probabilities of finding x in finite intervals have meaning. The distribution $P(x)$ is then a continuous density such that the probability of finding x between the interval x and $x + dx$ is $P(x)\,dx$.

4.1.1 Cumulative Distributions

Very often it is desired to know the probability of finding x between certain limits, e.g., $P(x_1 \leq x \leq x_2)$. This is given by the cumulative or integral distribution

$$P(x_1 \leq x \leq x_2) = \int_{x_1}^{x_2} P(x)\,dx\,, \tag{4.1}$$

where we have assumed $P(x)$ to be continuous. If $P(x)$ is discrete, the integral is replaced by a sum,

$$P(x_1 \leq x \leq x_2) = \sum_{i=1}^{2} P(x_i)\,. \tag{4.2}$$

By convention, also, the probability distribution is normalized to 1, i.e.,

$$\int P(x)\,dx = 1 \tag{4.3}$$

if x is continuous or

$$\sum_i P(x_i) = 1 \tag{4.4}$$

if x is discrete. This simply says that the probability of observing one of the possible outcomes in a given trial is defined as 1. It follows then that $P(x_i)$ or $\int P(x)\,dx$ cannot be greater than 1 or less than 0.

4.1.2 Expectation Values

An important definition which we will make use of later is the *expectation value* of a random variable or a random variable function. If x is a random variable distributed as $P(x)$, then

$$E[x] = \int x P(x)\,dx \tag{4.5}$$

is the *expected* value of x. The integration in (4.5) is over all admissible x. This, of course, is just the standard notion of an average value. For a discrete variable, (4.5) becomes a sum

$$E[x] = \sum_i x_i P(x_i)\,. \tag{4.6}$$

Similarly, if $f(x)$ is a function of x, then

$$E[f(x)] = \int f(x) P(x)\,dx \tag{4.7}$$

is the expected value of $f(x)$.

To simplify matters in the remainder of this section, we will present results assuming a continuous variable. Unless specified otherwise, the discrete case is found by replacing integrals with a summation.

4.1.3 Distribution Moments. The Mean and Variance

A probability distribution may be characterized by its *moments*. The rth moment of x about some fixed point x_0 is defined as the expectation value of $(x - x_0)^r$ where r is an integer. An analogy may be drawn here with the moments of a mass distribution in mechanics. In such a case, $P(x)$ plays the role of the mass density.

In practice, only the first two moments are of importance. And, indeed, many problems are solved with only a knowledge of these two quantities. The most important is the first moment about zero,

$$\mu = E[x] = \int x P(x)\, dx \,. \tag{4.8}$$

This can be recognized as simply the *mean* or *average* of x. If the analogy with mass moments is made, the mean thus represents the *"center of mass"* of the probability distribution.

It is very important here to distinguish the mean as defined in (4.8) from the mean which one calculates from a set of repeated measurements. The first refers to the theoretical mean, as calculated from the theoretical distribution, while the latter is an *experimental* mean taken from a sample. As we shall see in Sect. 4.4.2, the sample mean is an estimate of the theoretical mean. Throughout the remainder of this chapter, we shall always use the Greek letter μ to designate the theoretical mean.

The second characteristic quantity is the second moment about the mean (also known as the *second central moment*),

$$\sigma^2 = E[(x-\mu)^2] = \int (x-\mu)^2 P(x)\, dx \,. \tag{4.9}$$

This is commonly called the *variance* and is denoted as σ^2. The square root of the variance, σ, is known as the *standard deviation*. As can be seen from (4.9), the variance is the average squared deviation of x from the mean. The standard deviation, σ, thus measures the dispersion or width of the distribution and gives us an idea of how much the random variable x fluctuates about its mean. Like μ, (4.9) is the theoretical variance and should be distinguished from the sample variance to be discussed in Sect. 4.4.

Further moments, of course, may also be calculated, such as the third moment about the mean. This is known as the *skewness* and it gives a measure of the distribution's symmetry or asymmetry. It is employed on rare occasions, but very little information is generally gained from this moment or any of the following ones.

4.1.4 The Covariance

Thus far we have only considered the simple case of single variable probability distributions. In the more general case, the outcomes of a process may be characterized by several random variables, $x, y, z, \ldots$. The process is then described by a *multivariate* distribution $P(x, y, z, \ldots)$. An example is a playing card which is described by two variables: its denomination and its suit.

For multivariate distributions, the mean and variance of each separate random variable $x, y, \ldots$, are defined in the same way as before (except that the integration is over all variables). In addition a third important quantity must be defined:

$$\mathrm{cov}(x, y) = E[(x-\mu_x)(y-\mu_y)] \,, \tag{4.10}$$

where μ_x and μ_y are the means of x and y respectively. Equation (4.10) is known as the *covariance* of x and y and it is defined for each pair of variables in the probability density. Thus, if we have a trivariate distribution $P(x, y, z)$, there are three covariances: $\mathrm{cov}(x, y)$, $\mathrm{cov}(x, z)$ and $\mathrm{cov}(y, z)$.

The covariance is a measure of the linear correlation between the two variables. This is more often expressed as the *correlation coefficient* which is defined as

$$\rho = \frac{\mathrm{cov}(x, y)}{\sigma_x \sigma_y},\tag{4.11}$$

where σ_x and σ_y are the standard deviations of x and y. The correlation coefficient varies between -1 and $+1$ where the sign indicates the sense of the correlation. If the variables are perfectly correlated linearly, then $|\rho| = 1$. If the variables are independent[1] then $\rho = 0$. Care must be taken with the converse of this last statement, however. If ρ is found to be 0, then x and y can only be said to be *linearly* independent. If can be shown, in fact, that if x and y are related parabolically, (e.g., $y = x^2$), then $\rho = 0$.

4.2 Some Common Probability Distributions

While there are many different probability distributions, a large number of problems in physics are described or can be approximately described by a surprisingly small group of theoretical distributions. Three, in particular, the binomial, Poisson and Gaussian distributions find a remarkably large domain of application. Our aim in this section is to briefly survey these distributions and describe some of their mathematical properties.

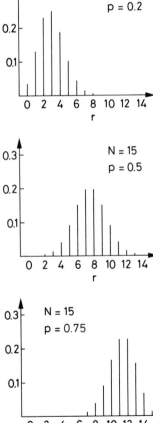

4.2.1 The Binomial Distribution

Many problems involve repeated, independent trials of a process in which the outcome of a single trial is dichotomous, for example, *yes* or *no*, *heads* or *tails*, *hit* or *miss*, etc. Examples are the tossing of a coin N times, the number of boys born to a group of N expectant mothers, or the number of hits scored after randomly throwing N balls at a small, fixed target.

More generally, let us designate the two possible outcomes as *success* and *failure*. We would then like to know the probability of r successes (or failures) in N tries regardless of the order in which they occur. If we assume that the probability of success does not change from one trial to the next, then this probability is given by the *binomial* distribution,

$$P(r) = \frac{N!}{r!\,(N-r)!}\,p^r(1-p)^{N-r},\tag{4.12}$$

where p is the probability of success in a single trial.

Equation (4.12) is a discrete distribution and Fig. 4.1 shows its form for various values of N and p. Using (4.8) and (4.9), the mean and variance many be calculated to yield

$$\mu = \sum_r rP(r) = Np \quad \text{and}\tag{4.13}$$

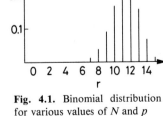

Fig. 4.1. Binomial distribution for various values of N and p

---

[1] The mathematical definition of independence is that the joint probability is a separable function, i.e., $P(x, y) = P_1(x)P_2(y)$.

$$\sigma^2 = \sum_r (r-\mu)^2 P(r) = Np(1-p) \,. \tag{4.14}$$

It can be shown that (4.12) is normalized by summing $P(r)$ from $r = 0$ to $r = N$. Here it will be noticed that $P(r)$ is nothing but the rth term of the binomial expansion (whence the name!), so that

$$\sum_r \frac{N!}{r!\,(N-r)!} p^r (1-p)^{N-r} = [(1-p)+p]^N = 1 \,. \tag{4.15}$$

Finding the cumulative distribution between limits other than 0 and N is somewhat more complicated, however, as no analytic form for the sum of terms exist. If there are not too many, the individual terms may be calculated separately and then summed. Otherwise, tabulations of the cumulative binomial distribution may be used.

In the limit of large N and not too small p, the binomial distribution may be approximated by a Gaussian distribution with mean and variance given by (4.13) and (4.14). For practical calculations, using a Gaussian is usually a good approximation when N is greater than about 30 and $p \geq 0.05$. It is necessary, of course, to ignore the discrete character of the binomial distribution when using this approximation (although there are corrections for this). If p is small (≤ 0.05), such that the product Np is finite, then the binomial distribution is approximated by the Poisson distribution discussed in the next section.

4.2.2 The Poisson Distribution

The *Poisson* distribution occurs as the limiting form of the binomial distribution when the probability $p \to 0$ and the number of trials $N \to \infty$, such that the mean $\mu = Np$, remains finite. The probability of observing r events in this limit then reduces to

$$P(r) = \frac{\mu^r e^{-\mu}}{r!} \,. \tag{4.16}$$

Like (4.12), the Poisson distribution is discrete. It essentially describes processes for which the single trial probability of success is very small but in which the number of trials is so large that there is nevertheless a reasonable rate of events. Two important examples of such processes are radioactive decay and particle reactions.

To take a concrete example, consider a typical radioactive source such as ^{137}Cs which has a half-life of 27 years. The probability per unit time for a single nucleus to decay is then $\lambda = \ln 2/27 = 0.026\text{ /year} = 8.2 \times 10^{-10}\text{ s}^{-1}$. A small probability indeed! However, even a 1 µg sample of ^{137}Cs will contain about 10^{15} nuclei. Since each nucleus constitutes a *trial*, the mean number of decays from the sample will be $\mu = Np = 8.2 \times 10^5$ decays/s. This satisfies the limiting conditions described above, so that the probability of observing r decays is given by (4.16). Similar arguments can also be made for particle scattering.

Note that in (4.16), only the mean appears so that knowledge of N and p is not always necessary. This is the usual case in experiments involving radioactive processes or particle reactions where the mean counting rate is known rather than the number of nuclei or particles in the beam. In many problems also, the mean per unit dimension λ, e.g. the number of reactions per second, is specified and it is desired to know the proba-

Fig. 4.2. Poisson distribution for various values of μ

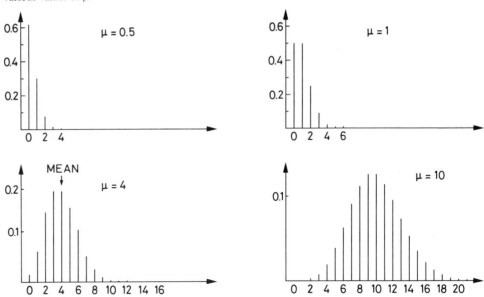

bility of observing r events in t units, for example, $t = 3$ s. An important point to note is that the mean in (4.16) refers to the mean number in t units. Thus, $\mu = \lambda t$. In these types of problems we can rewrite (4.16) as

$$P(r) = \frac{(\lambda t)^r e^{-\lambda t}}{r!}.$$ (4.17)

An important feature of the Poisson distribution is that it depends on only one parameter: μ. [That μ is indeed the mean can be verified by using (4.8)]. From (4.9), we also the find that

$$\sigma^2 = \mu,$$ (4.18)

that is the variance of the Poisson distribution is equal to the mean. The standard deviation is then $\sigma = \sqrt{\mu}$. This explains the use of the square roots in counting experiments (see Examples 1.1 and 1.2, on p. 11 and 12).

Figure 4.2 plots the Poisson distribution for various values of μ. Note that the distribution is *not* symmetric. The peak or maximum of the distribution does not, therefore, correspond to the mean. However, as μ becomes large, the distribution becomes more and more symmetric and approaches a Gaussian form. For $\mu \geq 20$, a Gaussian distribution with mean μ and variance $\sigma^2 = \mu$, in fact, becomes a relatively good approximation and can be used in place of the Poisson for numerical calculations. Again, one must neglect the fact that we are replacing a discrete distribution by a continuous one.

4.2.3 The Gaussian or Normal Distribution

The *Gaussian* or *normal* distribution plays a central role in all of statistics and is the most ubiquitous distribution in all the sciences. Measurement errors, and in particular,

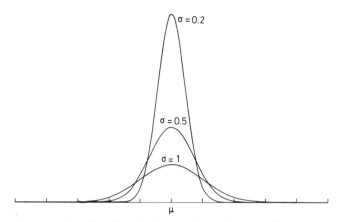

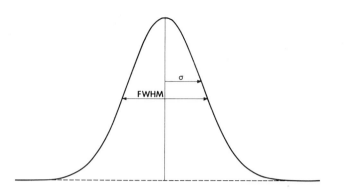

Fig. 4.3. The Gaussian distribution for various σ. The standard deviation determines the width of the distribution

Fig. 4.4. Relation between the standard deviation σ and the full width at half-maximum (FWHM)

instrumental errors are generally described by this probability distribution. Moreover, even in cases where its application is not strictly correct, the Gaussian often provides a good approximation to the true governing distribution.

The Gaussian is a continuous, symmetric distribution whose density is given by

$$P(x) = \frac{1}{\sigma\sqrt{2\pi}} \exp\left(-\frac{(x-\mu)^2}{2\sigma^2}\right). \tag{4.19}$$

The two parameters μ and σ^2 can be shown to correspond to the mean and variance of the distribution by applying (4.8) and (4.9).

The shape of the Gaussian is shown in Fig. 4.3. which illustrates this distribution for various σ. The significance of σ as a measure of the distribution width is clearly seen. As can be calculated from (4.19), the standard deviation corresponds to the half width of the peak at about 60% of the full height. In some applications, however, the full width at half maximum (FWHM) is often used instead. This is somewhat larger than σ and can easily be shown to be

$$\text{FWHM} = 2\sigma\sqrt{2\ln 2} = 2.35\,\sigma. \tag{4.20}$$

This is illustrated in Fig. 4.4. In such cases, care should be taken to be clear about which parameter is being used. Another width parameter which is also seen in the literature is the full-width at one-tenth maximum (FWTM).

The integral distribution for the Gaussian density, unfortunately, cannot be calculated analytically so that one must resort to numerical integration. Tables of integral values are readily found as well. These are tabulated in terms of a *reduced* Gaussian distribution with $\mu = 0$ and $\sigma^2 = 1$. All Gaussian distributions may be transformed to this reduced form by making the variable transformation

$$z = \frac{x-\mu}{\sigma}, \tag{4.21}$$

where μ and σ are the mean and standard deviation of the original distribution. It is a trivial matter then to verify that z is distributed as a reduced Gaussian.

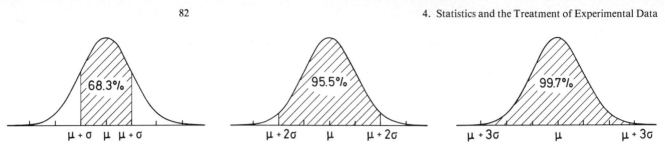

Fig. 4.5. The area contained between the limits $\mu \pm 1\sigma$, $\mu \pm 2\sigma$ and $\mu \pm 3\sigma$ in a Gaussian distribution

An important practical note is the area under the Gaussian between integral intervals of σ. This is shown in Fig. 4.5. These values should be kept in mind when interpreting measurement errors. The presentation of a result as $x \pm \sigma$ signifies, in fact, that the true value has $\simeq 68\%$ probability of lying between the limits $x - \sigma$ and $x + \sigma$ or a 95% probability of lying between $x - 2\sigma$ and $x + 2\sigma$, etc. Note that for a 1σ interval, there is almost a 1/3 probability that the true value is outside these limits! If two standard deviations are taken, then, the probability of being outside is only $\simeq 5\%$, etc.

4.2.4 The Chi-Square Distribution

As we will see in Sect. 4.7, the *chi-square* distribution is particulary useful for testing the *goodness-of-fit* of theoretical formulae to experimental data. Mathematically, the chi-square is defined in the following manner. Suppose we have a set of n independent random variables, x_i, distributed as Gaussian densities with theoretical means μ_i and standard deviations σ_i, respectively. The sum

$$u = \sum_{i=1}^{n} \left(\frac{x_i - \mu_i}{\sigma_i} \right)^2 \tag{4.22}$$

is then known as the *chi-square*. This is more often designated by the Greek letter χ^2; however, to avoid confusion due to the exponent we will use $u = \chi^2$ instead. Since x_i is a random variable, u is also a random variable and it can be shown to follow the distribution

$$P(u) = \frac{(u/2)^{(\nu/2)-1} \exp(-u/2)}{2\,\Gamma(\nu/2)}, \tag{4.23}$$

where ν is an integer and $\Gamma(\nu/2)$ is the *gamma function*. The integer ν is known as the *degrees of freedom* and is the sole parameter of the distribution. Its value thus determines the form of the distribution. The *degrees of freedom* can be interpreted as a parameter related to the number of independent variables in the sum (4.22).

Figure 4.6 plots the chi-square distribution for various values of ν. The mean and variance of (4.23) can also be shown to be

$$\mu = \nu, \quad \sigma^2 = 2\nu. \tag{4.24}$$

To see what the chi-square represents, let us examine (4.22) more closely. Ignoring the exponent for a moment, each term in the sum is just the deviation of x_i from its theoretical mean divided by its expected dispersion. The chi-square thus characterizes

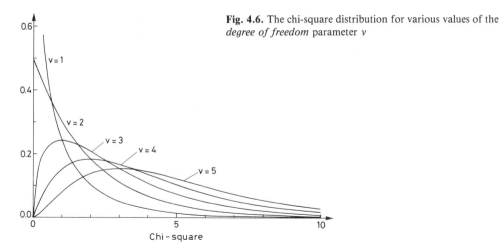

Fig. 4.6. The chi-square distribution for various values of the *degree of freedom* parameter v

the fluctuations in the data x_i. If indeed the x_i are distributed as Gaussians with the parameters indicated, then on the average, each ratio should be about 1 and the chi-square, $u = v$. For any given set of x_i, of course, there will be a fluctuation of u from this mean with a probability given by (4.23). The utility of this distribution is that it can be used to test hypotheses. By forming the chi-square between measured data and an assumed theoretical mean, a measure of the *reasonableness* of the fluctuations in the measured data about this hypothetical mean can be obtained. If an improbable chi-square value is obtained, one must then begin questioning the theoretical parameters used.

4.3 Measurement Errors and the Measurement Process

Measurements of any kind, in any experiment, are always subject to uncertainties or *errors*, as they are more often called. We will argue in this section that the measurement process is, in fact, a random process described by an abstract probability distribution whose parameters contain the information desired. The results of a measurement are then samples from this distribution which allow an estimate of the theoretical parameters. In this view, measurement errors can be seen then as sampling errors.

Before going into this argument, however, it is first necessary to distinguish between two types of errors: *systematic* and *random*.

4.3.1 Systematic Errors

Systematic errors are uncertainties in the bias of the data. A simple example is the zeroing of an instrument such as a voltmeter. If the voltmeter is not correctly zeroed before use, then all values measured by the voltmeter will be biased, i.e., offset by some constant amount or factor. However, even if the utmost care is taken in setting the instrument to zero, one can only say that it has been zeroed to within some value. This value may be small, but it sets a limit on the degree of certainty in the measurements and thus to the conclusions that can be drawn.

An important point to be clear about is that a systematic error implies that all measurements in a set of data taken with the same instrument are shifted in the same direction by the same amount − in unison. This is in sharp contrast to random errors where each individual measurement fluctuates independently of the others. Systematic errors, therefore, are usually most important when groups of data points taken under the same conditions are being considered. Unfortunately, there is no consistent method by which systematic errors may be treated or analyzed. Each experiment must generally be considered individually and it is often very difficult just to identify the possible sources let alone estimate the magnitude of the error. Our discussion in the remainder of this chapter, therefore, will not be concerned with this topic.

4.3.2 Random Errors

In contrast to systematic errors, random errors may be handled by the theory of statistics. These uncertainties may arise from instrumental imprecisions, and/or, from the inherent statistical nature of the phenomena being observed. Statistically, both are treated in the same manner as uncertainties arising from the finite sampling of an infinite population of events. The measurement process, as we have suggested, is a sampling process much like an opinion poll. The experimenter attempts to determine the parameters of a population or distribution too large to measure in its entirety by taking a random sample of finite size and using the sample parameters as an estimate of the true values.

This point of view is most easily seen in measurements of statistical processes, for example, radioactive decay, proton-proton scattering, etc. These processes are all governed by the probabilistic laws of quantum mechanics, so that the number of disintegrations or scatterings in a given time period is a random variable. What is usually of interest in these processes is the mean of the theoretical probability distribution. When a measurement of the number of decays or scatterings per unit time is made, a sample from this distribution is taken, i.e., the variable x takes on a value x_1. Repeated measurements can be made to obtain x_2, x_3, etc. This, of course, is equivalent to tossing a coin or throwing a pair of dice and recording the result. From these data, the experimenter may estimate the value of the mean. Since the sample is finite, however, there is an uncertainty on the estimate and this represents our measurement error. Errors arising from the measurement of inherently random processes are called *statistical* errors.

Now consider the measurement of a quantity such as the length of a table or the voltage between two electrodes. Here the quantities of interest are well-defined numbers and not random variables. How then do these processes fit into the view of measurement as a sampling process? What distribution is being sampled?

To take an example, consider an experiment such as the measurement of the length of a table with say, a simple folding ruler. Let us make a set of repeated measurements reading the ruler as accurately as possible. (The reader can try this himself!). It will then be noticed that the values fluctuate about and indeed, if we plot the frequency of the results in the form of a histogram, we see the outlines of a definite distribution beginning to take form. The differing values are the result of many small factors which are not controlled by the experimenter and which may change from one measurement to the next, for example, play in the mechanical joints, contractions and expansions due to temperature changes, failure of the experimenter to place the zero at exactly the same point each time, etc. These are all sources of *instrumental error*, where the term *instrument* also includes the observer! The more these factors are taken under control, of course, the smaller will be the magnitude of the fluctuations. The instrument is then said to be more *precise*. In the limit of an ideal, perfect instrument, the distribution then becomes a δ-function centered at the true value of the measured quantity. In reality, of course, such is never the case.

The measurement of a fixed quantity, therefore, involves taking a sample from an abstract, theoretical distribution determined by the imprecisions of the instrument. In almost all cases of instrumental errors, it can be argued that the distribution is Gaussian. Assuming no systematic error, the mean of the Gaussian should then be equal to the true value of the quantity being measured and the standard deviation proportional to the precision of the instrument.

Let us now see how sampled data are used to estimate the true parameters.

4.4 Sampling and Parameter Estimation.
The Maximum Likelihood Method

Sampling is the experimental method by which information can be obtained about the parameters of an unknown distribution. As is well known from the debate over opinion polls, it is important to have a representative and unbiased sample. For the experimentalist, this means *not* rejecting data because it does not "*look right*". The rejection of data, in fact, is something to be avoided unless there are overpowering reasons for doing so.

Given a data sample, one would then like to have a method for determining the *best* value of the true parameters from the data. The *best* value here is that which minimizes the variance between the estimate and the true value. In statistics, this is known as *estimation*. The estimation problem consists of two parts: (1) determining the best estimate and (2) determining the uncertainty on the estimate. There are a number of different principles which yield formulae for combining data to obtain a best estimate. However, the most widely accepted method and the one most applicable to our purposes is the principle of *maximum likelihood*. We shall very briefly demonstrate this principle in the following sections in order to give a feeling for how the results are derived. The reader interested in more detail or in some of the other methods should consult some of the standard tests given in the bibliography. Before treating this topic, however, we will first define a few terms.

4.4.1 Sample Moments

Let $x_1, x_2, x_3, \ldots, x_n$ be a sample of size n from a distribution whose theoretical mean is μ and variance σ^2. This is known as the sample population. The *sample mean*, $\bar{x}$, is then defined as

$$\bar{x} = \frac{1}{n} \sum_{i=1}^{n} x_i, \tag{4.25}$$

which is just the arithmetic average of the sample. In the limit $n \to \infty$, this can be shown to approach the theoretical mean,

$$\mu = \lim_{n \to \infty} \frac{1}{n} \sum_{i=1}^{n} x_i .$$ (4.26)

Similarly, the *sample variance*, which we denote by s^2 is

$$s^2 = \frac{1}{n} \sum_{i=1}^{n} (x_i - \bar{x})^2 ,$$ (4.27)

which is the average of the squared deviations. In the limit $n \to \infty$, this also approaches the theoretical variance σ^2.

In the case of multivariate samples, for example, (x_1, y_1), (x_2, y_2), ..., the sample means and variances for each variable are calculated as above. In an analogous manner, the *sample covariance* can be calculated by

$$\mathrm{cov}(x, y) = \frac{1}{n} \sum_{i=1}^{n} (x_i - \bar{x})(y_i - \bar{y}) .$$ (4.28)

In the limit of infinite n, (4.28), not surprisingly, also approaches the theoretical covariance (4.10).

4.4.2 The Maximum Likelihood Method

The method of maximum likelihood is only applicable if the form of the theoretical distribution from which the sample is taken is known. For most measurements in physics, this is either the Gaussian or Poisson distribution. But, to be more general, suppose we have a sample of n independent observations x_1, x_2, ..., x_n, from a theoretical distribution $f(x|\theta)$ where θ is the parameter to be estimated. The method then consists of calculating the *likelihood* function,

$$L(\theta|x) = f(x_1|\theta)f(x_2|\theta) \ldots f(x_n|\theta) ,$$ (4.29)

which can be recognized as the probability for observing the sequence of values x_1, x_2, ..., x_n. The principle now states that this probability is a maximum for the observed values. Thus, the parameter θ must be such that L is a maximum. If L is a regular function, θ can be found by solving the equation,

$$\frac{dL}{d\theta} = 0 .$$ (4.30)

If there is more than one parameter, then the partial derivatives of L with respect to each parameter must be taken to obtain a system of equations. Depending on the form of L, it may also be easier to maximize the logarithm of L rather than L itself. Solving the equation

$$\frac{d(\ln L)}{d\theta} = 0$$ (4.31)

then yields results equivalent to (4.30). The solution, $\hat{\theta}$, is known as the maximum likelihood *estimator* for the parameter θ. In order to distinguish the estimated value from the true value, we have used a caret over the parameter to signify it as the estimator.

It should be realized now that $\hat{\theta}$ is also a random variable, since it is a function of the x_i. If a second sample is taken, $\hat{\theta}$ will have a different value and so on. The estimator is thus also described by a probability distribution. This leads us to the second half of the estimation problem: What is the error on $\hat{\theta}$? This is given by the standard deviation of the estimator distribution. We can calculate this from the likelihood function if we recall that L is just the probability for observing the sampled values $x_1, x_2, \ldots, x_n$. Since these values are used to calculate $\hat{\theta}$, L is related to the distribution for $\hat{\theta}$. Using (4.9), the variance is then

$$\sigma^2(\hat{\theta}) = \int (\hat{\theta} - \theta)^2 L(\theta|x)\, dx_1\, dx_2 \ldots dx_n .\tag{4.32}$$

This is a general formula, but, unfortunately, only in a few simple cases can an analytic result be obtained. An easier, but only approximate method which works in the limit of large numbers, is to calculate the inverse second derivative of the log-likelihood function evaluated at the maximum,

$$\sigma^2(\hat{\theta}) \simeq -\left(\frac{d^2 \ln L}{d\theta^2} \right)^{-1} .\tag{4.33}$$

If there is more than one parameter, the matrix of the second derivatives must be formed, i.e.,

$$U_{ij} = -\frac{\partial^2 \ln L}{\partial \theta_i\, \partial \theta_j} .\tag{4.34}$$

The diagonal elements of the inverse matrix then give the approximate variances,

$$\sigma^2(\hat{\theta}_i) \simeq (U^{-1})_{ii} .\tag{4.35}$$

A technical point which must be noted is that we have assumed that the mean value of $\hat{\theta}$ is the theoretical θ. This is a desirable, but not essential property for an estimator, guaranteed by the maximum likelihood method only for infinite n. Estimators which have this property are *non-biased*. We will see one example in the following sections in which this is not the case. Equation (4.32), nevertheless, remains valid for all $\hat{\theta}$, since the error desired is the deviation from the true mean irrespective of the bias.

Another useful property of maximum likelihood estimators is invariance under transformations. If $u = f(\theta)$, then the best estimate of u can be shown to be $\hat{u} = f(\hat{\theta})$.

Let us illustrate the method now by applying it to the Poisson and Gaussian distributions.

4.4.3 Estimator for the Poisson Distribution

Suppose we have n measurements of samples, $x_1, x_2, x_3, \ldots, x_n$ from a Poisson distribution with mean μ. The likelihood function for this case is then

$$L(\mu|x) = \prod_{i=1}^{n} \frac{\mu^{x_i}}{x_i!} \exp(-\mu) = \exp(-n\mu) + \prod_{i=1}^{n} \frac{\mu^{x_i}}{x_i!} .\tag{4.36}$$

To eliminate the product sign, we take the logarithm

$$L^* = \ln L = -n\mu + \sum x_i \ln \mu - \sum \ln x_i! \ . \tag{4.37}$$

Differentiating and setting the result to zero, we when find

$$\frac{dL^*}{d\mu} = -n + \frac{1}{\mu} \sum x_i = 0 \ , \tag{4.38}$$

which yields the solution

$$\hat{\mu} = \frac{1}{n} \sum x_i = \bar{x} \ . \tag{4.39}$$

Equation (4.39), of course, is just the sample mean. This is of no great surprise, but it does confirm the often unconscious use of (4.39).

The variance of $\bar{x}$ can be found by using (4.33); however, in this particular case, we will use a different way. From (4.9) we have the definition

$$\sigma^2(\bar{x}) = E[(\bar{x} - \mu)^2] \ . \tag{4.40}$$

Applying this to the sample mean and rearranging the terms, we thus have

$$E[(\bar{x} - \mu)^2] = E\left[\left(\frac{1}{n} \sum x_i - \mu\right)^2\right] = \frac{1}{n^2} E[(\sum x_i - n\mu)^2] = \frac{1}{n^2} E[\{\sum (x_i - \mu)\}^2] \ . \tag{4.41}$$

Expanding the square of the sum, we find

$$[\sum (x_i - \mu)]^2 = \sum_i (x_i - \mu)^2 + \sum_{i \neq j} \sum (x_i - \mu)(x_j - \mu) \ . \tag{4.42}$$

If now the expectation value is taken, the cross term vanishes, so that

$$\sigma^2(\bar{x}) = \frac{1}{n^2} E[\sum (x_i - \mu)^2] = \frac{1}{n^2} \sum E[(x_i - \mu)^2] = \frac{n\sigma^2}{n^2} = \frac{\sigma^2}{n} \ . \tag{4.43}$$

As the reader may have noticed, (4.43) was derived without reference to the Poisson distribution, so that (4.43) is, in fact, a general result: the variance of the sample mean is given by the variance of the parent distribution, whatever it may be, divided by the sample size.

For a Poisson distribution, $\sigma^2 = \mu$, so that the error on the estimated Poisson mean is

$$\sigma(\hat{\mu}) = \sqrt{\frac{\mu}{n}} \simeq \sqrt{\frac{\hat{\mu}}{n}} = \sqrt{\frac{\bar{x}}{n}} \tag{4.44}$$

where have substituted the estimated value $\hat{\mu}$ for the theoretical μ.

4.4.4 Estimators for the Gaussian Distribution

For a sample of n points, all taken from the same Gaussian distribution, the likelihood function is

$$L = \prod_{i=1}^{n} \frac{1}{\sigma\sqrt{2\pi}} \exp\left[-\frac{(x_i-\mu)^2}{2\sigma^2}\right]. \tag{4.45}$$

Once again, taking the logarithm,

$$L^* = \ln L = -\frac{n}{2}\ln(2\pi\sigma^2) - \frac{1}{2}\sum\frac{(x_i-\mu)^2}{\sigma^2}. \tag{4.46}$$

Taking the derivatives with respect to μ and σ^2 and setting them to 0, we then have

$$\frac{\partial L^*}{\partial\mu} = \sum\frac{x_i-\mu}{\sigma^2} = 0 \tag{4.47}$$

and

$$\frac{\partial L^*}{\partial\sigma^2} = -\frac{n}{2\sigma^2} + \frac{1}{2}\sum\left(\frac{x_i-\mu}{\sigma}\right)^2\frac{1}{\sigma^2} = 0. \tag{4.48}$$

Solving (4.47) first yields

$$\hat{\mu} = \frac{\sum x_i}{n} = \bar{x}. \tag{4.49}$$

The best estimate of the theoretical mean for a Gaussian is thus the sample mean, which again comes as no great surprise.

From the general result in (4.43), the uncertainty on the estimator is thus

$$\sigma(\bar{x}) = \frac{\sigma}{\sqrt{n}}. \tag{4.50}$$

This is usually referred to as the *standard error of the mean*. Note that the error depends on the sample number as one would expect. As n increases, the estimate $\bar{x}$ becomes more and more precise. When only one measurement is made, $n = 1$, $\sigma(\bar{x})$ reduces to σ. For a measuring device, σ thus represents the *precision* of the instrument.

For the moment, however, σ is still unknown. Solving (4.48) for σ^2 yields the estimator

$$\hat{\sigma}^2 = \frac{1}{n}\sum(x_i-\mu)^2 \approx \frac{1}{n}\sum(x_i-\bar{x})^2 = s^2, \tag{4.51}$$

where we have replaced μ by its solution in (4.50). This, of course, is just the sample variance.

For finite values of n, however, the sample variance turns out to be a *biased estimator*, that is the expectation value of s^2 does not equal the true value, but is offset from it by a constant factor. It is not hard to show, in fact, that $E[s^2] = \sigma^2 - \sigma^2/n = (n-1)\sigma^2/n$. Thus for n very large, s^2 approaches the true variance as desired; however, for small n, σ^2 is underestimated by s^2. The reason is quite simple: for small samples, the occurrence of large values far from the mean is rare, so the sample variance tends to be weighted more towards smaller values. For practical use, a somewhat better estimate therefore, would be to multiply (4.51) by the factor $n/(n-1)$,

$$\hat{\sigma}^2 = \frac{\sum (x_i - \bar{x})^2}{n-1} . \tag{4.52}$$

Equation (4.52) is unbiased, however, it is no longer the *best* estimate in the sense that its average deviation from the true value is somewhat greater than that for (4.51). The difference is small however, so that (4.52) still provides a good estimate. Equation (4.52) then is the recommended formula for estimating the variance. Note that unlike the mean, it is impossible to estimate the standard deviation from one measurement because of the $(n-1)$ term in the denominator. This makes sense, of course, as it quite obviously requires more than one point to determine a dispersion!

The variance of $\hat{\sigma}^2$ in (4.52) may also be shown to be

$$\sigma^2(\hat{\sigma}^2) = \frac{2\sigma^4}{n-1} \simeq \frac{2\,\hat{\sigma}^4}{n-1} \tag{4.53}$$

and standard deviation of $\hat{\sigma}$

$$\sigma(\hat{\sigma}) = \frac{\sigma}{\sqrt{2(n-1)}} \simeq \frac{\hat{\sigma}}{\sqrt{2(n-1)}} . \tag{4.54}$$

4.4.5 The Weighted Mean

We have thus far discussed the estimation of the mean and standard deviation from a series of measurements of the same quantity with the same instrument. If often occurs, however, that one must combine two or more measurements of the same quantity with differing errors. A simple minded procedure would be to take the average of the measurements. This, unfortunately, ignores the fact that some measurements are more precise than others and should therefore be given more importance. A more valid method would be to weight each measurement in proportion to its error. The maximum likelihood method allows us to determine the weighting function to use.

From a statistics point of view, we thus have a sample $x_1, x_2, \ldots, x_n$, where each value is from a Gaussian distribution having the same mean μ but a different standard deviation σ_i. The likelihood function is thus the same as (4.45), but with σ replaced by σ_i. Maximizing this we then find the *weighted mean*

$$\hat{\mu} = \frac{\sum x_i/\sigma_i^2}{\sum 1/\sigma_i^2} . \tag{4.55}$$

Thus the weighting factor is the inverse square of the error, i.e., $1/\sigma_i^2$. This corresponds to our logic as the smaller the σ_i, the larger the weight and vice-versa.

Using (4.33), the error on the weighted mean can now be shown to be

$$\sigma^2(\hat{\mu}) = \frac{1}{\sum 1/\sigma_i^2} . \tag{4.56}$$

Note that if all the σ_i are the same, the weighted mean reduces to the normal formula in (4.49) and the error on the mean to (4.50).

4.5 Examples of Applications

4.5.1 Mean and Error from a Series of Measurements

Example 4.1 Consider the simple experiment proposed in Sect. 4.3.2 to measure the length of an object. The following results are from such a measurement:

17.62	17.62	17.615	17.62	17.61
17.61	17.62	17.625	17.62	17.62
17.61	17.615	17.61	17.605	17.61

What is the best estimate for the length of this object?

Since the errors in the measurement are instrumental, the measurements are Gaussian distributed. From (4.49), the best estimate for the mean value is then

$$\bar{x} = 17.61533$$

while (4.52) gives the standard deviation

$$\hat{\sigma} = 5.855 \times 10^{-3} .$$

This can now be used to calculate the standard error of the mean (4.50),

$$\sigma(\bar{x}) = \hat{\sigma}/\sqrt{15} = 0.0015 .$$

The best value for the length of the object is thus

$$x = 17.616 \pm 0.002 .$$

Note that the uncertainty on the mean is given by the standard error of the mean and *not* the standard deviation!

4.5.2 Combining Data with Different Errors

Example 4.2 It is necessary to use the lifetime of the muon in a calculation. However, in searching through the literature, 7 values are found from different experiments:

2.198 ± 0.001 µs	2.203 ± 0.004 µs	2.202 ± 0.003 µs
2.197 ± 0.005 µs	2.198 ± 0.002 µs	2.1966 ± 0.0020 µs
2.1948 ± 0.0010 µs		

What is the best value to use?

One way to solve this problem is to take the measurement with the smallest error; however, there is no reason for ignoring the results of the other measurements. Indeed, even though the other experiments are less precise, they still contain valid information on the lifetime of the muon. To take into account all available information we must take the weighted mean. This then yields then mean value

$$\tau = 2.19696$$

with an error

$$\sigma(\tau) = 0.00061.$$

Note that this value is smaller than the error on any of the individual measurements. The best value for the lifetime is thus

$$\tau = 2.1970 \pm 0.0006 \ \mu s \ .$$

4.5.3 Determination of Count Rates and Their Errors

Example 4.3   Consider the following series of measurements of the counts per minute from a detector viewing a ^{22}Na source,

2201	2145	2222	2160	2300

What is the decay rate and its uncertainty?

Since radioactive decay is described by a Poisson distribution, we use the estimators for this distribution to find

$$\hat{\mu} = \bar{x} = 2205.6 \quad \text{and}$$

$$\sigma(\hat{\mu}) = \sqrt{\frac{\bar{x}}{n}} = \sqrt{\frac{2205.6}{5}} = 21 \ .$$

The count rate is thus

Count Rate $= (2206 \pm 21)$ counts/min.

It is interesting to see what would happen if instead of counting five one-minute periods we had counted the total 5 minutes without stopping. We would have then observed a total of 11 028 counts. This constitutes a sample of $n = 1$. The mean count rate for 5 minutes is thus 11 208 and the error on this, $\sigma = \sqrt{11\,208} = 106$. To find the counts per minute, we divide by 5 (see the next section) to obtain 2206 ± 21, which is identical to what was found before. Note that the error taken was the square root of the count rate in 5 minutes. A common error to be avoided is to first calculate the rate per minute and then take the square root of this number.

4.5.4 Null Experiments. Setting Confidence Limits When No Counts Are Observed

Many experiments in physics test the validity of certain theoretical conservation laws by searching for the presence of specific reactions or decays forbidden by these laws. In such measurements, an observation is made for a certain amount of time T. Obviously, if one or more events are observed, the theoretical law is disproven. However, if no events are observed, the converse cannot be said to be true. Instead a limit on the lifetime of the reaction or decay is set.

Let us assume therefore that the process has some mean reaction rate λ. Then the probability for observing no counts in a time period T is

$$P(0 \mid \lambda) = \exp(-\lambda T) \ . \tag{4.57}$$

This, now, can also be interpreted as the probability distribution for λ when no counts are observed in a period T. We can now ask the question: What is the probability that λ is less λ_0? From (4.1),

$$P(\lambda \leq \lambda_0) = \int_0^{\lambda_0} T \exp(-\lambda T) \, d\lambda = 1 - \exp(-\lambda_0 T) , \qquad (4.58)$$

where we have normalized (4.57) with the extra factor T. This probability is known as the *confidence level* for the interval between 0 to λ_0. To make a strong statement we can choose a high confidence level (CL), for example, 90%. Setting (4.58) equal to this probability then gives us the value of λ_0,

$$\lambda_0 = -\frac{1}{T} \ln(1 - CL) . \qquad (4.59)$$

For a given confidence level, the corresponding interval is, in general, not unique and one can find other intervals which yield the same integral probability. For example, it might be possible to integrate (4.57) from some lower limit λ' to infinity and still obtain the same area under the curve. The probability that the true λ is *greater* than λ' is then also 90%. As a general rule, however, one should take those limits which cover the smallest range in λ.

Example 4.4   A 50 g sample of ^{82}Se is observed for 100 days for neutrinoless double beta decay, a reaction normally forbidden by lepton conservation. However, current theories suggest that this might occur. The apparatus has a detection efficiency of 20%. No events with the correct signature for this decay are observed. Set an upper limit on the lifetime for this decay mode.

Choosing a confidence limit of 90%, (4.59) yields

$$\lambda \leq \lambda_0 = -\frac{1}{100 \times 0.2} \ln(1 - 0.9) = 0.115 \text{ day}^{-1} ,$$

where we have corrected for the 20% efficiency of the detector. This limit must now be translated into a lifetime per nucleus. For 50 g, the total number of nuclei is

$$N = \frac{N_a}{82} \times 50 = 3.67 \times 10^{23} ,$$

which implies a limit on the decay rate per nucleus of

$$\lambda \leq \frac{0.115}{3.67 \times 10^{23}} = 3.13 \times 10^{-25} \text{ day}^{-1} .$$

The lifetime is just the inverse of λ which yields

$$\tau \geq 8.75 \times 10^{21} \text{ years} \quad 90\% \text{ CL} ,$$

where we have converted the units to years. Thus, neutrinoless double beta decay may exist but it is certainly a rare process!

4.5.5 Distribution of Time Intervals Between Counts

A distribution which we will make use of later is the distribution of time intervals between events from a random source. Suppose we observe a radioactive source with a mean rate λ. The probability of observing no counts in a period T is then given by (4.57). In a manner similar to Sect. 4.5.4, we can interpret (4.57) as the probability density for the time interval T during which no counts are observed. Normalizing (4.57), we obtain the distribution

$$P(T) = \lambda \exp(\lambda T) \tag{4.60}$$

for the time T between counts. Equation (4.60) is just an exponential distribution and can, in fact, be measured.

4.6 Propagation of Errors

We have seen in the preceding sections how to calculate the errors on directly measured quantities. Very often, however, it is necessary to calculate other quantities from these data. Clearly, the calculated result will then contain an uncertainty which is carried over from the measured data.

To see how the errors are propagated, consider a quantity $u = f(x, y)$ where x and y are quantities having errors σ_x and σ_y, respectively. To simplify the algebra, we only consider a function of two variables here; however, the extension to more variables will be obvious. We would like then to calculate the standard deviation σ_u as a function of σ_x and σ_y. The variance σ_u^2 can be defined as

$$\sigma_u^2 = E[(u - \bar{u})^2] , \tag{4.61}$$

where $\bar{u} = f(\bar{x}, \bar{y})$, i.e., the mean value of u is obtained by substituting the mean values of x and y. To express the deviation of u in terms of the deviations in x and y, let us expand $(u - \bar{u})$ to first order

$$(u - \bar{u}) \simeq (x - \bar{x}) \left. \frac{\partial f}{\partial x} \right|_{\bar{x}} + (y - \bar{y}) \left. \frac{\partial f}{\partial y} \right|_{\bar{y}} , \tag{4.62}$$

where the partial derivatives are evaluated at the mean values. Squaring (4.62) and substituting into (4.61) then yields

$$E[(u - \bar{u})^2] \simeq E\left[(x - \bar{x})^2 \left(\frac{\partial f}{\partial x} \right)^2 + (y - \bar{y})^2 \left(\frac{\partial f}{\partial y} \right)^2 + 2(x - \bar{x})(y - \bar{y}) \frac{\partial f}{\partial x} \frac{\partial f}{\partial y} \right] . \tag{4.63}$$

Now taking the expectation value of each term separately and making use of the definitions (4.8, 9) and (4.10), we find

$$\sigma_u^2 \simeq \left(\frac{\partial f}{\partial x} \right)^2 \sigma_x^2 + \left(\frac{\partial f}{\partial y} \right)^2 \sigma_y^2 + 2 \operatorname{cov}(x, y) \frac{\partial f}{\partial x} \frac{\partial f}{\partial y} . \tag{4.64}$$

The errors therefore are added quadratically with a modifying term due to the covariance. Depending on its sign and magnitude, the covariance can increase or decrease the errors by dramatic amounts. In general most measurements in physics experiments are independent or should be arranged so that the covariance will be zero. Equation (4.64) then reduces to a simple sum of squares. Where correlations can arise, however, is when two or more parameters are extracted from the same set of measured data. While the raw data points are independent, the parameters will generally be correlated. One common example are parameters resulting from a fit. The correlations can be calculated in the fitting procedure and all good computer fitting programs should supply this information. An example is given in Sect. 4.7.2. If these parameters are used in a calculation, the correlation must be taken into account. A second example of this type which might have occurred to the reader is the estimation of the mean and variance from a set of data. Fortunately, it can be proved that the estimators (4.49) and (4.52) are statistically independent so that $\rho = 0$!

4.6.1 Examples

As a first example let us derive the formulas for the sum, difference, product and ratio of two quantities x and y with errors σ_x and σ_y.

i) Error of a Sum: $u = x + y$

$$\sigma_u^2 = \sigma_x^2 + \sigma_y^2 + 2 \operatorname{cov}(x, y) . \tag{4.65}$$

ii) Error of a Difference: $u = x - y$

$$\sigma_u^2 = \sigma_x^2 + \sigma_y^2 - 2 \operatorname{cov}(x, y) . \tag{4.66}$$

If the covariance is 0, the errors on both a sum and difference then reduce to the same sum of squares. The relative error, σ_u/u, however, is much larger for the case of a difference since u is smaller. This illustrates the disadvantage of taking differences between two numbers with errors. If possible, therefore, a difference should always be directly measured rather than calculated from two measurements!

iii) Error of a Product: $u = xy$

$$\sigma_u^2 \simeq y^2 \sigma_x^2 + x^2 \sigma_y^2 + 2 \operatorname{cov}(x, y) \, xy .$$

Dividing the left side by u^2 and the right side by $x^2 y^2$,

$$\frac{\sigma_u^2}{u^2} \simeq \frac{\sigma_x^2}{x^2} + \frac{\sigma_y^2}{y^2} + 2 \frac{\operatorname{cov}(x, y)}{xy} . \tag{4.67}$$

iv) Error of a Ratio: $u = x/y$

$$\sigma_u^2 \simeq y^{-2} \sigma_x^2 + x^2 y^{-4} \sigma_y^2 - 2 \operatorname{cov}(x, y) \, xy^{-3} .$$

Dividing both sides by u^2 as in (iii), we find

$$\frac{\sigma_u^2}{u^2} \simeq \frac{\sigma_x^2}{x^2} + \frac{\sigma_y^2}{y^2} - 2 \frac{\operatorname{cov}(x, y)}{xy} , \tag{4.68}$$

which, with the exception of the sign of the covariance term is identical to the formula for a product. Equation (4.68) is generally valid when the relative errors are not too large. For ratios of small numbers, however, (4.68) is inapplicable and some additional considerations are required. This is treted in detail by *James* and *Roos* [4.1].

Example 4.5   The classical method for measuring the polarization of a particle such as a proton or neutron is to scatter it from a suitable *analyzing* target and to measure the asymmetry in the scattered particle distribution. One can, for example, count the number of particles scattered to the left of the beam at certain angle and to the right of the beam at the same corresponding angle. If R is the number scattered to the right and L the number to the left, the asymmetry is then given by

$$\varepsilon = \frac{R-L}{R+L} .$$

Calculate the error on ε as a function of the counts R and L.

This is a straight forward application of (4.64). Taking ther derivatives of ε, we thus find

$$\frac{\partial \varepsilon}{\partial R} = \frac{1}{R+L} + \frac{R-L}{(R+L)^2} = \frac{2L}{N_{tot}^2}$$

$$\frac{\partial \varepsilon}{\partial L} = \frac{1}{R+L} + \frac{R-L}{(R+L)^2} = \frac{-2R}{N_{tot}^2} ,$$

where the total number of counts $N_{tot} = R + L$. The error is thus

$$\sigma^2(\varepsilon) \simeq \frac{4L^2}{N_{tot}^4} \sigma_R^2 + \frac{4R^2}{N_{tot}^4} \sigma_L^2 .$$

The covariance is obviously 0 here since the measurements are independent. The errors on R and L are now given by the Poisson distribution, so that $\sigma_R^2 = R$ and $\sigma_L^2 = L$. Substituting into the above, then yields

$$\sigma^2(\varepsilon) \simeq 4 \frac{(L^2 R + R^2 L)}{N_{tot}^4} = 4 \frac{RL}{N_{tot}^3} .$$

If the asymmetry is small such that $R \simeq L \simeq N_{tot}/2$, we have the result that

$$\sigma(\varepsilon) \simeq \sqrt{\frac{1}{N_{tot}}} .$$

4.7 Curve Fitting

In many experiments, the functional relation between two or more variables describing a physical process, $y = f(x_1, x_2, \ldots)$, is investigated by measuring the value of y for various of $x_1, x_2, \ldots$. It is then desired to find the parameters of a theoretical curve which best describe these points. For example, to determine the lifetime of a certain radioactive source, measurements of the count rates, $N_1, N_2, \ldots, N_n$, at various times, $t_1, t_2, \ldots, t_n$, could be made and the data fitted to the expression

$$N(t) = N_0 \exp(-t/\tau) . \tag{4.69}$$

Since the count rate is subject to statistical fluctuations, the values N_i will have uncertainties $\sigma_i = \sqrt{N_i}$ and will not all lie along a smooth curve. What then is the best curve or equivalently, the best values for τ and N_0 and how do we determine them? The method most useful for this is the method of *least squares*.

4.7.1 The Least Squares Method

Let us suppose that measurements at n points, x_i, are made of the variable y_i with an error σ_i ($i = 1, 2, \ldots, n$), and that it is desired to fit a function $f(x; a_1, a_2, \ldots, a_m)$ to these data where $a_1, a_2, \ldots, a_m$ are unknown parameters to be determined. Of course, the number of points must be greater than the number of parameters. The method of least squares states that the best values of a_j are those for which the sum

$$S = \sum_{i=1}^{n} \left[\frac{y_i - f(x_i; a_j)}{\sigma_i} \right]^2 \tag{4.70}$$

is a minimum. Examining (4.70) we can see that this is just the sum of the squared deviations of the data points from the curve $f(x_i)$ weighted by the respective errors on y_i. The reader might also recognize this as the chi-square in (4.22). For this reason, the method is also sometimes referred to as *chi-square minimization*. Strictly speaking this is not quite correct as y_i must be Gaussian distributed with mean $f(x_i; a_j)$ and variance σ_i^2 in order for S to be a true chi-square. However, as this is almost always the case for measurements in physics, this is a valid hypothesis most of the time. The least squares method, however, is totally general and does not require knowledge of the parent distribution. If the parent distribution is known the method of maximum likelihood may also be used. In the case of Gaussian distributed errors this yields identical results.

To find the values of a_j, one must now solve the system of equations

$$\frac{\partial S}{\partial a_j} = 0 . \tag{4.71}$$

Depending on the function $f(x)$, (4.71) may or may not yield on analytic solution. In general, numerical methods requiring a computer must be used to minimize S.

Assuming we have the best values for a_j, it is necessary to estimate the errors on the parameters. For this, we form the so-called covariance or *error matrix*, V_{ij},

$$(\underline{V}^{-1})_{ij} = \frac{1}{2} \frac{\partial^2 S}{\partial a_i \partial a_j} , \tag{4.72}$$

where the second derivative is evaluated at the minimum. (Note the second derivatives form the *inverse* of the error matrix). The diagonal elements V_{ij} can then be shown to be the variances for a_i, while the offdiagonal elements V_{ij} represent the covariances between a_i and a_j. Thus,

$$\underline{V} = \begin{pmatrix} \sigma_1^2 & \mathrm{cov}(1,2) & \mathrm{cov}(1,3) & \ldots \\ \cdot & \sigma_2^2 & \mathrm{cov}(2,3) & \ldots \\ \cdot & \cdot & \sigma_3^2 & \ldots \\ \cdot & \cdot & \cdot & \end{pmatrix} \tag{4.73}$$

and so on.

4.7.2 Linear Fits. The Straight Line

In the case of functions linear in their parameters a_j, i.e., there are no terms which are products or ratios of different a_j's, (4.71) can be solved analytically. Let us illustrate this for the case of a straight line

$$y = f(x) = ax + b , \qquad (4.74)$$

where a and b are the parameters to be determined. Forming S, we find

$$S = \sum \frac{(y_i - ax_i - b)^2}{\sigma_i^2} . \qquad (4.75)$$

Taking the partial derivatives with respect to a and b, we then have the equations

$$\frac{\partial S}{\partial a} = -2 \sum \frac{(y_i - ax_i - b)x_i}{\sigma_i^2} = 0 ,$$

$$\frac{\partial S}{\partial b} = -2 \sum \frac{(y_i - ax_i - b)}{\sigma_i^2} = 0 . \qquad (4.76)$$

To simplify the notation, let us define the terms

$$A = \sum \frac{x_i}{\sigma_i^2} \qquad B = \sum \frac{1}{\sigma_i^2}$$

$$C = \sum \frac{y_i}{\sigma_i^2} \qquad D = \sum \frac{x_i^2}{\sigma_i^2} \qquad (4.77)$$

$$E = \sum \frac{x_i y_i}{\sigma_i^2} \qquad F = \sum \frac{y_i^2}{\sigma_i^2} .$$

Using these definitions, (4.76) becomes

$$2(-E + aD + bA) = 0 ,$$

$$2(-C + aA + bB) = 0 . \qquad (4.78)$$

This then leads to the solution,

$$a = \frac{EB - CA}{DB - A^2} \qquad b = \frac{DC - EA}{DB - A^2} . \qquad (4.79)$$

Our work is not complete, however, as the errors on a and b must also be determined. Forming the inverse error matrix, we then have

$$V^{-1} = \begin{pmatrix} A_{11} & A_{12} \\ A_{21} & A_{22} \end{pmatrix} , \qquad \text{where} \qquad (4.80)$$

$$A_{11} = \frac{1}{2} \frac{\partial^2 S}{\partial a^2} , \qquad A_{22} = \frac{1}{2} \frac{\partial^2 S}{\partial b^2} , \qquad A_{12} = A_{21} = \frac{1}{2} \frac{\partial^2 S}{\partial a \, \partial b} .$$

Inverting (4.80), we find

$$V = \frac{1}{A_{11}A_{22} - A_{12}^2} \begin{pmatrix} A_{22} & -A_{12} \\ -A_{12} & A_{11} \end{pmatrix} \tag{4.81}$$

so that

$$\sigma^2(a) = \frac{A_{22}}{A_{11}A_{22} - A_{12}^2} = \frac{B}{BD - A^2},$$

$$\sigma^2(b) = \frac{A_{11}}{A_{11}A_{22} - A_{12}^2} = \frac{D}{BD - A^2}, \tag{4.82}$$

$$\text{cov}(a, b) = \frac{-A_{12}}{A_{11}A_{22} - A_{12}^2} = \frac{-A}{BD - A^2}.$$

To complete the process, now, it is necessary to also have an idea of the *quality of the fit*. Do the data, in fact, correspond to the function $f(x)$ we have assumed? This can be tested by means of the chi-square. This is just the value of S at the minimum. Recalling Sect. 4.2.4, we saw that if the data correspond to the function and the deviations are Gaussian, S should be expected to follow a chi-square distribution with mean value equal to the degrees of freedom, v. In the above problem, there are n independent data points from which m parameters are extracted. The degrees of freedom is thus $v = n - m$. In the case of a linear fit, $m = 2$, so that $v = n - 2$. We thus expect S to be close to $v = n - 2$ if the fit is *good*. A quick and easy test is to form the *reduced chi-square*

$$\frac{\chi^2}{v} = \frac{S}{v}, \tag{4.83}$$

which should be close to 1 for a *good fit*.

A more rigorous test is to look at the probability of obtaining a χ^2 value greater than S, i.e., $P(\chi^2 \geq S)$. This requires integrating the chi-square distribution or using cumulative distribution tables. In general, if $P(\chi^2 \geq S)$ is greater than 5%, the fit can be accepted. Beyond this point, some questions must be asked.

An equally important point to consider is when S is very small. This implies that the points are *not* fluctuating enough. Barring falsified data, the most likely cause is an overestimation of the errors on the data points. If the reader will recall, the error bars represent a 1σ deviation, so that about 1/3 of the data points should, in fact, be expected to fall outside the fit!

Example 4.6 Find the best straight line through the following measured points

x	0	1	2	3	4	5
y	0.92	4.15	9.78	14.46	17.26	21.90
σ	0.5	1.0	0.75	1.25	1.0	1.5

Applying (4.75) to (4.82), we find

$$a = 4.227 \qquad b = 0.878$$

$$\sigma(a) = 0.044 \qquad \sigma(b) = 0.203 \qquad \text{and}$$

$$\text{cov}(a, b) = -0.0629.$$

To test the *goodness-of-fit*, we must look at the chi-square

$$\chi^2 = 2.078$$

for 4 degrees of freedom. Forming the reduced chi-square, $\chi^2/\nu \simeq 0.5$, we can see already that his is a good fit. If we calculate the probability $P(\chi^2 > 2.07)$ for 4 degrees of freedom, we find $P \simeq 97.5\%$ which is well within acceptable limits.

Example 4.7 For certain nonlinear functions, a *linearization* may be effected so that the method of linear least squares becomes applicable. One case is the example of the exponential, (4.69), which we gave at the beginning of this section. Consider a decaying radioactive source whose activity is measured at intervals of 15 seconds. The total counts during each period are given below.

t [s]	1	15	30	45	60	75	90	105	120	135
N[cts]	106	80	98	75	74	73	49	38	37	22

What is the lifetime for this source?

The obvious procedure is to fit (4.69) to these data in order to determine τ. Equation (4.69), of course, is nonlinear, however it can belinearized by taking the logarithm of both sides. This then yields

$$\ln N = \frac{-t}{\tau} + \ln N_0 .$$

Setting $y = \ln N$, $a = -1/\tau$ and $b = \ln N_0$, we see that this is just a straight line, so that our linear least-squares procedure can be used. One point which we must be careful about, however, is the errors. The statistical errors on N, of course, are Poissonian, so that $\sigma(N) = \sqrt{N}$. In the fit, however, it is the logarithm of N which is being used. The errors must therefore be transformed using the propagation of errors formula; we then have

$$\sigma^2(\ln N) = \left(\frac{\partial \ln N}{\partial N}\right)^2 \sigma^2(N) = \frac{1}{N^2} N = N^{-1} .$$

Using (4.75) to (4.82) now, we find

$$a = -1/\tau = -0.008999 \qquad \sigma(a) = 0.001$$
$$b = \ln N_0 = 4.721 \qquad \sigma(b) = 0.064.$$

The lifetime is thus

$$\tau = 111.1 \pm 0.1 \text{ s}.$$

The chi-square for this fit is $\chi^2 = 15.6$ with 8 degrees of freedom. The reduced chi-square is thus $15.6/(8 \simeq 1.96$, which is somewhat high. If we calculate the probability $P(\chi^2 > 15) \simeq 0.05$, however, we find that the fit is just acceptable. The data and the best straight line are sketched in Fig. 4.7 on a semi-log plot.

While the above fit is acceptable, the relatively large chi-square should, nevertheless, prompt some questions. For example, in the treatment above, background counts

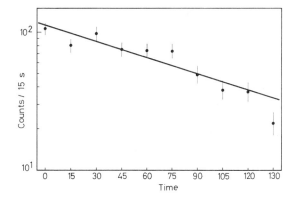

Fig. 4.7. Fit to data of Example 4.7. Note that the error bars of about 1/3 of the points do not touch the fitted line. This is consistent with the Gaussian nature of the measurements. Since the region defined by the errors bars ($\pm 1\sigma$) comprises 68% of the Gaussian distribution (see Fig. 4.5), there is a 32% chance that a measurement will exceed these limits!

were ignored. An improvement in our fit might therefore be obtained if we took this into account. If we assume a constant background, then the equation to fit would be

$$N(t) = N_0 \exp(-t/\tau) + C.$$

Another hypothesis could be that the source has more than one decay component in which case the function to fit would be a sum of exponentials. These forms unfortunately cannot be linearized as above and recourse must be made to nonlinear methods. In the special case described above, a non-iterative procedure [4.2 – 6] exists which may also be helpful.

4.7.3 Linear Fits When Both Variables Have Errors

In the previous examples, it was assumed that the independent variables x_i were completely free of errors. Strictly speaking, of course, this is never the case, although in many problems the errors on x are small with respect to those on y so that they may be neglected. In cases where the errors on both variables are comparable, however, ignoring the errors on x leads to incorrect parameters and an underestimation of their errors. For these problems the *effective variance method* may be used. Without deriving the result which is discussed by *Lybanon* [4.7] and *Orear* [4.8] for Gaussian distributed errors, the method consists of simply replacing the variance σ_i^2 in (4.70) by

$$\sigma_i^2 \rightarrow \sigma_y^2 + \left(\frac{df}{dx}\right)^2 \sigma_x^2, \tag{4.84}$$

where σ_x and σ_y are the errors on x and y respectively. Since the derivative is normally a function of the parameters a_j, S is nonlinear and numerical methods must be used to minimize S.

4.7.4 Nonlinear Fits

As we have already mentioned, nonlinear fits generally require a numerical procedure for minimizing S. Function minimization or maximization[2] is a problem unto itself and

---

[2] A minimization can be turned into a maximization by simply adding a minus sign in front of the function and vice-versa.

a number of methods have been developed for this purpose. However, no one method can be said to be applicable to all functions, and, indeed, the problem often is to find the right method for the function or functions in question. A discussion of the different methods available, of course, is largely outside the scope of this book. However, it is nevertheless worthwhile to briefly survey the most used methods so as to provide the reader with a basis for more detailed study and an idea of some of the problems to be expected. For practical purposes, a computer is necessary and we strongly advise the reader to find a ready-made program rather than attempt to write it himself. More detailed discussions may be found in [4.9 – 11].

Function Minimization Techniques. Numerical minimization methods are generally iterative in nature, i.e., repeated calculations are made while varying the parameters in some way, until the desired minimum is reached. The criteria for selecting a method, therefore, are speed and stability against divergences. In general, the methods can be classified into two broad categories: *grid* searches and *gradient* methods.

The grid methods are the simplest. The most elementary procedure is to form a grid of equally spaced points in the variables of interest and evaluate the function at each of these points. The point with the smallest value is then the approximate minimum. Thus, if $F(x)$ is the function to minimize, we would evaluate F at x_0, $x_0 + \Delta x$, $x_0 + 2\Delta x$, etc. and choose the point x' for which F is smallest. The size of the grid step, Δx, depends on the accuracy desired. This method, of course, can only be used over a finite range of x and in cases where x ranges over infinity it is necessary to have an approximate idea of where the minimum is. Several ranges may also be tried.

The elementary grid method is intrinsically stable, but it is quite obviously inefficient and time consuming. Indeed, in more than one dimension, the number of function evaluations becomes prohibitively large even for a computer! (In contrast to the simple grid method is the *Monte Carlo* or *random* search. Instead of equally spaced points, random points are generated according to some distribution, e.g., a uniform density.)

More efficient grid searches make use of variable stepping methods so as to reduce the number of evaluations while converging onto the minimum more rapidly. A relatively recent technique is the *simplex* method [4.12]. A simplex is the simplest geometrical figure in n dimensions having $n+1$ vertices. In $n = 2$ dimensions, for example, the simplex is a triangle while in $n = 3$ dimensions, the simplex is a tetrahedron, etc. The method takes on the name simplex because it uses $n+1$ points at each step. As an illustration, consider a function in two dimensions, the contours of which are shown in Fig. 4.8. The method begins by choosing $n+1 = 3$ points in some way or another, perhaps at random. A simplex is thus formed as shown in the figure. The point with the highest value is denoted as P_H, while the lowest is P_L. The next step is to replace P_H with a better point. To do this, P_H is reflected through the center of gravity of all points except P_H, i.e., the point

$$\bar{P} = \sum_i \frac{P_i - P_H}{n} .$$ (4.85)

This yields the point $P^* = \bar{P} + (\bar{P} - P_H)$. If $F(P^*) < F(P_L)$, a new minimum has been found and an attempt is made to do even better by trying the point $P^{**} = \bar{P} + 2(\bar{P} - P_H)$. The best point is then kept. If $F(P^*) > F(P_H)$ the reflection is brought backwards to $P^{**} = \bar{P} - \frac{1}{2}(\bar{P} - P_H)$. If this is not better than P_H, a new

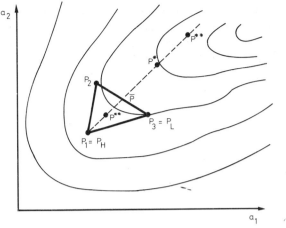

Fig. 4.8. The simplex method for function minimization

simplex is formed with points at $P_i = (P_i + P_L)/2$ and the procedure restarted. In this manner, one can imagine the triangle in Fig. 4.8 "*falling*" to the minimum. The simplex technique is a good method since it is relatively insensitive to the type of function, but it can also be rather slow.

Gradient methods are techniques which make use of the derivatives of the function to be minimized. These can either be calculated numerically or supplied by the user if known. One obvious use of the derivatives is to serve as guides pointing in the direction of decreasing F. This idea is used in techniques such as the method of *steepest descent*. A more widely used technique, however, is *Newton's method* which uses the derivatives to form a second-degree Taylor expansion of the function about the point x_0,

$$F(x) \simeq F(x_0) + \frac{\partial F}{\partial x}\bigg|_{x_0} (x - x_0) + \frac{1}{2} \frac{\partial^2 F}{\partial x^2}\bigg|_{x_0} (x - x_0)^2 \,. \tag{4.86}$$

In n dimensions, this is generalized to

$$F(x) \simeq F(x_0) + g^T(x - x_0) + \tfrac{1}{2}(x - x_0)^T G(x - x_0) \,, \tag{4.87}$$

where g_i is the vector of first derivatives $\partial F/\partial x_i$ and G_{ij} the matrix of second derivatives $\partial^2 F/\partial x_i \partial x_j$. The matrix G is also called the *hessian*. In essence, the method approximates the function around x_0 by a quadratic surface. Under this assumption, it is very easy to calculate the minimum of the n dimensional parabola analytically,

$$x_{\min} = x_0 - G^{-1} g \,. \tag{4.88}$$

This, of course, is not the true minimum of the function; but by forming a new parabolic surface about $x_{\min}$ and calculating its minimum, etc., a convergence to the true minimum can be obtained rather rapidly. The basic problem with the technique is that it requires G to be everywhere positive definite, otherwise at some point in the iteration a maximum may be calculated rather than a minimum, and the whole process diverges. This is more easily seen in the one-dimensional case in (4.86). If the second derivative is negative, then, we clearly have an inverted parabola rather than the desired well-shape figure.

Despite this defect, Newton's method is quite powerful and algorithms have been developed in which the matrix G is artificially altered whenever it becomes negative. In this manner, the iteration continues in the right direction until a region of positive-definiteness is reached. Such variations are called *quasi-Newton* methods.

The disadvantage of the Newton methods is that each iteration requires an evaluation of the matrix G and its inverse. This, of course, becomes quite costly in terms of time. This problem has given rise to another class of methods which either avoid calculating G or calculate it once and then "*update*" G with some correcting function after each iteration. These methods are described in more detail by *James* [4.11].

In the specific case of least squares minimization, a common procedure used with Newton's method is to *linearize* the fitting function. This is equivalent to approximating the hessian in the following manner. Rewriting (4.70) as

$$S = \sum_k s_k^2, \tag{4.89}$$

where $s_k = [y_k - f(x_k)]/\sigma_k$, the hessian becomes

$$\frac{\partial^2 S}{\partial x_i \partial x_j} = \frac{\partial}{\partial x_i} \frac{\partial}{\partial x_j} \sum_k s_k^2$$

$$\frac{\partial^2 S}{\partial x_i \partial x_j} = \frac{\partial}{\partial x_i} \sum_k 2 s_k \frac{\partial s_k}{\partial x_j} = 2 \sum_k \frac{\partial s_k}{\partial x_i} \frac{\partial s_k}{\partial x_j} + s_k \frac{\partial^2 s_k}{\partial x_i \partial x_j}. \tag{4.90}$$

The second term in the sum can be considered as a second order correction and is set to zero. The hessian is then

$$G_{ij} \simeq 2 \sum \frac{\partial s_k}{\partial x_i} \frac{\partial s_k}{\partial x_j}. \tag{4.91}$$

This approximation has the advantage of ensuring positive-definiteness and the result converges to the correct minimum. However, the covariance matrix will not in general converge to the correct covariance values, so that the errors as determined by this matrix may not be correct.

As the reader can see, it is not a trivial task to implement a nonlinear least squares program. For this reason we have advised the use of a ready-made program. A variety of routines may be found in the NAG library [4.13], for example. A very powerful program allowing the use of a variety of minimization methods such a simplex, Newton, etc., is *Minuit* [4.14] which is available in the CERN program library. This library is distributed to many laboratories and universities.

Local vs Global Minima. Up to now we have assumed that the function F contains only one minimum. More generally, of course, an arbitrary function can have many local minima in addition to a global, absolute minimum. The methods we have described are all designed to locate a local minimum without regard for any other possible minima. It is up to the user to decide if the minimum obtained is indeed what he wants. It is important, therefore, to have some idea of what the true values are so as to start the search in the right region. Even in this case, however, there is no guarantee that the process will converge onto the closest minimum. A good technique is to fix those parameters which are approximately known and vary the rest. The result can then be used to start a second search in which all the parameters are varied. Other problems which can arise are the occurrence of overflow or underflow in the computer. This occurs very often with exponential functions. Here good starting values are generally necessary to obtain a result.

Errors. While the methods we have discussed allow us to find the parameter values which minimize the function S, there is no prescription for calculating the errors on the parameters. A clue, however, can be taken from the linear one-dimensional case. Here we saw the variance of a parameter θ was given by the inverse of the second derivative (4.72),

$$\sigma^2 = \left| \frac{1}{2} \frac{\partial^2 S}{\partial \theta^2} \right|^{-1} . \tag{4.92}$$

If we expand S in Taylor series about the minimum

$$S(\theta) = S(\theta^*) + \frac{1}{2} \frac{\partial^2 S}{\partial \theta^2} (\theta - \theta^*)^2$$

$$= S(\theta^*) + \frac{1}{\sigma^2} (\theta - \theta^*)^2 . \tag{4.93}$$

At the point $\theta = \theta^* + \sigma$, we thus find that

$$S(\theta^* + \sigma) = S(\theta^*) + 1 . \tag{4.94}$$

Thus the error on θ corresponds to the distance between the minimum and where the S distribution increases by 1.

This can be generalized to the nonlinear case where the S distribution is not generally parabolic around the minimum. Finding the errors for each parameter then implies finding those points for which the S value changes by 1 from the minimum. If S has a complicated form, of course, this is not always easy to determine and once again, a numerical method must be used to solve the equation. If the form of S can be approximated by a quadratic surface, then, the error matrix in (4.73) can be calculated and inverted as in the linear case. This should then give an estimate of the errors and covariances.

4.8 Some General Rules for Rounding-off Numbers for Final Presentation

As a final remark in this chapter, we will suggest here a few general rules for the rounding off of numerical data for their final presentation.

The number of digits to be kept in a numerical result is determined by the errors on that result. For example, suppose our result after measurement and analysis is calculated to be $x = 17.615334$ with an error $\sigma(x) = 0.0233$. The error, of course, tells us that the result is uncertain at the level of the second decimal place, so that all following digits have absolutely no meaning. The result therefore should be rounded-off to correspond with the error.

Rounding off also applies to the calculated error. Only the first significant digit has any meaning, of course, but it is generally a good idea to keep two digits (but not more) in case the results are used in some other analysis. The extra digit then helps avoid a cumulative round-off error. In the example above, then, the error is rounded off to $\sigma = 0.0233 \rightarrow 0.023$; the result, x, should thus be given to three decimal places.

A general method for rounding off numbers is to take all digits to be rejected and to place a decimal point in front. Then

1) if the fraction thus formed is less than 0.5, the least significant digit is kept as is,
2) if the fraction is greater than 0.5, the least significant digit is increased by 1,
3) if the fraction is exactly 0.5, the number is increased if the least significant digit is odd and kept if it is even.

In the example above, three decimal places are to be kept. Placing a decimal point in front of the rejected digits then yields 0.334. Since this is less than 0.5, the rounded result is $x = 17.615 \pm 0.023$.

One thing which should be avoided is rounding off in steps of one digit at a time. For example, consider the number 2.346 which is to be rounded-off to one decimal place. Using the method above, we find $2.346 \rightarrow 2.3$. Rounding-off one digit at a time, however, yields

$$2.346 \rightarrow 2.35 \rightarrow 2.4\,!$$

5. General Characteristics of Detectors

As an introduction to the following chapters on detectors, we will define and describe here some general characteristics common to detectors as a class of devices. For the reader without detector experience, these characteristics will probably take on more significance when examples of specific detectors are treated. He should not hesitate to continue on, therefore, and return at a later time if he has not fully understood the contents of this chapter.

While the history of nuclear and elementary particle physics has seen the development of many different types of detectors, all are based on the same fundamental principle: the transfer of part or all of the radiation energy to the detector mass where it is converted into some other form more accessible to human perception. As we have seen in Chap. 2, charged particles transfer their energy to matter through direct collisions with the atomic electrons, thus inducing excitation or ionization of the atoms. Neutral radiation, on the other hand, must first undergo some sort of reaction in the detector producing charged particles, which in turn ionize and excite the detector atoms. The form in which the converted energy appears depends on the detector and its design. The gaseous detectors discussed in the next chapter, for example, are designed to directly collect the ionization electrons to form a current signal, while in scintillators, both the excitation and ionization contribute to inducing molecular transitions which result in the emission of light. Similarly, in photographic emulsions, the ionization induces chemical reactions which allow a track image to be formed, and so on.

Modern detectors today are essentially electrical in nature, i.e., at some point along the way the information from the detector is transformed into electrical impulses which can be treated by electronic means. This, of course, takes advantage of the great progress that has been made in electronics and computers to provide for faster and more accurate treatment of the information. Indeed, most modern detectors cannot be exploited otherwise. When discussing "*detectors*", therefore, we will also take this to mean the electronics as well. This, of course, is not to say that only electrical detectors are used in modern experiments, and indeed there are many other types which are employed. However, if an electrical detector can be used, it is generally preferred for the reasons already mentioned. Our discussion in the following sections, therefore, will only be concerned with this type.

5.1 Sensitivity

The first consideration for a detector is its sensitivity, i.e., its capability of producing a usable signal for a given type of radiation and energy. No detector can be sensitive to all types of radiation at all energies. Instead, they are designed to be sensitive to certain types of radiation in a given energy range. Going outside this region usually results in an unusable signal or greatly decreased efficiency.

Detector sensitivity to a given type of radiation of a given energy depends on several factors:

1) the cross section for ionizing reactions in the detector
2) the detector mass
3) the inherent detector noise
4) the protective material surrounding the sensitive volume of the detector.

The cross-section and detector mass determine the probability that the incident radiation will convert part or all its energy in the detector into the form of ionization. (We assume here that the properties of the detector are such that the ionization created will be efficiently used.) As we saw in Chap. 2, charged particles are highly ionizing, so that most detectors even of low density and small volume will have some ionization produced in their sensitive volume. For neutral particles, this is much less the case, as they must first undergo an interaction which produces charged particles capable of ionizing the detector medium. These interaction cross sections are usually much smaller so that a higher mass density and volume are necessary to ensure a reasonable interaction rate, otherwise the detector becomes essentially transparent to the neutral radiation. The mass required, depends on the type of radiation and the energy range of interest. In the case of the neutrino, for example, detector masses on the order of tons are usually necessary!

Even if ionization is produced in the detector, however, a certain minimum amount is necessary in order for the signal to be usable. This lower limit is determined by the noise from the detector and the associated electronics. The noise appears as a fluctuating voltage or current at the detector output and is always present whether there is radiation or not. Obviously, the ionization signal must be larger than the average noise level in order to be usable. For a given radiation type in a given energy range, the total amount of ionization produced is determined by the sensitive volume.

A second limiting factor is the material covering the entrance window to the sensitive volume of the detector. Because of absorption, only radiation with sufficient energy to penetrate this layer can be detected. The thickness of this material thus sets a lower limit on the energy which can be detected.

5.2 Detector Response

In addition to detecting the presence of radiation, most detectors are also capable of providing some information on the energy of the radiation. This follows since the amount of ionization produced by radiation in a detector is proportional to the energy it loses in the sensitive volume. If the detector is sufficiently large such that the radiation is completely absorbed, then this ionization gives a measure of the energy of the radiation. Depending on the design of the detector, this information may or may not be preserved as the signal is processed, however.

In general, the output signal of electrical detectors is in the form of a current pulse[1]. The amount of ionization is then reflected in the electrical charge contained in this

---

[1] Detectors may also be operated in a continuous mode in which the signal is a continuous current or voltage varying in time with the intensity of the radiation. This can be performed by electrically integrating the number of pules over a certain period of time.

signal, i.e., the integral of the pulse with respect to time. Assuming that the shape of the pulse does not change from one event to the other, however, this integral is directly proportional to the amplitude or *pulse height* of the signal, so that this characteristic may be used instead. The relation between the radiation energy and the total charge or pulse height of the output signal is referred to as the *response* of the detector.

Ideally, of course, one would like this relation to be linear although it is not absolutely necessary. It does, however, simplify matters greatly when transforming the measured pulse height to energy. For many detectors, the response is linear or approximately so over a certain range of energies. In general, however, the response is a function of the particle type and energy, and it does not automatically follow that a detector with a linear response for one type of radiation will be linear for another. A good example is organic scintillator. As will be seen later, the response is linear for electrons down to a very low energies but is nonlinear for heavier particles such as the proton, deuteron, etc. This is due to the different reaction mechanisms which are triggered in the medium by the different particles.

5.3 Energy Resolution. The Fano Factor

For detectors which are designed to measure the energy of the incident radiation, the most important factor is the energy resolution. This is the extent to which the detector can distinguish two close lying energies. In general, the resolution can be measured by sending a monoenergetic beam of radiation into the detector and observing the resulting spectrum. Ideally, of course, one would like to see a sharp delta-function peak. In reality, this is never the case and one observes a peak structure with a finite width, usually Gaussian in shape. This width arises because of fluctuations in the number of ionizations and excitations produced.

The resolution is usually given in terms of the *full width at half maximum* of the peak (FWHM). Energies which are closer than this interval are usually considered unresolvable. This is illustrated in Fig. 5.1. If we denote this width as ΔE, then the relative resolution at the energy E is

$$\text{Resolution} = \Delta E/E . \tag{5.1}$$

Equation (5.1) is usually expressed in percent. A NaI detector has about a 8% or 9% resolution for γ-rays of about 1 MeV, for example, while germanium detectors have resolutions on the order of 0.1%!

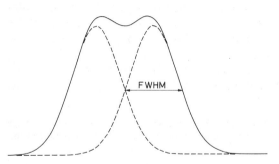

Fig. 5.1. Definition of energy resolution. Two peaks are generally considered to be resolved if they are separated by a distance greater than their full widths at half maximum (FWHM). The *solid line* shows the sum of two identical Gaussian peaks separated by just this amount

In general, the resolution is a function of the energy deposited in the detector, with the ratio (5.1) improving with higher energy. This is due to the Poisson or Poisson-like statistics of ionization and excitation. Indeed, it is found that the average energy required to produce an ionization is a fixed number, w, dependent only on the material. For an energy E, therefore, one would expect on the average, $J = E/w$ ionizations. Thus as energy increases, the number of the ionization events also increases resulting in smaller relative fluctuations.

To calculate the fluctuations it is necessary to consider two cases. For a detector in which the radiation energy is not totally absorbed, for example, a thin transmission detector which just measures the dE/dx loss of a passing particle, the number of signal-producing reactions is given by a Poisson distribution. The variance is then given by

$$\sigma^2 = J, \qquad\qquad (5.2)$$

where J is the mean number of events produced. The energy dependence of the resolution can then be seen to be

$$R = 2.35 \frac{\sqrt{J}}{J} = 2.35 \sqrt{\frac{w}{E}}, \qquad\qquad (5.3)$$

where the factor 2.35 relates the standard deviation of a Gaussian to its FWHM. Thus the resolution varies inversely as the square root of the energy.

If the full energy of the radiation is absorbed as is the case for detectors used in spectroscopy experiments, the naive assumption of Poisson statistics is incorrect. And indeed, it is observed that the resolution of many such detectors is actually smaller than that calculated from Poisson statistics. The difference here is that the total energy deposited is a fixed, constant value, while in the previous case, the energy deposited can fluctuate. The total number of ionizations which can occur and the energy lost in each ionization are thus constrained by this value. Statistically, this means that the ionization events are not all independent so that Poisson statistics is not applicable. *Fano* [5.1] was the first to calculate the variance under this condition and found

$$\sigma^2 = FJ, \qquad\qquad (5.4)$$

where J is the mean ionization produced and F is a number known as the Fano factor.

The factor F is a function of all the various fundamental processes which can lead to an energy transfer in the detector. This includes all reactions which do not lead to ionization as well, for example, phonon excitations, etc. It is thus an intrinsic constant of the detecting medium. Theoretically, F is very difficult to calculate accurately as it requires a detailed knowledge of all the reactions which can take place in the detector. From (5.4), the resolution is then given by

$$R = 2.35 \frac{\sqrt{FJ}}{J} = 2.35 \sqrt{\frac{Fw}{E}}. \qquad\qquad (5.5)$$

If $F = 1$, the variance is the same as that for a Poisson distribution and (5.5) reduces to (5.3). This seems to be the case for scintillators, however, for many detectors such as semiconductors or gases, $F < 1$. This, of course, greatly increases the resolution of these types of detectors.

In addition to the fluctuations in ionization, a number of external factors can affect the overall resolution of a detector. This includes effects from the associated electronics such as noise, drifts, etc. Assuming all these sources are independent and distributed as Gaussians, the total resolution E is then given by (4.64), i.e.,

$$(\Delta E)^2 = (\Delta E_{\text{det}})^2 + (\Delta E_{\text{elect}})^2 + \ldots . \tag{5.6}$$

5.4 The Response Function

For the measurement of energy spectra, an important factor which must be considered is the *response function* of the detector for the type of radiation being detected. This is the spectrum of pulse heights observed from the detector when it is bombarded by a monoenergetic beam of the given radiation. Up to now, we have assumed that this response spectrum is a Gaussian peak. If we ignore the finite width for a moment, this essentially corresponds to a Dirac delta function, i.e., for a fixed incident energy the output signal has a single, fixed amplitude. Then, if the response is linear, the spectrum of pulse heights measured from the detector corresponds directly to the energy spectrum of the incident radiation. This is the ideal case. Unfortunately, a Gaussian peak response is not always realized particularly in the case of neutral radiation.

The response function of a detector at a given energy is determined by the different interactions which the radiation can undergo in the detector and its design and geometry. To take an example, consider monoenergetic charged particles, say electrons, incident on a detector thick enough to stop the particles. Assuming all the electrons lose their energy by atomic collisions, it is clear that the spectrum of pulse heights will be a Gaussian peak. In reality, however, some of the electrons will scatter out of the detector before fully depositing their energy. This produces a low energy tail. Similarly some electrons will emit bremsstrahlung photons which may escape from the detector. This again gives rise to events at a lower energy than the peak. The response function thus consists of a Gaussian peak with a low energy tail determined by the amount of scattering and bremsstrahlung energy loss. If the tail is small, however, this can still be a reasonable approximation to the ideal Gaussian response depending on the precision desired. Moreover, the response function can be improved by changing the design and geometry of the detector. A material of lower atomic number Z can be chosen, for example, to minimize backscattering and bremsstrahlung. Similarly if the detector is made to surround the source, backscattering electrons will be captured thus decreasing the escape of these particles, etc.

To see how the response function can change with radiation type, consider the same detector with gamma rays instead. As we have already mentioned, gamma rays must first convert into charged particles in order to be detected. The principal mechanisms for this are the photoelectric effect, Compton scattering and pair production. In the photoelectric effect, the gamma ray energy is transferred to the photoelectron which is then stopped in the detector. Since the energy of all the photoelectrons is the same, this results in a sharp peak in the pulse height spectrum, which is the desired Gaussian response. However, some gamma rays will also suffer Compton scatterings. As given by (2.109), the Compton electrons are distributed continuously in energy so that a distribution, similar to Fig. 2.24, also appears in the response function. This, of course,

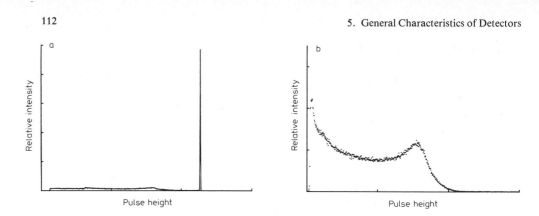

Fig. 5.2a, b. The response functions of two different detectors for 661 keV gamma rays. **(a)** shows the response of a germanium detector which has a large photoelectric cross section relative to the Compton scattering cross section at this energy. A large photopeak with a relatively small continuous Compton distribution is thus observed. **(b)** is the response of an organic scintillator detector. Since this material has a low atomic number Z, Compton scattering is predominant and only this distribution is seen in the response function

immediately destroys the ideal delta-function response. In a similar manner, those events interacting via the pair production mechanism will also contribute a structure to the function. One such total response function is sketched in Fig. 5.2. The observed pulse height spectrum, therefore, simply reflects the different interactions which occur in the detector volume. Since the relative intensity of each structure in the spectrum is determined by the relative cross sections for each interaction mechanism, the response function will also be different at different energies and for different detector media.

If the detector is now used to measure a spectrum of gamma rays, the observed pulse height distribution will be a convolution of the gamma ray spectrum and the detector response, i.e.,

$$PH(E) = \int S(E')R(E, E')\, dE' , \tag{5.7}$$

where $R(E, E')$ is the response function at the incident energy E' and $S(E')$ is the spectrum of gamma ray energies. To determine the gamma ray spectrum $S(E')$, from the measured pulse height distribution then requires knowing $R(E, E')$ in order to invert (5.7). Here, of course, we see the utility of having $R(E, E') = \delta(E' - E)$!

5.5 Response Time

A very important characteristic of a detector is its response time. This is the time which the detector takes to form the signal after the arrival of the radiation. This is crucial to the timing properties of the detector. For good timing, it is necessary for the signal to be quickly formed into a sharp pulse with a rising flank as close to vertical as possible. In this way a more precise moment in time is *marked* by the signal.

The duration of the signal is also of importance. During this period, a second event cannot be accepted either because the detector is insensitive or because the second signal will *pile up* on the first. This contributes to the *dead* time of the detector and limits the count rate at which it can be operated. The effect of dead time is discussed in Sect. 5.7.

5.6 Detector Efficiency

Two types of efficiency are generally referred to when discussing radiation detection: *absolute* efficiency and *intrinsic* detection efficiency. The absolute or total efficiency of a detector is defined as that fraction of events emitted by the source which is actually registered by the detector, i.e.,

$$\mathcal{E}_{tot} = \frac{\text{events registered}}{\text{events emitted by source}} . \tag{5.8}$$

This is a function of the detector geometry and the probability of an interaction in the detector. As an example, consider a cylindrical detector with a point source at a distance d on the detector axis as shown in Fig. 5.3. If the source emits isotropically, then, the probability of the particle being emitted at an angle θ is

$$P(\theta)\,d\Omega = d\Omega/4\pi . \tag{5.9}$$

The probability that a particle hitting the detector will have an interaction in the detector is given by (2.7). Combining the two then yields

$$d\mathcal{E}_{tot} = \left[1 - \exp\left(\frac{-x}{\lambda} \right) \right] \frac{d\Omega}{4\pi} , \tag{5.10}$$

where x is the path length in the detector and λ is the mean free path for an interaction. The total efficiency is then found by integrating (5.10) over the volume of the detector.

In many cases, however, the value of x does not vary too much over the detector or the value of λ is so small that the exponential can be considered as zero. The absolute efficiency can then be factored into two parts: the *intrinsic* efficiency, $\mathcal{E}_{int}$, and the *geometrical* efficiency or *acceptance*, $\mathcal{E}_{geom}$. The total or absolute efficiency of the detector is then given by the product

$$\mathcal{E}_{tot} \simeq \mathcal{E}_{int}\,\mathcal{E}_{geom} . \tag{5.11}$$

The intrinsic efficiency is that fraction of events actually hitting the detector which is registered, i.e.,

$$\mathcal{E}_{int} = \frac{\text{events registered}}{\text{events impinging on detector}} . \tag{5.12}$$

This probability depends on the interaction cross sections of the incident radiation on the detector medium. The intrinsic efficiency is thus a function of the type of radiation,

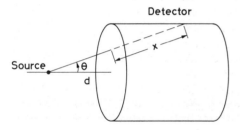

Fig. 5.3. Calculating the detection efficiency of a cylindrical detector for a point source

its energy and the detector material. For charged particles, the intrinsic efficiency is generally good for most detectors, since it is rare for a charged particle *not* to produce some sort of ionization. For heavier particles, though, quenching effects may be present in some materials which drain the ionization produced. The problem of efficiency is generally more important for neutral particles as they must first interact to create secondary charged particles. These interactions are much rarer in general, so that capturing a good fraction of the incident neutral radiation is not always assured. The dimensions of the detector become important as sufficient mass must be present in order to provide a good probability of interaction.

The geometric efficiency, in contrast, is that fraction of the source radiation which is geometrically intercepted by the detector. This, of course, depends entirely on the geometrical configuration of the detector and source. The angular distribution of the incident radiation must also be taken into account. For the cylindrical detector in Fig. 5.3, $\mathscr{E}_{geom}$ is simply the average solid angle fraction. For multidetector systems, where coincidence requirements are imposed, however, the calculations can be somewhat complicated and recourse to numerical simulation with Monte Carlo methods must be made.

5.7 Dead Time

Related to the efficiency is the *dead time* of the detector. This is the finite time required by the detector to process an event which is usually related to the duration of the pulse signal. Depending on the type, a detector may or may not remain sensitive to other events during this period. If the detector is insensitive, any further events arriving during this period are lost. If the detector retains its sensitivity, then, these events may *pile-up* on the first resulting in a distortion of the signal and subsequent loss of information from both events. These losses affect the observed count rates and distort the time distribution between the arrival of events. In particular, events from a random source will no longer have the Poissonian time distribution given by (4.60). To avoid large dead time effects, the counting rate of the detector must be kept sufficiently low such that the probability of a second event occurring during a dead time period is small. The remaining effect can then be corrected.

When calculating the effects of dead time, the entire detection system must be taken into account. Each element of a detector system has it own dead time and, indeed, it is often the electronics which account for the larger part of the effect. Moreover, when several elements have comparable dead times, combining the effects is also a difficult task and a general method does not exist for solving such problems.

As an illustration let us analyze the effect on count rate due to the dead time of simple element in the system. Suppose the element has a dead time τ and that τ is constant for all events. Two fundamental cases are usually distinguished: *extendable* or *non-extendable* dead times. These are also referred to as the *paralyzable* or *non-paralyzable* models. In the extendable case, the arrival of a second event during a dead time period extends this period by adding on its dead time τ starting from the moment of its arrival. This is illustrated in Fig. 5.4. This occurs in elements which remain sensitive during the dead time. In principle if the event rate is sufficiently high, events can arrive such that their respective dead time periods all overlap. This produces a prolonged period during which no event is accepted. The element is thus *paralyzed*. The non-

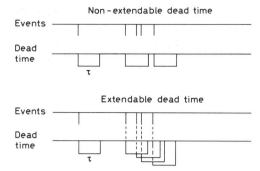

Fig. 5.4. Extendable (paralyzable) and non-extendable (non-paralyzable) dead time models

extendable case, in contrast, corresponds to a element which is insensitive during the dead time period. The arrival of a second event during this period simply goes unnoticed and after a time τ the element becomes active again.

Let us consider the non-extendable case first. Suppose m is the true count rate and the detector registers k counts in a time T. Since each detected count n engenders a dead time τ, a total dead time $k\tau$ is accumulated during the counting period T. During this dead period, a total of $mk\tau$ counts is lost. The true number of counts is therefore

$$mT = k + mk\tau . \tag{5.13}$$

Solving for m in terms of k, we find

$$m = \frac{k/T}{1 - (k/T)\tau} . \tag{5.14}$$

Thus (5.14) provides us with a formula for finding the true rate m from the observed rate k/T.

The extendable case is somewhat more difficult. Here, one realizes that only those counts which arrive at time intervals greater than τ are recorded. As given by (4.60), the distribution of time intervals between events decaying at a rate m, is

$$P(t) = \frac{1}{m} \exp(-mt) . \tag{5.15}$$

The probability that $t > \tau$ is then

$$P(t > \tau) = \frac{1}{m} \int_{\tau}^{\infty} \exp(-mt)\, dt = \exp(-m\tau) . \tag{5.16}$$

The number of counts observed in a time T, therefore, is just that fraction of the mT true events whose arrival times satisfy this condition,

$$k = mT \exp(-m\tau) . \tag{5.17}$$

To find the true value, m, (5.17) must be solved numerically. Figure 5.5 shows the behavior of (5.17). Note that the function first increases, goes through a maximum at $m = 1/\tau$ and then decreases once again. This means that for a given observed rate, k/T, there are two corresponding solutions for m. Care should be taken, therefore, to distinguish between two.

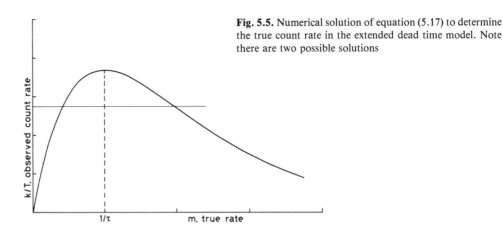

Fig. 5.5. Numerical solution of equation (5.17) to determine the true count rate in the extended dead time model. Note there are two possible solutions

The above results are generally adequate for most practical problems, however, they are only first order approximations. More rigorous treatments are given by *Foglio Parra* and *Mandelli Bettoni* [5.2] and a general discussion of dead time problems by *Muller* [5.3]. The case of a variable dead time is also treated by *Libert* [5.4].

Given the above results, the problem which often arises is to determine which class, extendable or non-extendable, is applicable. Indeed, many detectors systems are combinations of both, having some elements which are extendable and others which are not. And some may not be in either class. Moreover the dead time of the elements could be variable depending on the count rate, the pulse shapes, etc. A solution often used is to deliberately add in a blocking circuit element with a dead time larger than all other elements into the system such that the detector system can be treated by one of the fundamental models. This, of course, slows down the system but removes the uncertainty in the dead-time model. This should be done quite early in the system, however, in order to avoid pile-up problems later. See for example the *Inhibit* in Chap. 16.

5.7.1 Measuring Dead Time

The classical method of measuring dead time is the so-called *two-source* technique. In this procedure, the count rates of two different sources are measured separately and together. To illustrate the principle of the method, let us suppose n_1 and n_2 are the true count rates of the two sources and R_1, R_2 and R_{12} are the rates observed for the separate and combined sources. For simplicity also, let us assume that there is no background. In the non-extended case, we then have the relations

$$n_1 = \frac{R_1}{1-R_1\tau}, \quad n_2 = \frac{R_2}{1-R_2\tau} \quad \text{and}$$
$$n_1 + n_2 = \frac{R_{12}}{1-R_{12}\tau}. \tag{5.18}$$

Eliminating the n's, we have

$$\frac{R_{12}}{1-R_{12}\tau} = \frac{R_1}{1-R_1\tau} + \frac{R_2}{1-R_2\tau} \tag{5.19}$$

which yields the solution

$$\tau = \frac{R_1 R_2 - [R_1 R_2 (R_{12} - R_1)(R_{12} - R_2)]^{1/2}}{R_1 R_2 R_{12}} . \tag{5.20}$$

While conceptually the double source method is quite simple, it is in practice, a cumbersome and time consuming method which yields results to no better than $5 - 10\%$ [5.3]. This can be seen already in (5.20) which shows τ to be given by a difference between two numbers. From the point of view of statistical errors (see Sect. 4.6.1), of course, this is disadvantageous. Experimentally, care must also be taken to ensure the same positioning for the sources when they are measured separately and together. Even then, however, there may be scattering effects of one source on the other which may modify the combined rates, etc.

A number of other methods have been proposed, however. One technique is to replace one of the sources with a pulse generator [5.5] of frequency $f < (3\tau)^{-1}$. If R_s is the observed rate of the source alone and R_c is the observed combined rate of the source and generator, then it can be shown that the dead time in the non-extended case is

$$\tau \simeq \frac{1 - [(R_c - R_s)/f]^{1/2}}{R_s} . \tag{5.21}$$

Equation (5.21) is only approximate, but it should give results to better than 1% as long as the oscillator frequency condition stated above is met [5.3]. This method, of course, avoids the problem of maintaining a fixed source geometry but it does require an estimate of the dead time and a fast pulser to ensure the frequency condition above. A more general formula valid at all frequencies has been worked out by *Muller* [5.6] requiring, however, a long numerical calculation of a correction factor.

For an extended dead time, it can also be shown [5.3] that

$$(m + f - mf\tau) \exp(-m\tau) = R_c \tag{5.22}$$

where m is the true rate of the source. If (5.17) for the case of source alone is used, the relation

$$\frac{R_c - R_s}{f} = (1 - m\tau) \exp(-m\tau) \tag{5.23}$$

is found which can be solved for $m\tau$. If m is known then, of course, τ follows.

A very quick and accurate method which can be used for measuring the dead time of the electronics system alone is to inject pulses from two oscillators [5.3] of frequency f_1 and f_2 and to measure the mean frequency of the combined pulses, f_c. For the non-extended model, it can be shown then,

$$f_c = \begin{cases} f_1 + f_2 - 2f_1 f_2 \tau & \text{for} \quad 0 < \tau < T/2 \\ 1/T & \text{for} \quad T/2 < \tau < T \end{cases} \tag{5.24}$$

where T is the period of the faster oscillator, i.e., the smaller of $1/f_1$ and $1/f_2$. For the extended model, we have similarly

$$f_c = \begin{cases} f_1 + f_2 - 2f_1 f_2 \tau & \text{for} \quad 0 < \tau < T \\ 0 & \text{for} \quad \tau > T . \end{cases} \tag{5.25}$$

The expressions are thus identical if the frequency of the faster oscillator is chosen such that $f < (2\tau)^{-1}$; the dead time, *irrespective* of the model, is then given by

$$\tau = \frac{t_m}{2} \frac{R_1 + R_2 - R_c}{R_1 R_2} \qquad (5.26)$$

where R_1, R_2 and R_c are the total measured counts for the two oscillators separately and combined in a measurement period t_m. If f is chosen greater then $(2\tau)^{-1}$, then a determination of the model type can be made by comparing the results to the predictions in (5.24) and (5.25).

When using this method, of course, it is important to assure that the form of the pulses are close to those of true detector signals and that the frequencies of the oscillators are stable. In such cases, the two-oscillator method can yield quick and accurate results to a precision better than 10^{-3}.

6. Ionization Detectors

Ionization detectors were the first electrical devices developed for radiation detection. These instruments are based on the direct collection of the ionization electrons and ions produced in a gas by passing radiation. During the first half of the century, three basic types of detector were developed: the *ionization chamber,* the *proportional counter* and the *Geiger-Müller counter.* Except for specific applications, these particular devices are not in widespread use in modern nuclear and particle physics experiments today. They are, however, still very much employed in the laboratory as radiation monitors. They are cheap, simple to operate and easy to maintain. Their basic design and structure, in fact, have changed little since the late 1940's when the then newly-developed scintillation counter began taking the place of these instruments in nuclear research.

During the late 1960's, a renewed interest in gas ionization instruments was stimulated in the particle physics domain by the invention of the *multi-wire proportional chamber.* These devices were capable of localizing particle trajectories to less than a millimeter and were quickly adopted in high-energy experiments. Stimulated by this success, the following years saw the development of the *drift chamber* and, somewhat later, the *time projection chamber.* These devices operate on the same basic principles as the simple proportional counter, but otherwise bear little physical resemblance to their simpler predecessor. They are now used extensively in high energy particle physics experiments and require more sophisticated electronics as well as data acquisition by computer. There are also many variants of the above instruments which have been designed for more particular needs.

Because of their higher density, attention has also been focused on the use of liquids as an ionizing medium. The physics of ionization and transport in liquids is not as well understood as in gases, but much progress in this domain has been made and development is continuing in this area.

6.1 Gaseous Ionization Detectors

Because of the greater mobility of electrons and ions, a gas is the obvious medium to use for the collection of ionization from radiation. Many ionization phenomena arise in gases and over the years these have been studied and exploited in the detectors we will describe below. The three original gas devices, i.e., the ionization chamber, the proportional counter and the Geiger-Müller counter, serve as a good illustration of the application of gas ionization phenomena in this class of instruments. These detectors are actually the same device working under different operating parameters, exploiting different phenomena. The basic configuration (Fig. 6.1) consists of a container, which we will take to be a cylinder for simplicity, with conducting walls and a thin end window. The cylinder is filled with a suitable gas, usually a noble gas such as argon. Along

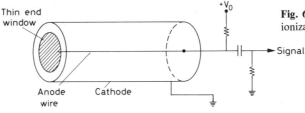

Fig. 6.1. Basic construction of a simple gas ionization detector

its axis is suspended a conducting wire to which a positive voltage, $+V_0$, relative to the walls is applied. A radial electric field

$$E = \frac{1}{r} \frac{V_0}{\ln(b/a)} \tag{6.1}$$

with r: radial distance from axis; b: inside radius of cylinder; a: radius of central wire is thereby established. If radiation now penetrates the cylinder, a certain number of electron-ion pairs will be created, either directly, if the radiation is a charged particle, or indirectly through secondary reactions if the radiation is neutral. The mean number of pairs created is proportional to the energy deposited in the counter. Under the action of the electric field, the electrons will be accelerated towards the anode and the ions toward the cathode where they are collected.

The current signal observed, however, depends on the field intensity. This is illustrated in Fig. 6.2 which plots the total charge collected as a function of V. At zero voltage, of course, no charge is collected as the ion-electron pairs recombine under their own electrical attraction. As the voltage is raised, however, the recombination forces

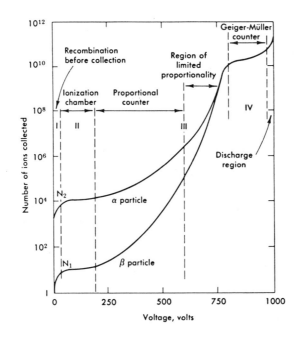

Fig. 6.2. Number of ions collected versus applied voltage in a single wire gas chamber (from *Melissinos* [6.1])

are overcome and the current begins to increase as more and more of the electron-ion pairs are collected before they can recombine. At some point, of course, all created pairs will be collected and further increases in voltage show no effect. This corresponds to the first flat region in Fig. 6.2. A detector working in this region (II) is called an *ionization chamber* since it collects the ionization produced directly by the passing radiation. The signal current, of course, is very small and must usually be measured with an electrometer. Ionization chambers are generally used for measuring gamma ray exposure and as monitoring instruments for large fluxes of radiation.

Returning to Fig. 6.2, if we now increase the voltage beyond region II we find that the current increases again with the voltage. At this point, the electric field is strong enough to accelerate freed electrons to an energy where they are also capable of ionizing gas molecules in the cylinder. The electrons liberated in these secondary ionizations, of course, are also accelerated to produce still more ionization and so on. This results in an ionization *avalanche* or *cascade*. Since the electric field is strongest near the anode, as seen from (6.1), this avalanche occurs very quickly and almost entirely within a few radii of this wire. The number of electron-ion pairs in the avalanche, however, is directly proportional to the number of primary electrons. What results then is a proportional amplification of the current, with a multiplication factor depending on the working voltage V. This factor can be as high as 10^6 so that the output signal is much larger than that from an ionization chamber, but still in proportion to the original ionization produced in the detector. This region of proportional multiplication extends up to point III and a detector operating in this domain is known as a *proportional chamber*. Because it is the basic model for the more sophisticated gas devices to be described later, we will treat the proportional counter in more detail later.

If the voltage is now increased beyond point III, the total amount of ionization created through multiplication becomes sufficiently large that the space charge created distorts the electric field about the anode. Proportionality thus begins to be lost. This is known as the region of limited proportionality. Increasing V still higher, the energy becomes so large that a discharge occurs in the gas. What happens physically is that instead of a single, localized avalanche at some point along the anode wire (as in a proportional counter), a *chain reaction* of many avalanches spread out along the entire length of the anode is triggered. These secondary avalanches are caused by photons emitted by deexciting molecules which travel to other parts of the counter to cause further ionizing events. The output current thus becomes completely saturated, always giving the same amplitude regardless of the energy of the initial event. In order to stop the discharge, a *quenching* gas must be present in the medium to absorb the photons and drain their energy into other channels. Detectors working in this voltage region are called *Geiger-Müller* or *breakdown* counters. The Geiger voltage region, in fact, is characterized by a plateau over which the count rate varies little. The width of the plateau depends on the efficacy of the quencher in the gas. In general, the working voltage of a Geiger counter is chosen to be in the middle of the plateau in order to minimize any variations due to voltage drift.

Finally, now, if the voltage is increased still further a continuous breakdown occurs with or without radiation. This region, of course, is to be avoided to prevent damage to the counter. In this illustration, we thus see how phenomena such as gas multiplication and discharge, in addition to gas ionization, can be used for radiation detection.

6.2 Ionization and Transport Phenomena in Gases

Because of the importance of ionization detectors in physics research much work has been and still is devoted to the ionization process and the movement of electrons and ions in gases. We shall, therefore, devote some time to reviewing some of these processes in the following sections.

6.2.1 Ionization Mechanisms

As we saw in Chap. 2, the energy loss of a charged particle in matter is essentially divided between two types of reaction: (1) excitation, and (2) ionization in which a free electron and ion are created. The excitation of an atom, X,

$$X + p \rightarrow X^* + p \ , \tag{6.2}$$

where p is a charged particle, is a *resonant* reaction which requires the correct amount of energy to be transferred. Typical cross-sections in noble gases at resonance [6.2] are on the order of $\sigma \simeq 10^{-17} \, cm^2$. While no free electrons or ions are created, the excited molecule or atom may participate in further reactions which do result in ionization. This is discussed later.

For an ionization,

$$X + p \rightarrow X^+ + p + e^- \ , \tag{6.3}$$

there is, of course, no exact energy requirement and, in fact, its cross-section is somewhat higher with $\sigma \simeq 10^{-16} \, cm^2$ [6.2]. However, the ionization process has a energy threshold which is relatively high, and since low energy transfers are more probable, the excitation reactions generally dominate.

The electrons and ions created by the incident radiation itself, (6.3), are known as *primary* ionization. In a number of these ionizations, however, a sufficiently large amount of energy is transferred to the electron (delta-rays) such that this electron also creates ion-electron pairs. This latter ionization is known as *secondary* ionization. If their energy is high enough, the secondary ionization electrons may also ionize and so on until the threshold for ionizing reactions is reached.

A second mechanism of ionization in gases is the *Penning Effect*. In certain atoms, metastable states are excited which, because of a large spin-parity difference, are unable to deexcite immediately to the ground state by the emission of a photon. In such atoms, a deexcitation may occur through a collision with a second atom resulting in the ionization of the latter. Common examples are molecular gases on noble gases and noble gases on noble gases, e.g.,

$$Ne^* + Ar \rightarrow Ne + Ar^+ + e^- \ . \tag{6.4}$$

A third important mechanism which occurs in noble gases is the formation of molecular ions. In this process, a positive gas ion interacts with a neutral atom of the same type to form a molecular ion, i.e.,

$$He^+ + He \rightarrow He_2^+ + e^- \ . \tag{6.5}$$

6.2.2 Mean Number of Electron-Ion Pairs Created

Since the occurrence of the ionizing reactions above is statistical in nature, two identical particles will not, in general, produce the same number of ion-electron pairs. We can ask, however: What is the average number of ion-electron pairs (from all mechanisms) created for a given energy loss? Note that this is *not* equal to the energy loss divided by the ionization potential, since some energy is also lost to excitation! For gases, this average turns out to be on the order of 1 ion-electron pair per 30 eV of energy lost, that is, for a 3 keV particle, an average of $3000/30 = 100$ ion-electron pairs will be created. Moreover, what is surprising is that this average value does not depend very strongly on particle type and only weakly on the type of gas. Table 6.1 gives a comparison of the measured values for this average for several types of gas used in ionization detectors.

Table 6.1. Excitation and ionization characteristics of various gases

	Excitation potential [eV]	Ionization potential [eV]	Mean energy for ion-electron pair creation [eV]
H_2	10.8	15.4	37
He	19.8	24.6	41
N_2	8.1	15.5	35
O_2	7.9	12.2	31
Ne	16.6	21.6	36
Ar	11.6	15.8	26
Kr	10.0	14.0	24
Xe	8.4	12.1	22
CO_2	10.0	13.7	33
CH_4		13.1	28
C_4H_{10}		10.8	23

Table 6.2. Measured Fano factors for various gas mixtures

Gas	F	Ref.
Ar 100%	$0.2^{+0.01}_{-0.02}$	[6.4]
	$<0.40 \pm 0.03$	[6.5]
Ar + 80% Xe	$<0.21 \pm 0.03$	[6.5]
Ar + 24% Xe	$<0.23 \pm 0.02$	[6.5]
Ar + 20% Xe	$<0.16 \pm 0.02$	[6.5]
Ar + 5% Xe	$<0.14 \pm 0.03$	[6.5]
Ar + 5% Kr	$<0.37 \pm 0.06$	[6.5]
Ar + 20% Kr	$<0.12 \pm 0.02$	[6.5]
Ar + 79% Kr	$<0.13 \pm 0.02$	[6.5]
Xe 100%	$<0.15 \pm 0.01$	[6.6]
	$<0.15 \pm 0.03$	[6.5]
Kr 100%	$<0.23 \pm 0.01$	[6.6]
	$<0.19 \pm 0.02$	[6.5]
Kr + 1.3% Xe	$<0.19 \pm 0.01$	[6.6]
Kr + 20% Xe	$<0.21 \pm 0.02$	[6.6]
Kr + 40% Xe	$<0.22 \pm 0.01$	[6.6]
Kr + 60% Xe	$<0.21 \pm 0.01$	[6.6]
Kr + 95% Xe	$<0.21 \pm 0.01$	[6.6]

The average energy, w, required for creating an electron-ion pair is important since it determines the efficiency and the energy resolution of the detector. From (5.6), the resolution for a particle of energy E is

$$R = 2.35 \sqrt{\frac{Fw}{E}} \, , \tag{6.6}$$

where F is the Fano factor for the gas medium. While the Fano factor is not well determined for most gases, it is clear that F is much less than 1. Table 6.2 gives some measured values for various noble gas mixtures.

6.2.3 Recombination and Electron Attachment

While the number of electron-ion pairs created is important for the efficiency and energy resolution of the detector, it is equally important that these pairs again remain in a free state long enough to be collected. Two processes, in particular, hinder this operation: *recombination* and *electron attachment*.

When there is no electric field, ion-electron pairs will generally recombine under the force of their electric attraction, emitting a photon in the process,

$$X^+ + e^- \rightarrow X + h\nu \, . \tag{6.7}$$

For molecular ions, a similar recombination reaction occurs

$$X^- + Y^+ \rightarrow XY + e^- \, . \tag{6.8}$$

In general, the rate of recombinations will depend on the concentrations of the positive and negative ions so that,

$$dn = b n^- n^+ dt \, , \tag{6.9}$$

where b is a constant dependent on the type of gas and n^+ and n^- are the positive and negative ion concentrations respectively. If we set $n^+ = n^- = n$, integration then yields the result

$$n = \frac{n_0}{1 + b n_0 t} \, , \tag{6.10}$$

where n_0 is the initial concentration at $t = 0$.

Electron attachment involves the capture of free electrons by electronegative atoms to form negative ions,

$$e^- + X \rightarrow X^- + h\nu \, . \tag{6.11}$$

These are atoms which have an almost full outer electron shell so that the addition of an extra electron actually results in the release of energy. The negative ion formed is consequently stable. The energy released in this capture is known as the *electron affinity*. Clearly, therefore, the presence of any electronegative gases in the detector will severely diminish the efficiency of electron-ion collection by trapping the electrons before they can reach the electrodes. Some well known electronegative gases are O_2, H_2O, CO_2, CCl_4 and SF_6. The noble gases He, Ne, Ar, in contrast, have negative electron affinities.

6.3 Transport of Electrons and Ions in Gases

For ionization detectors, an understanding of the motion of the electrons and ions in gases is extremely important as these factors influence many operating characteristics of the detector. For the most part, this motion is described by the classical kinetic theory of gases. Two phenomena are of particular importance: diffusion, and drift in an electric field.

6.3.1 Diffusion

In the absence of an electric field, electrons and ions liberated by passing radiation diffuse uniformly outward from their point of creation. In the process they suffer multiple collisions with the gas molecules and lose their energy. They thus come quickly into thermal equilibrium with the gas and eventually recombine. At thermal energies, the velocities of the charges are described by the Maxwell distribution which gives a mean speed of

$$v = \sqrt{\frac{8\,kT}{\pi m}} \; ,$$

(6.12)

where k is Boltzmann's constant, T the temperature and m the mass of the particle. Quite obviously, the average speed of the electrons is much greater than that of the ions due their smaller mass. At room temperature, the electron speed is a few times 10^6 cm/s while the positive ion speeds are on the order of 10^4 cm/s.

From kinetic theory, the linear distribution of charges after diffusing a time t can be shown to be Gaussian,

$$\frac{dN}{dx} = \frac{N_0}{\sqrt{4\,\pi Dt}} \exp\left(-\frac{x^2}{4\,Dt}\right) ,$$

(6.13)

where N_0 is the total number of charges, x the distance from the point of creation and D the *diffusion coefficient*. The *rms* spread in x is thus

$$\sigma(x) = \sqrt{2\,Dt} \; .$$

(6.14)

If three dimensions are considered, the spherical spread is given by

$$\sigma(r) = \sqrt{6\,Dt} \; ,$$

(6.15)

where r is the radial distance. The radial spread of ions in air under normal conditions, for example, is about 1 mm after 1 second [6.3]. The diffusion coefficient is a parameter which can be calculated from kinetic theory and can be shown to be

$$D = \tfrac{1}{3} v \lambda \ , \tag{6.16}$$

where λ is the mean free path of the electron or ion in the gas. For a classical ideal gas, the mean free path is related to the temperature T, and the pressure p, by

$$\lambda = \frac{1}{\sqrt{2}} \frac{kT}{\sigma_0 p} \ , \tag{6.17}$$

where σ_0 is the total cross section for a collision with a gas molecule. Substituting (6.12) and (6.17) into (6.16) then gives the explicit expression

$$D = \frac{2}{3\sqrt{\pi}} \frac{1}{p \sigma_0} \sqrt{\frac{(kT)^3}{m}} \ . \tag{6.18}$$

The dependence of D on the various parameters of the gas now becomes evident.

6.3.2 Drift and Mobility

In the presence of an electric field, the electrons and ions freed by radiation are accelerated along the field lines towards the anode and cathode respectively. This acceleration is interrupted by collisions with the gas molecules which limit the maximum average velocity which can be attained by the charge along the field direction. The average velocity attained is known as the *drift velocity* of the charge and is superimposed upon its normal random movement. Compared to their thermal velocities, the drift speed of the ions is slow, however, for electrons this can be much higher since they are much lighter.

In kinetic theory, it is useful to define the *mobility* of a charge as

$$\mu = u/E \ , \tag{6.19}$$

where u is the drift velocity and E the electric field strength. For positive ions, the drift velocity is found to depend linearly on the ratio E/p, (also known as the *reduced* electric field), up to relatively high electric fields. At a constant pressure, this implies that the mobility μ is a constant. For a given E, it is also quite clear that μ varies as the inverse of the pressure p.

For ideal gases, in which the moving charges remain in thermal equilibrium, the mobility can be shown to be related to the diffusion constant by

$$D/\mu = kT/e \ . \tag{6.20}$$

This is the result of a classical argument and is known as the Einstein relation.

Unlike positive ions, the mobility for electrons is much greater and is found to be a function of E. Velocities as high as a few times 10^6 cm/s can generally be attained before saturation sets in. The electric fields at this point are generally on the order of 1 kV/cm-atm. Figure 6.3 shows some measured results for electrons in different gas mixtures.

The gain in velocity of the electrons may also affect the diffusion rate if the mean energy of the electrons exceeds thermal energies. The factor kT in (6.20) is then replaced by this mean energy. The diffusion constant D then increases accordingly causing a greater spread of the electron cloud as given by (6.14) and (6.15). This has impor-

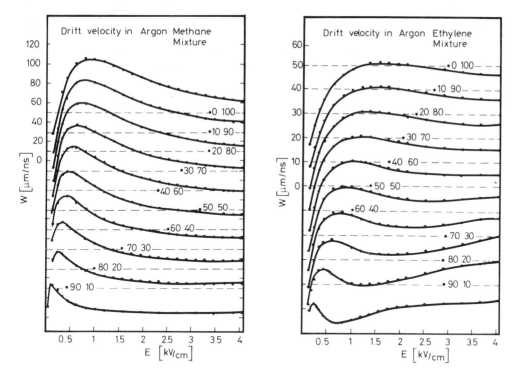

Fig. 6.3. Drift velocities of electrons in various gas mixtures as a function of electric field (from *Jean-Marie* et al. [6.7])

tant consequences for detectors such as the drift chambers which attempt to determine the position of a track by measuring the drift time of the ionization electrons. A more rigorous theory of electron transport in gases is given by *Palladino* and *Sadoulet* [6.9].

6.4 Avalanche Multiplication

Multiplication in gas detectors occurs when the primary ionization electrons gain sufficient energy from the accelerating electric field to also ionize gas molecules. The resulting secondary electrons then produce tertiary ionization and so on. This results in the formation of an avalanche. Because of the greater mobility of the electrons, the avalanche has the form of a liquid-drop with the electrons grouped near the head and the slower ions trailing behind as shown in Fig. 6.4.

If α is the mean free path of the electron for a secondary ionizing collision, then $1/\alpha$ is the probability of an ionization per unit path length. This is better known as the *first Townsend coefficient*. Figure 6.5 shows the coefficients for different gases. If there are n electrons, then in a path dx, there will be

$$dn = n\alpha\,dx \tag{6.21}$$

new electrons created. Integrating, this yields the total number of electrons created in a path x,

$$n = n_0 \exp(\alpha x) \;, \tag{6.22}$$

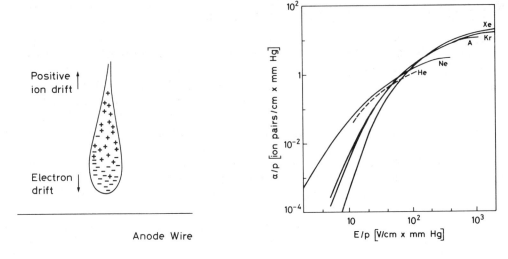

Anode Wire

▲

Fig. 6.4. Avalanche formation. Since the electrons are more mobile than the positive ions, the avalanche takes on the form of a liquid drop with the electrons at the head

Fig. 6.5. First Townsend avalanche coefficients for several gases (from *Brown* [6.9])

where n_0 is the original number of electrons. The multiplication factor is then

$$M = n/n_0 = \exp(\alpha x) \ . \tag{6.23}$$

More generally in the case of nonuniform electric fields such as (6.1), α is a function of x, in which case

$$M = \exp\left[\int_{r_1}^{r_2} \alpha(x)\,dx\right] \ . \tag{6.24}$$

While (6.24) can increase without limit, physically, the multiplication factor is limited to about $M < 10^8$ or $\alpha x < 20$ after which breakdown occurs. This is known as the Raether limit.

The multiplication factor or *gas gain* is of fundamental importance for the development of proportional counters. For this reason, various theoretical models have been developed for calculating α for different gases. A very early model by *Rose* and *Korff* [6.10], for example, gives

$$\frac{\alpha}{p} = A \exp\left(\frac{-Bp}{E}\right)$$

where A and B are constants depending on the gas. A short review of this and other models is given by *Kowalski* [6.11].

6.5 The Cylindrical Proportional Counter

As an introduction to the more sophisticated particle tracking detectors which will be discussed in the following sections, we will examine in more detail the simple proportional counter described in Sect. 6.1. The ionization chamber and Geiger counter, as we have mentioned, are not widely used in current physics experiments and are generally found as radiation monitoring and survey instruments. Good discussions of their construction and characteristics are given in the books by *Knoll* [6.12] and in the older works by *Rossi* and *Staub*, and *Wilkinson* [6.13].

The simple proportional counter is generally used for detecting low energy x-rays on the order of a few keV and very low energy electrons from sources which can be mixed with the counter gas. With a filling gas of high neutron capture cross section, such as BF_3 or ^3He, proportional counters can also be used for thermal or epithermal neutron detection. In general, the fill gases are at normal atmospheric pressure, however, to increase density and thus the efficiency, higher pressures may be used.

The basic feature of the proportional counter, of course, is the proportional gas multiplication which occurs. In order for this to be useful, however, the geometry of the electric field must be considered. Consider, for example, a planar detector consisting of parallel anode and cathode plates with a gas filling in between. The electric field is thus uniform and perpendicular to the plates. If a high enough voltage is applied to the electrodes, then the electrons created in an ionizing event will be accelerated towards the anode plate triggering avalanches along its path. It is not hard to see, however, that the total ionization produced will depend on the length of the path and thus on where the ionizing event occurs. For events with the same energy, therefore, the signal amplitude will vary with position and the relation between signal and energy is lost.

This problem can be solved by using a cylindrical geometry as we have already described in Sect. 6.1. The electric field, as we saw, is then given by (6.1). Because of the $1/r$ dependence, the field is relatively weak at large r but becomes intense very close to the surface of the anode wire. If the voltage is correctly chosen, ions and electrons created in the cylindrical volume will simply drift towards their respective electrodes. Only when the electrons are very close to the anode wire (a few wire diameters) does the electric field become intense enough for multiplication to occur. At this point, the avalanche occurs very quickly and the signal is generated. Regardless of the where the ionizing event occurs, therefore, all multiplications take place in a small region about the anode.

Figure 6.4 illustrates the development of an avalanche near the anode wire. The avalanche takes on a drop-like form with the electrons at the head and the positive ions in the rear. As the drop approaches the anode wire, diffusion of the charges causes the drop to surround the wire in the azimuthal direction. The avalanche, however, remains highly localized in the direction along the wire. The electrons are then collected very quickly (~ 1 ns) while the positive ions begin drifting towards the cathode. This ion drift is mainly responsible for the signal seen on the electrodes as we will see in the next section.

6.5.1 Pulse Formation and Shape

Contrary to what might be inferred from the brief description of ionization counters in Sect. 6.1, the pulse signal on the electrodes of ionization devices is formed by induction due to the movement of the ions and electrons as they drift towards the cathode and anode, rather than by the actual collection of the charges itself. Let us see how this occurs. For the cylindrical proportional counter, the electric field and potential can be written as

$$E(r) = \frac{CV_0}{2\pi\varepsilon}\frac{1}{r} \, ,$$

$$\varphi(r) = -\frac{CV_0}{2\pi\varepsilon}\ln\left(\frac{r}{a}\right) \, , \tag{6.25}$$

where r is the radial distance from the wire, V_0 the applied voltage, ε the dielectric constant of the gas, and

$$C = \frac{2\pi\varepsilon}{\ln(b/a)} \tag{6.26}$$

is the capacitance per unit length of this configuration.

Suppose that there is now a charge q located at a distance r from the central wire. The potential energy of the charge is then

$$E = q\varphi(r) \, . \tag{6.27}$$

If now the charge moves a distance dr, the change in potential energy is

$$dE = q\frac{d\varphi(r)}{dr}dr \, . \tag{6.28}$$

For a cylindrical capacitor, however, the electrostatic energy contained in the electric field is $E = \frac{1}{2}lCV_0^2$, where l is the length of the cylinder. If the movement of the charges is fast relative to the time that an external power supply can react to changes in the energy of the system, we can consider the system as closed. Energy is then conserved, so that

$$dE = lCV_0\,dV = q\frac{d\varphi(r)}{dr}dr \, . \tag{6.29}$$

Thus there is a voltage change,

$$dV = \frac{q}{lCV_0}\frac{d\varphi(r)}{dr}dr \tag{6.30}$$

induced across the electrodes by the displacement of the charge. Equation (6.30) is a general result, in fact, and can be used for any configuration.

For our cylindrical proportional counter, let us assume that an ionizing event has occurred and that multiplication takes place at a distance r' from the anode. The total induced voltage from the electrons is then

$$V^- = \frac{-q}{lCV_0}\int_{a+r'}^{a}\frac{d\varphi}{dr}dr = -\frac{q}{2\pi\varepsilon l}\ln\left(\frac{a+r'}{a}\right) \tag{6.31}$$

while that from the positive ions is

$$V^+ = \frac{q}{lCV_0} \int_{a+r'}^{b} \frac{d\varphi}{dr} \, dr = -\frac{q}{2\pi\varepsilon l} \ln\frac{b}{a+r'} \; . \qquad (6.32)$$

The sum of the two contributions is then $V = V^- + V^+ = -q/lC$ and their ratio of the contributions is

$$\frac{V^-}{V^+} = \frac{\ln\dfrac{a+r'}{a}}{\ln\dfrac{b}{a+r'}} \; . \qquad (6.33)$$

Since the multiplication region is limited to a distance of a few wire radii, it is easy to see that the contribution of the electrons is small compared to the positive ions. Taking some typical values of $a = 10$ μm, $b = 10$ mm and $r' = 1$ μm, V^- turns out to be less than 1% of V^+. The induced signal, therefore, is almost entirely due to the motion of the positive charges and one can ignore the motion of the electrons[1].

With this simplification we can now calculate the time development of the pulse. Thus,

$$V(t) = \int_{r(0)}^{r(t)} \frac{dV}{dr} \, dr = -\frac{q}{2\pi\varepsilon l} \ln\frac{r(t)}{a} \; . \qquad (6.34)$$

To find $r(t)$, we have the definition (6.19)

$$\frac{dr}{dt} = \mu E(r) = \frac{\mu CV_0}{\pi\varepsilon} \frac{1}{r} \qquad (6.35)$$

so that

$$r \, dr = \frac{\mu CV_0}{\pi\varepsilon} \, dt \; . \qquad (6.36)$$

Since the positive ions all come from the region close to the anode, we can set $r(0) = 0$ for simplicity. Integration then yields

$$r(t) = \left(a^2 + \frac{\mu CV_0}{\pi\varepsilon} t\right)^{1/2} \; . \qquad (6.37)$$

---

[1] The contribution of the electrons can be ignored only if they are all created near the anode. In some high gain gases, such as the magic gas to be discussed later, this is not always the case. Indeed, ultraviolet photons emitted in avalanches near the anode can extend the avalanche radially outward where the process is finally halted by the low field. In such cases the path length of the electrons is long and their contribution to the induced signal becomes significant [6.14].

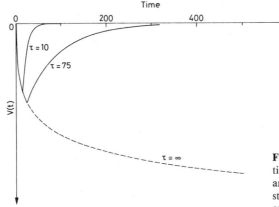

Fig. 6.6. Pulse signal from a cylindrical proportional counter. The pulse is usually cut short by an RC differentiating circuit with a time constant τ. The figure shows the effect of two different constants

Substituting into (6.34), we find

$$V(t) = -\frac{q}{4\,\pi\varepsilon l}\ln\left(1+\frac{\mu CV_0}{\pi\varepsilon a^2}t\right) = -\frac{q}{4\,\pi\varepsilon l}\ln\left(1+\frac{t}{t_0}\right)\;,\tag{6.38}$$

where $t_0 = \pi\varepsilon/\mu CV_0$. For this distance the total drift time T is

$$T = t_0(b^2 - a^2)\;.\tag{6.39}$$

This function is graphed in Fig. 6.6 for some typical values. Since it is not necessary to use the entire signal, the pulse is usually differentiated (see Sect. 14.23.2) to shorten its duration. In this manner only the faster rising part of the pulse is exploited. Depending on the time constant of the differentiator, the fall time of the resulting pulse will vary.

6.5.2 Choice of Fill Gas

The choice of a filling gas for proportional counters is governed by several factors: low working voltage, high gain, good proportionality and high rate capability. In general, these conditions are met by using a gas mixture rather than a pure one. For a minimum working voltage, noble gases are usually chosen since they require the lowest electric field intensities for avalanche formation. Because of its higher specific ionization and lower cost, argon is usually preferred. Pure argon as a filling gas, however, cannot be operated with gains of more than about $10^3 - 10^4$ without continuous discharge occurring [6.2]. This arises because of the high excitation energy (11.6 eV) for this element. Excited argon atoms formed in the avalanche thus deexcite giving rise to high energy photons capable of ionizing the cathode and causing further avalanches.

This problem can be remedied by the addition of a polyatomic gas, such as methane or alcohol. A few inorganic gases, such as CO_2, BF_3 can also be used. These molecules act as *quenchers* by absorbing the radiated photons and then dissipating this energy through dissociation or elastic collisions. A small amount of polyatomic gas already produces dramatic changes in counter operation. Indeed gains of up to 10^6 are obtained. In conventional proportional counters a commonly used mixture is 90% Ar and 10% methane (CH_4). This mixture is also known as *P10* gas. Another often used quencher is isobutane.

The gain can still be further increased by adding a judicious amount of electronegative gas such as freon (CF_3Br) or ethyl bromide. Apart from absorbing photons, these gases can also trap electrons extracted from the cathode before they can reach the anode to cause an avalanche. A gain of 10^7 can then be attained before the onset of Geiger-Müller operation.

The use of an organic quencher, unfortunately, results in further problems after high fluxes of radiation have been absorbed. In effect, the recombination of dissociated organic molecules results in the formation of solid or liquid polymers which accumulate on the anode and cathode of the detector. Positive ions reaching the cathode must then slowly diffuse through this layer to be neutralized. When a sufficiently large flux of radiation is present, however, the rate at which ions are produced is greater than the number leaking through to the cathode, so that a positive charge build-up occurs. This causes a continuous discharge in the counter which continues even after the radiation is removed. Only a complete cleaning can then regenerate the counter.

One possible solution to the problem is to use inorganic quenchers; however, they are much less efficient. The remedy instead is to add still another gas component. This time a small quantity of nonpolymerizing agent such as methylal or propylic alcohol. These agents then change the molecular ions at the cathode into a non-polymer species through an ion-exchange mechanism.

For sealed gas counters, an additional problem which arises is the large amount of quencher consumed in each detected event. At a gain of 10^6 and assuming 100 electron-ion pairs/event, about 10^8 molecules are dissociated per event. For a 10 cm^3 counter with a $90-10\%$ mixture at atmospheric pressure, then, changes in the operational characteristics will be observed after total 10^{10} events [6.2]. This, of course, is not a problem if a continuous gas flow is used.

6.6 The Multiwire Proportional Chamber (MWPC)

One of the basic requirements of experimental particle physics is the determination of particle trajectories. Up until about 1970, all tracking devices were, for the most part, optical in nature. Photographic emulsions, the cloud chamber, the bubble chamber, the spark chamber, etc. all required the recording of track information on film which was then analyzed frame by frame for events of interest. An all-electronic device, therefore, was greatly desired as it would allow more events to be treated more accurately. One possibility was to construct arrays of many proportional counters tubes; however, mechanically, this was not practical. The breakthrough occurred in 1968 with the invention of the multiwire proportional chamber by *Charpak* [6.15]. Charpak showed, in effect, that an array of many closely spaced anode wires in the same chamber could each act as independent proportional counters. Moreover, with transistorized electronics, each wire could have its own amplifier integrated onto the chamber frame to make a practical detector for position sensing. The MWPC was quickly adopted in high energy physics and stimulated a new generation of physics experiments. Since their invention, they have also found other applications as x-ray imaging devices in such diverse fields as astrophysics, crystallography, medecine, etc.

6.6.1 Basic Operating Principle

The basic MWPC consists of a plane of equally spaced anode wires centered between two cathode planes. Typical wire spacings are 2 mm with an anode-cathode gap width

Fig. 6.7. Basic configuration of a multiwire proportional chamber. Each wire acts as an independent proportional counter. The signal on the firing wire is negative while the signals on the neighboring wires are small and positive

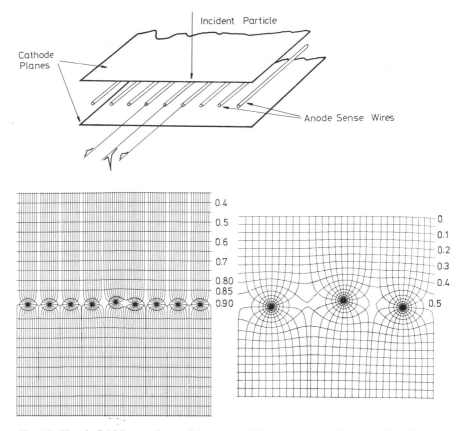

Fig. 6.8. Electric field lines and potentials in a multiwire proportional chamber. The effect of a slight wire displacement on the field lines is also shown (from *Charpak* et al. [6.16])

of 7 or 8 mm. Figure 6.7 illustrates this configuration schematically. If a negative voltage is applied to the cathode planes, an electric field configuration as shown in Fig. 6.8 arises. Except for the region very close to the anode wires, the field lines are essentially parallel and almost constant. If we assume an infinite anode plane with zero diameter wires, the potential is then given by

$$V(x, y) = -\frac{CV}{4\pi\varepsilon} \ln\left[4\left(\sin^2\frac{\pi x}{s} + \sinh^2\frac{\pi y}{s}\right)\right] ,$$ (6.40)

where V is the applied voltage, s the wire spacing, and C, the anode-cathode capacitance. If the $L \gg s \gg d$, this last quantity is given by

$$C = \frac{2\pi\varepsilon}{\dfrac{\pi L}{s} - \ln\dfrac{\pi d}{s}} ,$$ (6.41)

where L is the anode to cathode gap distance and d is the anode wire diameter. While the assumptions we have made are not met in a real chamber, (6.40) and (6.41) are usually good approximations for most purposes.

Near the anode wires the field takes on a $1/r$ dependence similar to the single wire cylindrical proportional chamber. If electrons and ions are now liberated in the constant field region they will drift along the field lines toward the nearest anode wire and opposing cathode. Upon reaching the high field region, the electrons will be quickly accelerated to produce an avalanche. The positive ions liberated in the multiplication process then induce a negative signal on the anode wire as we saw in the single proportional counter. The neighboring wires are also affected; however, the signals induced here are *positive* and of small amplitude as illustrated in Fig. 6.7. In a similar manner, a positive signal is induced on the cathode. There is thus no ambiguity as to which wire is closest to the ionizing event.

The signal from one anode plane, of course, only gives information on one coordinate of the ionizing event. The second coordinate may be obtained by using a second detector whose anode wires are oriented perpendicularly to the first. Usually both detectors are integrated into the same chamber to form an X-Y MWPC. This can be made even more sophisticated by adding diagonal planes of wires, etc., for additional information. (Another method for obtaining two-dimensional information is described in Sect. 6.6.5). To measure the trajectory of a particle, two or more aligned MWPCs may now be used to form a *telescope* as shown in Fig. 6.9. Reading the positions of the signaling wires then allows a reconstruction of the track.

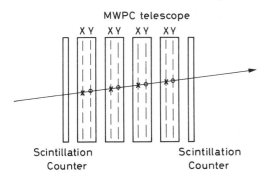

Fig. 6.9. A MWPC telescope for particle tracking. Each MWPC contains an X and Y wire plane. If the MWPC's are aligned, the measured coordinates allow a reconstruction of the particle trajectory. More than two planes of wires in a chamber may also be used

The spatial resolution of a MWPC depends on the anode wire spacing and is typically one-half this value. In a MWPC with typical 2 mm wire spacing, therefore, the spatial resolution is $\simeq \pm 1$ mm. This can be increased by using a finer spacing, however, going below 1 mm becomes difficult to work with.

An additional capability of the MWPC is multiple track resolution. Since each wire is a separate detector, two or more tracks can in principle be detected. However, this depends on their relative separation which must be at least as large as the wire spacing. Even before this point is reached, however, ambiguities in the reconstruction of the tracks may arise because of their proximity.

6.6.2 Construction

There are many methods for constructing a MWPC to which the abundant literature on this subject bears witness. The basic mechanical problem is to support the many wires and electrodes making up the chamber, which, for a high energy physics experiment, can be several square meters in area. A common technique is to stretch the anode wires out on a fiber-glass or epoxy frame and solder the ends to printed circuits which have been integrated on the frame. Tungsten wires, typically 20 – 40 μm in diameter, are generally used. The entire frame structure then has the appearance of a weaving loom. Similar frames are prepared for the cathode wire planes which can also be made of thin metal foils, wire mesh or thin strips of conducting material, and the detector windows which are usually made of mylar. The frames are then stacked and bolted together with the appropriate O-rings, etc., to assure gas tightness.

The alignment of the anode wires is, of course, critical for precise position measurement and it is important to consider their mechanical and electrical stability. When voltage is applied to the electrodes, there is an electrostatic repulsion between the wires which must be compensated by the mechanical tension in the wires. For a given voltage V and wire length L, the minimum mechanical tension [6.3] required is given by

$$T > \frac{1}{4\pi\varepsilon}\left(\frac{CVL}{s}\right)^2 , \tag{6.42}$$

where s is the wire spacing and C is the capacitance of the chamber. The maximum tension that can be applied, of course, depends on the wire thickness and elasticity. If the tension is insufficient, then the wires find a new equilibrium state in which they are alternatively displaced up and down, out of the plane of the anode.

A similar problem which arises is the attraction of the cathode towards the anode. In larger chambers, this can result in a *curving in* of the cathode near the center. This changes the gap width which then affects the multiplication factor of the detector. In such cases, a support structure of some kind must be placed in the gap. This, of course, modifies the electric field in the gap which then necessitates some sort of correcting electrode to restore the field.

For further details on chamber construction, we refer the reader to [6.2 – 3] and the references therein. Some more recent references are also given in [6.17].

6.6.3 Chamber Gas

The requirements for the fill gas of a multiwire chamber are identical to those for a simple proportional counter. A very high gain gas mixture which is widely used is the so-called *"magic gas"* consisting of Ar (75%), isobutane (24.5%) and freon-13B1 (0.5%), where the proportions are in volume. A small quantity of methylal may also be added to this mixture. The function of each of these components is described in Sect. 6.5.2. This gas was discovered early in the development of the MWPC and provides gains of close to 10^7. At this level of multiplication the signals are saturated and are thus independent of energy. However, they are fast because of the significant contribution of the electrons (see footnote in Sect. 6.5.1) and provide good timing resolution. The detector, moreover, is extremely efficient and the large signal amplitudes greatly simplify the readout electronics.

With modern low-noise, fast electronics, however, other gas mixtures perhaps with lower gain can be used to optimize other characteristics, for example, sensitivity to certain radiation type, energy resolution, high rate capability, etc.

6.6.4 Timing Resolution

The MWPC is a relatively fast detector and can also be used in timing applications. The timing resolution of the MWPC depends essentially on the drift time of the electrons.

The pulse on an anode wire, as we saw in Sect. 6.5.1, is induced by the positive ions in the avalanche as they drift towards the cathode. About 8% of this signal (see Fig. 6.6) is in the first 10 ns of the pulse so that the leading edge is quite fast. The timing uncertainty from this source is thus small. The limiting factor, instead, is the time spread between the arrival of the event and the occurrence of the avalanche at the anode. For a charged particle which traverses the chamber, ionization is distributed throughout the gap so that electrons will continue to arrive over a long period of time. The leading edge of the pulse thus rises in an approximately linear fashion. The first electrons to arrive, however, are those which pass closest to an anode wire. This distance cannot be more than a half a wire spacing so that the uncertainty in timing is not more than the time it takes for electrons to drift a half a wire spacing. For a typical two mm wire spacing, this is about $\simeq 25 - 30$ ns.

In the case of x-ray detection, the timing resolution is even better because of the short range of the conversion electrons. All electrons essentially arrive within a few nanoseconds of each other which greatly reduces the time spread.

6.6.5 Readout Methods

The information from a MWPC may be extracted in a number of different ways. The standard method is to consider each wire in the chamber as a separate detector connected to its own electronics. A typical pulse processing circuit for one wire is shown in Fig. 6.10. The signal is first amplified, then discriminated and shaped to a standard logic level. To allow selection or rejection of events, a gate preceded by an appropriate delay is added. A gating signal from the trigger or some other part of the electronics system can then be used to select only those events desired. Events passing through the gate are then stored in a memory from which the entire wire pattern of the chamber can later be readout by a computer. Besides serving to reject unwanted events the gate pulse can also be used to optimize detector performance by filtering events with multiple firing of the anode wires. This is discussed in the next section.

In addition to the separate wire readout method, a number of analog methods have been developed for obtaining one and two-dimensional information from one plane of anode wires only. These methods make use of the fact that the avalanche at the anode is also highly localized along the length of the wire. Information on this point can be obtained in a number of ways.

The *center of gravity* of method exploits the signals induced on the cathode of the MWPC. If the cathodes are arranged as a series of strips with one plane oriented parallel to the anode wires and other orthogonally, then the situation in Fig. 6.11 arises. The induced signals are largest on the strip closest to the avalanche and diminish proportionately with distance from the avalanche point. If y_i is the coordinate of the ith strip and Q_i is the measured charge on that strip, then the avalanche point can be estimated by calculating the center of gravity,

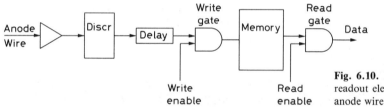

Fig. 6.10. Block diagram of the readout electronics for a MWPC anode wire

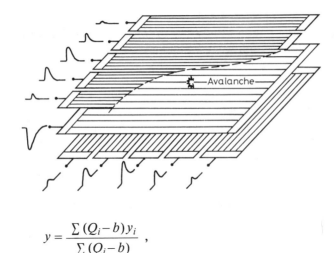

Fig. 6.11. Bidimensional readout using cathode strips (from *Breskin* et al. [6.18]). The signals from each strip are stored and the center of gravity of the distribution calculated to yield the position of the avalanche

$$y = \frac{\sum (Q_i - b) y_i}{\sum (Q_i - b)} \; , \tag{6.43}$$

where b is a small bias which is subtracted from each Q_i in order to correct for the dispersive effects of noise [6.19]. The x coordinate is obtained in a similar manner from the opposite cathode plane. While this should in principle give the same result as the anode signal, a better resolution in this coordinate can actually be obtained in some cases. This depends on the symmetry of the avalanche around the anode wire and consequently on the type of radiation being detected. With soft x-rays, a resolution of 35 μm has been obtained [6.20] in the y-direction along the anode wire, and a somewhat worse resolution in x. With a high energy charged particle beam, however, a resolution of closer to 100 μm is more typical.

The method of *charge division* relies on the fact that the charge collected at either end of a resistive anode wire is divided in proportion to the length of wire from the point at which the charge is injected. This is illustrated in Fig. 6.12. If Q_A and Q_B are the charges collected then the coordinate along the wire is

$$y = L \frac{Q_A}{Q_A + Q_B} \; , \tag{6.44}$$

where L is the length of the wire. Accuracies as high as 0.4% of the wire length have been obtained [6.21].

One of earliest and simplest analog methods is the delay line technique which was developed before the current sophisticated electronics for MWPCs became available. In this technique, external delay lines are capacitively coupled to the cathode or anode

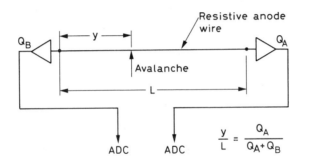

Fig. 6.12. Charge division method for coordinate readout on MWPC's

planes of the chamber [6.21]. Using the anode signal or some other triggering signal as a start, the time difference between the arrival of the signals at the ends of the delay line are measured. This then yields the two coordinates of the avalanche.

6.6.6 Track Clusters

Up to now, we have assumed that only one wire fires per event; this is usually not the case under real conditions. Particles traversing the chamber at an angle to the anode plane generally produce a cluster of firings as they cross over several wires as shown in Fig. 6.13. Moreover, even with perpendicularly incident particles, the creation of high energy ionization electrons (delta rays) will also produce multiple firings a fraction of the time. Because of the differing distances to the anode, the wires signals are spread out over a time period corresponding to the drift times of the electrons. The desired signal, of course, is the one which is closest to the event and the one which arrives first. One way to limit the cluster size is by adjusting the width of the gating signal in the readout electronics so that the late arriving electrons are eliminated. This can usually limit cluster size to 1 or 2 wires depending on the track angle.

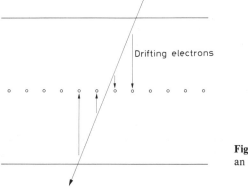

Fig. 6.13. Track clusters caused by particles arriving at an angle to the anode plane

The optimum gate width can be found by placing the MWPC in a well collimated beam and varying the width while recording the number of single and multiple firings. The point at which the single firings are at a maximum and the multiple firings at a minimum then determines the optimum gate width. In typical chambers with 2 mm wire spacing and 8 mm gaps, the minimum gate time is about 30 ns. With smaller gate widths, good events begin to be rejected and the detection efficiency drops.

The cluster size for an event may also be controlled by increasing the amount of electronegative gas mixed with the chamber gas. Electrons produced in the far regions will then have a smaller probability of reaching the anodes thus limiting the number of wires producing a signal. The amount of electronegative gas that can be added, of course, is limited by the requirement for a good overall efficiency of the chamber.

6.6.7 MWPC Efficiency

The intrinsic efficiency of the MWPC depends on the number of electron-ion pairs produced and collected in the chamber. As such it is dependent on the dE/dx of the fill gas, the width of the gap, the pressure of the gas, the amount of electronegative gases, the high voltage applied, the threshold set on the electronics, the gate width on the readout, etc.

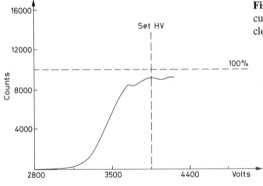

Fig. 6.14. A high voltage single track efficiency curve for an MWPC. In a good chamber this can be close to 100%

Assuming that the dimensions of the chamber gap and the stopping power and composition of the fill gas are adequate, the determining factors for the efficiency of a given chamber are the high voltage and the electronics. For the typical chambers we have been describing, the efficiency for charged particles can be very high, in fact, with values on the order of 98 – 99% or more. Figure 6.14 shows the measured efficiency for single tracks in a typical MWPC filled with *magic* gas as a function of the applied voltage. This measurement was performed by placing the chamber in a proton beam along with two beam-defining scintillation counters sandwiching the chamber. Using the coincidence between the counters as a trigger, the number of single tracks observed versus the number of triggers then yields the efficiency. As can be seen, a plateau is obtained at which the efficiency is close to 100%. The working voltage is then usually set at some point near the middle of this plateau.

The threshold on the electronics and the gate width on the readout are also important as we have mentioned. The threshold, of course, must be low enough such that all valid signals will be accepted but high enough so that noise will be rejected. In general, a threshold of about 1/10 of the peak amplitude is sufficient [6.3] for 100% detection efficiency of minimum ionizing particles. In typical MWPCs this is about 0.5 mV on 1 kΩ. The width of the gate on the readout electronics must also be sufficiently large so as not to lose events. The minimum width as we have indicated is about 30 ns.

The efficiency of a chamber is also dependent on the count rate. At a fixed voltage, the efficiency will generally decrease as the incident flux is increased. This is the result of a space charge build-up around the chamber wires due to the positive ions released in avalanches. Since their drift velocities are low, an accumulation of ions occurs as the number of avalanches increases. The effect of this charge is to alter the electric field and lower the gain around the wire. The pulse height spectrum is then shifted downward so that part of it is lost below the threshold thereby causing the efficiency to drop. This loss can be recovered to some extent by raising the voltage, however, the plateau will generally be narrower. The threshold may also be lowered, if possible. As a general rule, the maximum flux rate for per unit length of MWPC wire is about 10^4/s per millimeter.

6.7 The Drift Chamber

Early in the development of the MWPC, it was realized that spatial information could also be obtained by measuring the drift time of the electrons coming from an ionizing event. If a trigger is available to signal the arrival of a particle and the drift velocity is known, then the distance from the sensing wire to the origin of the electrons is

$$x = \int_{t_0}^{t_1} u \, dt \; , \tag{6.45}$$

where t_0 is the arrival time of the particle and t_1 is the time at which the pulse appears at the anode. In practice, of course, it is highly desirable to have a constant drift velocity, u, and hence a constant electric field, so as to have a linear relationship between time and distance. Drift chambers exploiting this aspect of electron transport were built shortly after the MWPC and have been used extensively in particle physics ever since.

Figure 6.15 schematically illustrates the basic operation of a drift chamber. The drift *cell* is defined at one end by a high voltage electrode and at the other end by the anode of a simple proportional counter. In order to create a constant electric field, a series of cathode *field* wires individually held at appropriate voltages line the drift region. To signal the arrival of a particle, a scintillation counter covering the entire sensitive area is placed before or after the chamber. A particle traversing the chamber and scintillator, now, liberates electrons in the gas which then begin drifting towards the anode. At the same time, the fast signal from the scintillator starts a timer. The signal created at the anode as the drifting electrons arrive then stops the timer to yield the drift time.

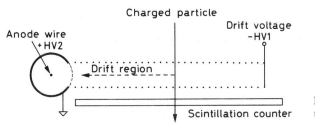

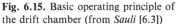

Fig. 6.15. Basic operating principle of the drift chamber (from *Sauli* [6.3])

While drift paths as long as 50 cm have been used with this simple structure, the usual drift region is about $5 - 10$ cm. Shorter pathlengths minimize the effect of diffusion and avoid the use of very high voltages. With typical drift velocities of about 5 cm/μs, this then yields drift times of 1 or 2 μs. This is also known as the *memory* time of the chamber. To cover a wider surface area, many adjacent drift cells can be used. Drift chambers several meters long have been constructed in this manner. To obtain several points on a track, several drift chambers with different wire orientations may also be stacked together.

In principle, the chamber structure used for a MWPC may also be employed for drift chambers as well. The wire spacings, of course, will be somewhat larger so as to have a more reasonable drift time. A problem, however, is the nonuniformity of the field in the inter-anode wire gap (see Fig. 6.8). To correct this, additional field wires are usually added in the space between anode wires. One design optimized for high-resolution [6.22] is shown in Fig. 6.16. Here, the potential on the cathode wires is not constant, but instead, uniformly graded downwards from 0 (ground) on the wire facing the

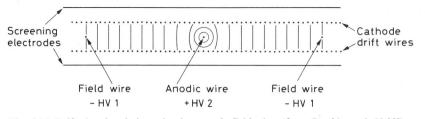

Fig. 6.16. Drift chamber design using interanode field wires (from *Breskin* et al. [6.22])

anode sense wire to a high negative voltage on the cathode wires facing the field wires on either side of the anode. The resulting equipotential lines are also shown in the figure.

While Fig. 6.16 shows a planar design, drift chambers can also be made in cylindrical form. Such chambers then give information on the r and ϕ coordinates of a particle trajectory. These are very much used at collider machines for direct visualization of outgoing particle tracks [6.23] without having to mathematically fit the detected coordinates. This requires many points on a track to be measured and thus a high density of wires. This also results in a better resolution of vertices and multiple tracks.

The advantage of drift chambers is the relatively small amount of wires and electronics required and the large surface areas which can be covered. They are generally easier to operate, however, much more attention must be given to the fill gas and the field uniformity if good resolution is desired. Like the MWPC, the maximum counting rates are also limited to about 10^4/s-mm per wire. Since there are less wires in a drift chamber, however, the maximum allowable flux on a drift chamber is generally less than for an MWPC.

6.7.1 Drift Gases

Since a precise knowledge of the drift velocity is necessary to operate drift chambers, the choice of a fill gas is of utmost importance. The basic criteria for choosing a drift gas are generally the same as those for a MWPC, however, particular attention must be given to the drift properties.

The purity of the gas, of course, is very important. In particular, if electronegative gases are present, electrons will be captured as they drift to the anodes. The allowable level for these impurities depends on the length of the drift path: the longer this path, the higher the required purity level.

To maximize operational stability, a gas exhibiting drift velocity saturation at not too high electric fields should be chosen. Some examples can be seen in Fig. 6.3 which shows the drift velocities of electrons in some gases flattening out over a range of electric field intensities. When operated in this region, the drift velocity is then less sensitive to field inhomogeneities, changes in the operating voltage, temperature, etc.

The magnitude of the drift velocity must also be considered. If the chamber is to operate at high count rates, then the drift velocity must be high so as to minimize dead time. If, instead, high spatial resolution is desired, a slower drift velocity is required to minimize timing errors. High drift velocities can be obtained with CF_4 and a hydrocarbon quencher [6.24], for example, while slow velocities are observed in gases such as dimethylether (DME) [6.25], CO_2, or $He - C_2H_6$.

6.7.2 Spatial Resolution

The spatial resolution of a drift chamber depends on how well the relation between drift time and space coordinate is known and the amount of diffusion suffered by the electrons as they drift. The latter factor depends on the length of the drift path. If we assume a uniform drift velocity, then it can be seen from (6.14), that the spread in the electron cloud after a distance x is simply

$$\sigma = \sqrt{\frac{2Dx}{\mu E}} \ . \tag{6.46}$$

The width of the diffusion distribution thus goes as the square root of the drift path. (This is not the error in localizing the center of the distribution, however. Recalling Sect. 4.4.4, the error is $\sigma/\sqrt{n}$ where n is the number of electrons in the cloud). To obtain higher spatial resolutions, therefore, a smaller drift length is necessary. With a 5 cm drift path, resolutions on the order of 100 μm can be obtained. The intrinsic accuracy of the drift chamber can be much better, of course, and has been measured to be as low as 50 μm over a drift space of about 5 mm [6.3].

In more recent years, much effort has been put into designing very high resolution chambers with spatial resolutions of 50 μm or less for the next generation of high-energy experiments. These efforts have included searches for new low-diffusion, low drift velocity gases and new chamber designs and concepts. Operation at high pressure has also been investigated as a means of reducing diffusion as well as a new means of timing by triggering on the scintillation light emitted by the avalanche. A review of some of these efforts up to the time of this writing may found in [6.20] and [6.26].

6.7.3 Operation in Magnetic Fields

A common occurrence in particle physics experiments is the location of the detector in or near the field of a magnet. In such cases, it is quite obvious that the path of the drifting electrons and the drift velocity will be altered by the Lorentz force. A precise knowledge of the magnetic field is then necessary in order to correlate the drift time with position. In some cases also, it may be possible to adjust the electric field direction so as to compensate the effects of the magnetic field.

6.8 The Time Projection Chamber (TPC)

The most sophisticated of the current ionization detectors is the *time projection chamber* or TPC. This device is essentially a three-dimensional tracking detector capable of providing information on many points of a particle track along with information on the specific energy loss, *dE/dx*, of the particle. For this reason, the TPC has been referred to as an *electronic bubble chamber*. It has played an essential role in physics experiments at high-energy electron-positron colliders and more recently has been proposed for use in many other types of experiments [6.27, 28].

The TPC makes use of ideas from both the MWPC and drift chamber. Its basic structure is sketched in Fig. 6.17. The detector is a essentially a large gas-filled cylinder

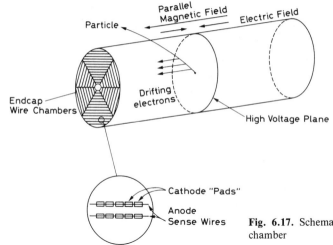

Fig. 6.17. Schematic diagram of a time projection chamber

with a thin high voltage electrode at the center. At high energy colliders, the diameter and length of the cylinder can be as large as two meters. When voltage is applied, a uniform electric field directed along the axis is created. A parallel magnetic field, the function of which will be explained in a moment, is also applied. The ends of the cylinder are covered by sector arrays of proportional anode wires arranged as shown. Parallel to each wire is a cathode strip cut up into rectangular segments. These segments are also known as cathode *pads*.

At a collider machine, the detector is positioned so that its center is at the interaction point. The TPC thus subtends a solid angle close to 4π. Particles emanating from this point pass through the cylinder volume producing free electrons which drift towards the endcaps where they are detected by the anode wires as in a MWPC. This yields the position of a space point projected onto the endcap plane. One coordinate is given by the position of the firing anode wire while the second is obtained from the signals induced on the row of cathode pads along the anode wire. Using the center-of-gravity method described in Sect. 6.6.5, this locates the position of the avalanche along the firing anode wire. The third coordinate, along the cylinder axis, is now given by the drift time of the ionization electrons. Since all ionization electrons created in the sensitive volume of the TPC will drift towards the endcap, each anode wire over which the particle trajectory crosses will sample that portion of the track. This yields many space points for each track allowing a full reconstruction of the particle trajectory. This is illustrated in more detail in Fig. 6.18.

Because of the relatively long drift distance, diffusion, particularly in the lateral direction, becomes a problem. This is remedied by the parallel magnetic field which confines the electrons to helical trajectories about the drift direction. This turns out to reduce diffusion by as much as a factor of 10. In order to avoid deviating the trajectories of the drifting electrons, the magnetic and electric fields must be in perfect alignment and uniform over the volume of the drift zone down to about one part in 10^4.

A problem which arises during operation is the accumulation of a space charge in the drift volume due to positive ions from avalanches drifting back towards the central cathode. These ions are sufficiently numerous that a distortion of the electric field in the drifting volume occurs. This is prevented by placing a grid at ground potential just

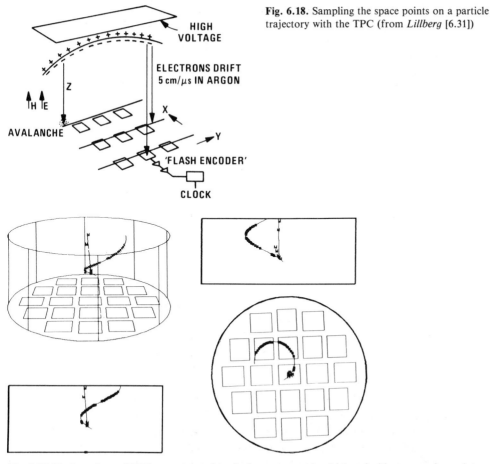

Fig. 6.18. Sampling the space points on a particle trajectory with the TPC (from *Lillberg* [6.31])

Fig. 6.19. Various views of TPC reconstructed tracks from an event in which an incident muon decays into a positron. The muon track is denoted by a *M* (from *Lillberg* [6.31])

before the anode wires. Positive ions are then captured as this grid rather than drifting back into the sensitive volume. The grid also serves to separate the drift region from the avalanche zone and allows an independent control of each.

Since the charge collected at the endcaps is proportional to the energy loss of the particle, the signal amplitudes from the anode also provide information on the dE/dx of the particle. If the momentum of the particle is known from the curvature of its trajectory in the magnetic field, for example, then this information can be used to identify the particle[2]. In order for this method to work, however, sufficient resolution in the dE/dx measurement must be obtained. This is much more difficult to realize as many factors must be considered, e.g., electron loss due to attachment, wire gain variations in position and time, calibration of the wires, saturation effects, choice of gas and operating pressure, etc., all of which require careful thought!

Because of the very large amount of data produced for each event, an important consideration is the readout and data acquisition system for a TPC. The original TPC

---

[2] A general review of the particle identification method using dE/dx measurements is given in [6.29].

used at the PEP electron-positron collider employed *charged-coupled devices* (CCDs) to store the information [6.30] from the sense wires for later readout by slower ADCs. Charge-coupled devices are essentially analog shift registers capable of storing time and pulse height information. The CCD continuously samples information from the TPC wires at a rate determined by an external clock (about 15 MHz) and dumps this information as long as there is not a trigger signifying a valid event. If a trigger signal arrives indicating a valid event, however, the clock rate is slowed down by about a factor of 100 and the CCD contents read and digitized by ADCs. The information is then passed to a computer for track reconstruction.

A second approach which has been used in later TPCs is to use *flash* ADCs (see Sect. 14.11) directly coupled to the sense wires. These ADCs are sufficiently fast such that several wires can be multiplexed into one ADC. A description of one such readout and data acquisition system is given in [6.31]. Figure 6.19 shows a reconstructed event from a TPC.

6.9 Liquid Ionization Detectors

With the ever increasing requirements of nuclear and high-energy experiments, physicists have for some time now looked at the possibility of using liquids as an ionizing medium. Liquids, of course, present many advantages such as higher density and lower diffusion. Per unit length of liquid, therefore, a greater number of ion-electron pairs will be created. This, for example, would allow MWPCs with smaller gap and wire spacings to be constructed. The smaller gap thickness, in turn, would also result in smaller wire clusters. Similarly, lower diffusion would result leading to less track broadening and a higher spatial resolution.

Liquids, unfortunately, suffer from a number of problems. Relative to gases, for example, ionization and transport phenomena in liquids are much less well understood. The principal problem, however, is technical: the presence of electronegative impurities such as oxygen which can attach drifting electrons. This requires purification to very high levels which is not always realizable for a given liquid. This situation has essentially restricted the choice of liquid ionization media to the noble elements, particularly argon and xenon, and a few hydrocarbons in which attaining a high level of purity would appear to be technically feasible. Since the boiling points of these elements are at low temperatures, this also implies an additional cryogenic system for these detectors.

The use of LID's dates back to the early 1950s when *Marshall* [6.32] was the first to build a small centimeter-size liquid detector and to actually use it in an experiment. Because of the successful development of semiconductor detectors at that time, however, LID research remained relatively inactive until the late 1960s when Luis Alvarez' group at Berkeley began investigating the use of liquified noble gases in wire chambers as a means of increasing spatial resolution. So as to keep the electronics simple, it was hoped that multiplication could be induced in the liquid to provide a larger signal. The experiments, however, were only partially successful. A type of Geiger-Müller multiplication was, in fact, observed in a simple cylindrical counter filled with liquid argon (LA) [6.33], however, the detection efficiency was very low (<20%). Better results were obtained with liquid xenon (LXe) in which proportional multiplication and an almost 100% detection efficiency were observed [6.34]; however, an impurity level of better than a few parts per million was necessary which required a long, painstaking

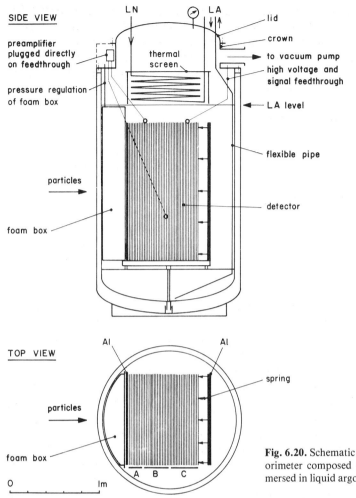

Fig. 6.20. Schematic diagram of a liquid argon calorimeter composed of a stack of steel plates immersed in liquid argon (from *Delfosse* et al. [6.39])

purification process. Moreover, very thin wires (3.5 μm) were required to obtain sufficient gain. Based on this experience, a small multi-wire LXe ionization chamber with 12 μm wires was constructed and demonstrated [6.35] to have a spatial resolution of ±15 μm. Given the thickness of the wires, no proportional multiplication was observed. Despite this success, however, developments in this direction have been very slow probably because of the many technical difficulties involved and the lack of understanding of ionization and charge transport processes in liquids. More recently, a prototype micro-strip ionization chamber [6.36] has been operated with a resolution of less than 8.5 μm. Again, however, many technical problems still remain.

At present, the only liquid ionization detector currently in routine use is the liquid argon calorimeter first introduced by *Willis* and *Radeka* [6.37] for electromagnetic shower detection in high energy physics experiments. This instrument essentially consists of a horizontal stack of equally-spaced steel plates immersed in liquid argon as shown in Fig. 6.20. By applying a voltage between the steel plates, this forms a series of ionization chambers in which the plates act as electrodes and the LA as an ionizing medium. Since the gaps are small on the order of 2 cm, the required purity level for the LA is only a few parts in a million, which is readily achieved for this liquid. High energy photons incident on the calorimeter face convert on the steel plates creating an electromagnetic shower which is absorbed by the following layers of steel and LA. The ionization created in each layer of argon allows one to sample the energy deposited by the shower as it develops. Since the entire shower is absorbed, the total ionization collected is proportional to the total energy of the shower. Such calorimeters have since become standard devices in high-energy experiments and similar instruments have been developed for hadron showers as well. A review of the calorimetric technique in general is given by *Fabjan* and *Ludlam* [6.38].

Despite the many technical problems, the attractiveness of liquids has nevertheless led to new proposals for more sophisticated detectors such as the liquid argon drift chamber [6.40] and the liquid argon TPC [6.41]. Because drift paths of a few tens of centimeters will be called for in these detectors, a purity level of less than 1 part in a billion will be necessary. Moreover, this purity must be maintained in the detector for long periods. This is to be contrasted to the LA calorimeter where purities of 1 ppm are generally used. Recent measurements of electron drift in a liquid argon [6.42] test chamber using a relatively simple purification system have been encouraging, however, yielding a purity of 0.2 ppb of oxygen equivalent and a measured extrapolated attenuation length of 7 m at an electric field strength of 1 kV/cm.

As a final remark, we note the recent research into new ionization liquids. While LA can be purified to very high levels, the cryogenic system required proves to be cumbersome and constraining. For this reason, some physicists have begun searching for a suitable room-temperature liquid with the required drift properties and purity. Nonpolar organic liquids such as tetramethylsilane or 2,2,4,4-tetramethylpentane appear to show some of the desired properties and are now currently under investigation. Ionization due to radiation has recently been observed in tetramethylsilane [6.43]. A general review of the physics and chemistry of these liquids is given in [6.44].

7. Scintillation Detectors

The *scintillation* detector is undoubtedly one of the most often and widely used particle detection devices in nuclear and particle physics today. It makes use of the fact that certain materials when struck by a nuclear particle or radiation, emit a small flash of light, i.e. a scintillation. When coupled to an amplifying device such as a photomultiplier, these scintillations can be converted into electrical pulses which can then be analyzed and counted electronically to give information concerning the incident radiation.

Probably the earliest example of the use of scintillators for particle detection was the spinthariscope invented by Crookes in 1903. This instrument consisted of a ZnS screen which produced weak scintillations when struck by α-particles. When viewed by a microscope in a darkened room, they could be discerned with the naked eye, although some practice was necessary. It was tedious to use, therefore, and thus never very popular, even though it was spectacularly employed by Geiger and Marsden in their famous α scattering experiments. Indeed, with the invention of the gaseous ionization instruments, the optical scintillation counter fell into quick disuse.

In 1944, not quite a half century later, Curran and Baker resuscitated the instrument by replacing the human eye with the then newly developed photomultiplier tube. The weak scintillations could now be counted with an efficiency and reliability equal to that of the gaseous ionization instruments. Thus was born the modern electronic scintillation detector. New developments and improvements followed rapidly so that by the mid-1950's scintillation detectors were among the most reliable and convenient available. This is still true today. In this chapter, we will survey the existing materials and current techniques in use as well as describe their basic underlying principles.

7.1 General Characteristics

The basic elements of a scintillation detector are sketched below in Fig. 7.1. Generally, it consists of a scintillating material which is optically coupled to a photomultiplier either directly or via a light guide. As radiation passes through the scintillator, it excites the atoms and molecules making up the scintillator causing light to be emitted. This

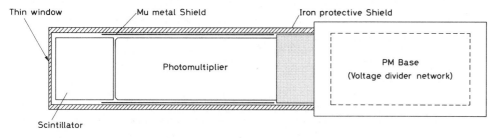

Fig. 7.1. Schematic diagram of a scintillation counter

light is transmitted to the photomultiplier (PM or PMT for short) where it is converted into a weak current of photoelectrons which is then further amplified by an electron-multiplier system. The resulting current signal is then analyzed by an electronics system.

In general, the scintillator signal is capable of providing a variety of information. Among its most outstanding features are:

1) *Sensitivity to Energy.* Above a certain minimum energy, most scintillators behave in a near linear fashion with respect to the energy deposited, i.e., the light output of a scintillator is directly proportional to the exciting energy. Since the photomultiplier is also a linear device, (when operated properly!), the amplitude of the final electrical signal will also be proportional to this energy. This makes the scintillator suitable as an energy spectrometer although it is not the ideal instrument for this purpose.

2) *Fast Time Response.* Scintillation detectors are fast instruments in the sense that their response and recovery times are short relative to other types of detectors. This faster response allows timing information, i.e., the time difference between two events, to be obtained with greater precision, for example. This and its fast recovery time also allow scintillation detectors to accept higher count rates since the dead time, i.e., the time that is lost while waiting for the scintillator to recover, is reduced.

3) *Pulse Shape Discrimination.* With certain scintillators, it is possible to distinguish between different types of particles by analyzing the shape of the emitted light pulses. This is due to the excitation of different fluorescence mechanisms by particles of different ionizing power. The technique is known as *pulse-shape discrimination* and is discussed in more detail later in this chapter.

Scintillator materials exhibit the property known as *luminescence.* Luminescent materials, when exposed to certain forms of energy, for example, light, heat, radiation, etc., absorb and reemit the energy in the form of visible light. If the reemission occurs immediately after absorption or more precisely within 10^{-8} s (10^{-8} s being roughly the time taken for atomic transitions), the process is usually called *fluorescence.* However, if reemission is delayed because the excited state is metastable, the process is called *phosphorescence* or *afterglow.* In such cases, the delay time between absorption and reemission may last anywhere from a few microseconds to hours depending on the material.

As a first approximation, the time evolution of the reemission process may be described as a simple exponential decay (Fig. 7.2)

$$N = \frac{N_0}{\tau_d} \exp\left(\frac{-t}{\tau_d}\right) , \tag{7.1}$$

where N is the number of photons emitted at time t, N_0 the total number of photons emitted, and τ_d the decay constant. The finite rise time from zero to the maximum in most materials is usually much more rapid than the decay time and has been taken as zero here for simplicity.

While this simple representation is adequate for most purposes, some, in fact, exhibit a more complex decay. A more accurate description, in these cases, may be given by a two-component exponential

$$N = A \exp\left(\frac{-t}{\tau_f}\right) + B \exp\left(\frac{-t}{\tau_s}\right) , \tag{7.2}$$

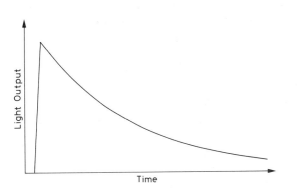

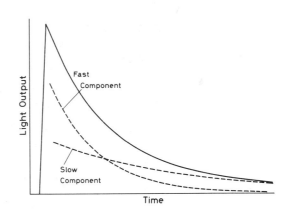

Fig. 7.2. Simple exponential decay of fluorescent radiation. The rise time is usually much faster than the decay time

Fig. 7.3. Resolving scintillation light into *fast* (prompt) and *slow* (delayed) components. The *solid line* represents the total light decay curve

where τ_s and τ_f are the decay constants. For most scintillators, one component is generally much faster than the other so that it has become customary to refer to them as the *fast* and *slow* components (hence the subscripts f and s), or the *prompt* and *delayed* components. Their relative magnitudes, A and B, vary from material to material, although it is the fast component which generally dominates. Figure 7.3 shows the relation between these components. As will be seen in a later section, the existence of these two components forms the basis for the technique of pulse shape discrimination.

While many scintillating materials exist, not all are suitable as detectors. In general, a good detector scintillator should satisfy the following requirements:

1) high efficiency for conversion of exciting energy to fluorescent radiation
2) transparency to its fluorescent radiation so as to allow transmission of the light
3) emission in a spectral range consistent with the spectral response of existing photomultipliers
4) a short decay constant, τ.

At present, six types of scintillator materials are in use: organic crystals, organic liquids, plastics, inorganic crystals, gases and glasses. In the following sections we will briefly describe each category. Their basic properties are summarized in Table 7.1.

7.2 Organic Scintillators

The organic scintillators are aromatic hydrocarbon compounds containing linked or condensed benzene-ring structures. Their most distinguishing feature is a very rapid decay time on the order of a few nanoseconds or less.

Scintillation light in these compounds arises from transitions made by the *free* valence electrons of the molecules. These delocalized electrons are not associated with any particular atom in the molecule and occupy what are known as the *π-molecular orbitals*. A typical energy diagram for these orbitals is shown in Fig. 7.4, where we have distinguished the spin singlet states from the spin triplet states. The ground state is a singlet state which we denote by S_0. Above this level are the excited singlet states (S^*, S^{**}, ...) and the lowest triplet state (T_0) and its excited levels (T^*, T^{**}, ...). Also associat-

Table 7.1. Physical properties of various commercial scintillators (data from Nuclear Enterprises scintillator catalog [7.1])

	Scintillator	Type	Density	Refractive index	Melting softening or boiling point C[a]	Light output (% Anthracene)	Decay constant, main component [ns]	Wavelength of maximum emission [nm]	Content of loading element (% by wt.)	H/C No. of H atoms/No. of C atoms	Principal applications
Plastic	NE 102A	Plastic	1.032	1.581	75	65	2.4	423		1.104	γ, α, β, fast n
	NE 104	Plastic	1.032	1.581	75	68	1.9	406		1.100	ultra-fast counting
	NE 104B	Plastic	1.032	1.58	75	59	3.0	406		1.107	with BBQ light guides
	NE 105	Plastic	1.037	1.58	75	46		423		1.098	dosimetry
	NE 110	Plastic	1.032	1.58	75	60	3.3	434		1.104	γ, α, β, fast n etc.
	NE 111A	Plastic	1.032	1.58	75	55	1.6	370		1.103	ultra-fast timing
	NE 114	Plastic	1.032	1.58	75	50	4.0	434		1.109	as for NE 110
	NE 160	Plastic	1.032	1.58	80	59	2.3	423		1.105	use at high temperatures
	Pilot U	Plastic	1.032	1.58	75	67	1.36	391		1.100	ultra fast timing
	Pilot 425	Plastic	1.19	1.49	100			425		1.6	Cherenkov detector
Liquid	NE 213	Liquid	0.874	1.508	141	78	3.7	425		1.213	fast n (P.S.D.)
	NE 216	Liquid	0.885	1.523	141	78	3.5	425		1.171	α, β (internal counting)
	NE 220	Liquid	1.036	1.442	104	65	3.8	425	O 29%	1.669	internal counting, dosimetry
	NE 221	Gel	1.08	1.442	104	55	4	425		1.669	α, β (internal counting)
	NE 224	Liquid	0.877	1.505	169	80	2.6	425		1.330	γ, fast n
	NE 226	Liquid	1.61	1.38	80	20	3.3	430		0	γ, insensitive to n
	NE 228	Liquid	0.71	1.403	99	45		385		2.11	n
	NE 230	Deuterated liquid	0.945	1.50	81	60	3.0	425	D 14.2%	0.984	(D/C) special applications
	NE 232	Deuterated liquid	0.89	1.43	81	60	4	430	D 24.5%	1.96	(D/C) special applications
	NE 233	Liquid	0.874	1.506	117	74	3.7	425		1.118	α, β (internal counting)
	NE 235	Liquid	0.858	1.47	350	40	4	420		2.0	large tanks
	NE 250	Liquid	1.035	1.452	104	50	4	425	O 32%	1.760	internal counting, dosimetry
Loaded liquid	NE 311 & 311A	B loaded liquid	0.91	1.411	85	65	3.8	425	B 5%	1.701	n, β
	NE 313	Gd loaded liquid	0.88	1.506	136	62	4.0	425	Gd 0.5%	1.220	n
	NE 316	Sn loaded liquid	0.93	1.496	148.5	35	4.0	425	Sn 10%	1.411	γ, x-rays
	NE 323	Gd loaded liquid	0.879	1.50	161	60	3.8	425	Gd 0.5%	1.377	n

Neutron (ZnS-type) and glass	NE 422 & 426	^{6}Li-ZnS(Ag)	2.36		110	300	200	450	Li 5%	0.715	slow n
	NE 451	ZnS(Ag) plastic	1.443		110	300	200	450		0.858	fast n
	NE 901, 902, 903	Glass	2.64	1.58	c. 1200	28	20 & 60	395	Li 2.3%		n, β
	NE 904, 905, 906	Glass	2.5	1.55	c. 1200	25	20 & 58	395	Li 6.6%		n
	NE 907, 908	Glass	2.42	1.566	c. 1200	20	18 & 62	399	Li 7.5%		n
	NE 912, 913	Glass	2.3	1.55	c. 1200	25	18 & 55	397	Li 7.7%		n, β (low background)
Crystal	Anthracene	Crystal	1.25	1.62	217	100	30	447			γ, α, β, fast n
	Stilbene	Crystal	1.16	1.626	125	50	4.5	410			fast n (P.S.D.), γ, etc.
	NaI(Tl)	Crystal	3.67	1.775	650	230	230	413			γ, x-rays
	NaI(pure)	Crystal	3.67	1.775	651	440[b]	60[b]	303[b]			γ, x-rays (fast counting)
	LiI(Eu)	Crystal	4.06	1.955	445	75	1200	475			n
	CsI(Tl)	Crystal	4.51	1.788	620	95	1100	580			heavy particles, γ (P.S.D.)
	CsI(Na)	Crystal	4.51	1.787	621	150.190	650	420			heavy particles, γ (P.S.D.)
	CsI(pure)	Crystal	4.51	1.788	621	500[b]	600[b]	c. 400[b]			heavy particles, γ (low energy)
	CaF$_2$(Eu)	Crystal	3.17	1.443	1418	110	1000	435			β, x-rays etc.
	CaWO$_4$	Crystal	6.1	1.92	1535	36	6000	430			γ (seldom used)
	ZnS(Ag)	Multi-crystal	4.09	2.356	1850	300	200	450			α
	ZnO(Ga)	Multi-crystal	5.61	2.02	1975	90	1.48	385			α

[a] Although NE 160 begins to soften very slightly at approximatley 80°C, it retains its shape up to at least 150°C unlike other plastic scintillators as NE 102A.
[b] At liquid nitrogen temperature.

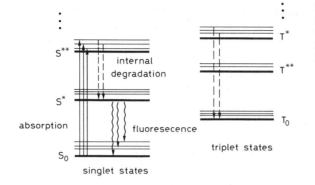

Fig. 7.4. Energy level diagram of an organic scintillator molecule. For clarity, the singlet states (denoted by S) are separated from the triplet states (denoted by T)

ed with each electron level is a fine structure which corresponds to excited vibrational modes of the molecule. The energy spacing between electron levels is on the order of a few eV whereas that between vibrational levels is of the order of a few tenths of eV.

Ionization energy from penetrating radiation excites both the electron and vibrational levels as shown by the solid arrows. The singlet excitations generally decay immediately (≤ 10 ps) to the S^* state without the emission of radiation, a process which is known as *internal degradation*. From S^*, there is generally a high probability of making a radiative decay to one of the vibrational states of the ground state S_0 (wavy lines) within a few nanoseconds time. This is the normal process of fluorescence which is described by the *prompt* exponential component in (7.2). The fact that S^* decays to excited vibrational states of S_0, with emission of radiation energy less than that required for the transition $S_0 \rightarrow S^*$ also explains the transparency of the scintillators to their own radiation.

For the triplet excited states, a similar internal degradation process occurs which brings the system to the lowest triplet state. While transitions from T_0 to S_0 are possible, they are, however, highly forbidden by multipole selection rules. The T_0 state, instead, decays mainly by interacting with another excited T_0 molecule,

$$T_0 + T_0 \rightarrow S^* + S_0 + \text{phonons} \qquad (7.3)$$

to leave one of the molecules in the S^* state. Radiation is then emitted by the S^* as described above. This light comes after a delay time characteristic of the interaction between the excited molecule and is the *delayed* or *slow* component of scintillator light. The contribution of this slow component to the total light output is only significant in certain organic materials, however.

Because of the molecular nature of luminescence in these materials, organics can be used in many physical forms without the loss of their scintillating properties. As detectors, they have been used in the form of pure crystals and as mixtures of one or more compounds in liquid and solid solutions. A brief description of these types is given below.

7.2.1 Organic Crystals

The most common crystals are anthracene ($C_{14}H_{10}$), trans-stilbene ($C_{14}H_{12}$) and naphthalene ($C_{10}H_8$). With the exception of anthracene which has a decay time of $\simeq 30$ ns, these crystals have a fast time response on the order of a few nanoseconds. However, due to channeling effects their amplitude response is anisotropic, that is, for a constant source of radiation the response varies with the orientation of the crystal. Obtaining a good energy resolution with a noncollimated source, then, can become a difficult problem.

They are hard crystals and thus very durable, although stilbene tends to be brittle and more sensitive to thermal shock than anthracene. For this reason also, the cutting of such crystals to desired forms and shapes is often a difficult task. This and other disadvantages, unfortunately, have caused anthracene and stilbene to fall into disuse in the past years.

Anthracene, nevertheless, has the distinction of having the highest light output of all the organic scintillators. For this reason, it is chosen as the reference to which the light outputs of other scintillators are compared. These outputs are thus usually expressed as percent of anthracene light.

7.2.2 Organic Liquids

These materials are liquid solutions of one or more organic scintillators in an organic solvent. While the scintillation process here is still the same as that described above, the mechanism of energy absorption is different. In solutions, the ionization energy seems to be absorbed mainly by the solvent and then passed on to the scintillation solute. This transfer usually occurs very quickly and efficiently, although the precise details of the mechanism are still not clear.

Some of the organic scintillators most commonly used as solutes are p-Terphenyl[1], PBD[2], PPO[3] and POPOP[4]. Among the solvents, the most successful seem to be xylene, toluene, benzene, phenylcyclohexane, triethylbenzene and decaline. Measurements have shown that the efficiency of liquid scintillators increases with solute concentration although a broad maximum is reached just before saturation of the solution. Typical concentrations are on the order of 3 g of solute per liter of solvent.

The response of liquid scintillators is generally quite fast with decay times on the order of 3 to 4 ns. They have a particular advantage in that they can be easily *loaded* with other materials so as to increase efficiency for a particular application. For example, Boron-11, which has a high neutron cross-section, can be added to increase efficiency for neutron detection. Similarly, *wavelength shifters*, i.e. materials which absorb light of one frequency and reemit it at another, can also be added to make the spectrum of emitted light more compatible with a photomultiplier cathode. Loading, however, usually causes a lengthening of the decay time and a drop in light output because of a quenching effect which is produced by these additives. It has been found, though, that by adding naphthalene, biphenyl and other compounds to the solvent, much of the quenching effect can be removed.

As a general rule, liquid scintillators are extremely sensitive to impurities in the solvent. It is not uncommon, in fact, to find two different samples of the same liquid scintillator with pulse heights differing by as much as a factor of 2 because of contaminating impurities. Dissolved oxygen, in particular, seems to have a large effect, although

---

[1] $C_{18}H_{14}$
[2] 2-phenyl,5-(4-biphenylyl)-1,3,4-oxadiazole ($C_{20}H_{14}N_2O$)
[3] 2,5-diphenyloxazole ($C_{15}H_{11}NO$)
[4] 1,4-Bis-[2-(5-phenyloxazolyl)]-benzene ($C_{24}H_{16}N_2O_2$)

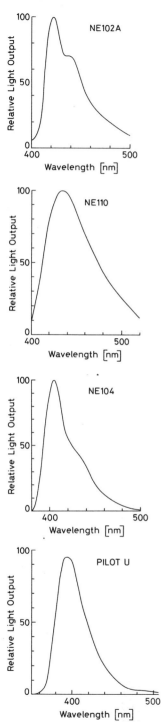

Fig. 7.5. Light emission spectra for several different plastic scintillators (from *Nuclear Enterprises* catalog [7.1])

this problem can be remedied to some extent by bubbling oxygen-free nitrogen through the liquid scintillator.

7.2.3 Plastics

In nuclear and particle physics, plastic scintillators are probably the most widely used of the organic detectors today. Like the organic liquids, plastic scintillators are also solutions of organic scintillators but in a solid plastic solvent. The most common and widely used plastics are polyvinyltoluene, polyphenylbenzene and polystyrene. Some common primary solutes are PBD, p-Terphenyl and PBO, which are dissolved in concentrations typically on the order of 10 g/l. Very often a secondary solute such as POPOP is also added for its wavelength shifting properties, but in a very much smaller proportion. The light emission spectra of several commercial plastics is shown in Fig. 7.5.

Plastics offer an extremely fast signal with a decay constant of about $2-3$ ns and a high light output. Because of this fast decay, the finite rise time cannot be ignored in the description of the light pulse as was done in (7.1). The best mathematical description, as shown by *Bengston* and *Moszynski* [7.2], appears to be the convolution of a Gaussian with an exponential,

$$N(t) = N_0 f(\sigma, t) \exp\left(\frac{-t}{\tau}\right) , \qquad (7.4)$$

where $f(\sigma, t)$ is a Gaussian with a standard deviation σ. Table 7.2 gives some fitted values of these parameters for a few common plastics.

Table 7.2. Gaussian and exponential parameters for light pulse description from several plastic scintillators (from *Bengston* and *Moszynski* [7.2])

Scintillator	σ [ns]	τ [ns]
NE102A	0.7	2.4
NE111	0.2	1.7
Naton 136	0.5	1.87

One of the major advantages of plastics is their flexibility. They are easily machined by normal means and shaped to desired forms. They are produced commercially in a wide variety of sizes and forms, ranging from thin films, a few $\mu g/cm^2$ thick, to large sheets, blocks and cylinders, and are relatively cheap. Moreover, various types of plastics are made offering differences in light transmission, speed, etc.

While they are generally quite rugged, plastics are easily attacked by organic solvents such as acetone and other aromatic compounds. They are, however, resistant to water pure methyal, silicon grease and lower alcohols. When handling unprotected plastic, it is generally advisable to wear cotton or terylene gloves as the body acids from one's hands can cause a cracking of the plastic (often referred to as *craze*) after a period of time.

7.3 Inorganic Crystals

The inorganic scintillators are mainly crystals of alkali halides containing a small activator impurity. By far, the most commonly used material is NaI(Tl), where Thallium (Tl) is the impurity activator. Somewhat less common, but in active use is CsI(Tl), also with Tl as an impurity activator. Others crystals include CsF_2, CsI(Tl), CsI(Na), KI(Tl) and LiI(Eu). Among the non-alkali materials are $Bi_4Ge_3O_{12}$ (bismuth germanate or BGO), BaF_2, ZnS(Ag), ZnO(Ga), $CaWO_4$ and $CdWO_4$ among others (see Table 7.1).

The spectrum of emitted light from some of the more commonly used crystals is shown in Fig. 7.6. In general, inorganic scintillators are 2 – 3 orders of magnitude slower (~ 500 ns) in response than organic scintillators due to phosphorescence. (The one exception is CsF which has a decay time of 5 ns!) However, the time evolution of emission is, in most cases, well described by the simple one or two exponential decay forms.

A major disadvantage of certain inorganic crystals is hygroscopicity. NaI, in particular, is a prime example. To protect it from moisture in the air, it must be housed in an air tight protective enclosure. Other hygroscopic crystals are CsF, LiI(Eu) and KI(Tl). BGO and BaF_2, on the other hand, are non-hygroscopic and can be handled without protection, while CsI(Tl) is only slightly hygroscopic but can generally be handled without protection.

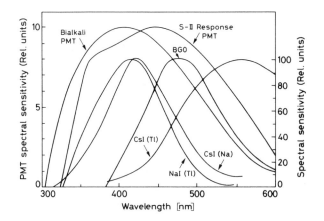

Fig. 7.6. Light emission spectra for different inorganic crystals (from *Harshaw Catalog* [7.3])

The advantage of inorganic crystals lies in their greater stopping power due to their higher density and higher atomic number. Among all the scintillators, they also have some of the highest light outputs, which results in better energy resolution. This makes them extremely suitable for the detection of gamma-rays and high-energy electrons and positrons.

While NaI has generally been the standard for these purposes, two new materials have recently drawn much attention from high-energy and nuclear physicists. These are $Bi_4Ge_3O_{12}$ (*Bismuth Germanate* or BGO) and BaF_2 (Barium Fluoride).

BGO is particularly interesting because of its very high-Z and greater efficiency for the photoelectric conversion of γ-rays. Relative to NaI, for example, it is 3 to 5 times more efficient and nonhygroscopic, as well. Its light output is lower than NaI, however, so that its resolution is about a factor two worse. Moreover, it is still relatively expensive and difficult to obtain in large quantities. Nevertheless, at very high energies it

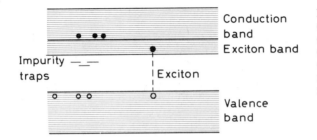

Fig. 7.7. Electronic band structure of inorganic crystals. Besides the formation of free electrons and holes, loosely coupled electron-hole pairs known as excitons are formed. Excitons can migrate through the crystal and be captured by impurity centers

presents an enormous advantage over NaI. BaF$_2$, on the other hand, has been discovered to have a very fast light component in the ultra-violet region. Decay times on the order of 500 ps have been measured in preliminary tests. This would make it twice as fast as the fastest plastics. The total light output of this component is low, however, and further development is necessary before its real usefulness can be determined.

Whereas the scintillation mechanism in organic materials is molecular in nature, that in inorganic scintillators is clearly characteristic of the electronic band structure found in crystals (see Fig. 7.7). When a nuclear particle enters the crystal, two principal processes can occur. It can ionize the crystal by exciting an electron from the valence band to the conduction band, creating a free electron and a free hole. Or it can create an *exciton* by exciting an electron to a band (the exciton band) located just below the conduction band. In this state the electron and hole remain bound together as a pair. However, the pair can move freely through the crystal. If the crystal now contains impurity atoms, electronic levels in the forbidden energy gap can be locally created. A migrating free hole or a hole from an exciton pair which encounters an impurity center, can then ionize the impurity atom. If now a subsequent electron arrives, it can fall into the opening left by the hole and make a transition from an excited state to the ground state, emitting radiation if such a deexcitation mode is allowed. If the transition is radiationless the impurity center becomes a *trap* and the energy is lost to other processes.

7.4 Gaseous Scintillators

These consist mainly of the noble gases: xenon, krypton, argon and helium, along with nitrogen. In these scintillators the atoms are individually excited and returned to their ground states within about 1 ns, so that their response is extremely rapid. However, the emitted light is generally in the ultraviolet, a wavelength region at which most photomultipliers are inefficient. One method of overcoming this difficulty has been to coat the walls of the container with a wavelength shifter such as diphenylstilbene (DPS). These materials strongly absorb light in the ultraviolet and emit in the blue-green region where photomultiplier cathodes are more efficient. In some cases, the PM windows are also coated with a thin layer of wavelength shifter as well.

Gas scintillators have generally been used in experiments with heavy charged particles or fission fragments. Here, mixtures of several gases (for example, 90% ^{3}He, 10% Xe) under pressures as high as 200 atm have been used to increase detection efficiency. In recent years, gas scintillators have been proposed as detectors in space physics as well.

Experiments have also been performed with solid and liquid xenon and liquid helium which have been found to scintillate.

7.5 Glasses

Glass scintillators are cerium activated lithium or boron silicates. However, boron glasses have light outputs some ten times lower than lithium so they are not very often employed today. Glass detectors are used primarily for neutron detection although they are also sensitive to β and γ radiation. They are most noteworthy for their resistance to all organic and inorganic reagents except for hydrofluoric acid. In addition, they have high melting points and are extremely resistant. These physical and chemical characteristics make them especially useful in extreme environmental conditions.

Their speed of response is between that of plastics and inorganic crystals, typically on the order of a few tens of nanoseconds. Light output, however, is low, never reaching more than 25 – 30% of that for anthracene.

For low-energy neutron detection, it is also possible to increase sensitivity by enriching the lithium component with ^{6}Li. Separation of neutron events from γ radiation events can then be made by using pulse height discrimination techniques.

7.6 Light Output Response

The light output of a scintillator refers more specifically to its efficiency for converting ionization energy to photons. This is an extremely important quantity, as it determines the efficiency and resolution of the scintillator. In general the light output is different for different types of particles at the same energy. Moreover, for a given particle type, it does not always vary linearly with energy.

As with ionization of gases, we can define an average energy loss required for the creation of a photon. Table 7.3 gives a brief list of this efficiency for several materials when electrons are the exciting particles. In general, this efficiency decreases for heavier particles. This behavior is seem in Fig. 7.8 for the case of plastic. Traditionally the light output of scintillators is referred to anthracene and is given as percent of anthracene light output. A more complete list of scintillator light outputs is given in Table 7.1.

It should be kept in mind that when considering the efficiency of a scintillation detector, the efficiency of the photomultiplier must also be taken into account, since they are inseparably coupled. A typical efficiency for the latter, as will be seen in the next chapter, is about 30%. Thus, assuming that all the emitted photons are collected, only 30% of these photons will ever be detected.

Table 7.3. Average energy loss per scintillator photon for electrons

Material	ε [eV/photon]
Anthracene	60
NaI	25
Plastic	100
BGO	300

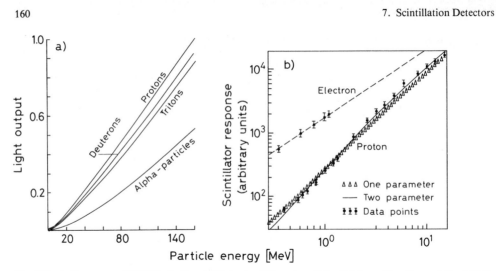

Fig. 7.8a, b. Response of NE 102 plastic scintillator to different particles ((**a**) from *Gooding* and *Pugh* [7.6]; (**b**) from *Craun* and *Smith* [7.7])

7.6.1 Linearity

Up until now, we have assumed that scintillators respond in a linear fashion with respect to the exciting energy, that is the fluorescent light emitted, L, is directly proportional to the energy, E, deposited by the ionizing particle,

$$L \propto E \ . \tag{7.5}$$

Strictly speaking, this linear relation is not true, although for many applications it can be considered as a good approximation. In reality, the response of scintillators is a complex function of not only energy but the type of particle and its specific ionization.

In organic materials, non-linearities are readily observed for electrons at energies below 125 keV, although they are small [7.4]. For heavier particles, however, the deviations are more pronounced and become very noticeable at lower energies, with the higher ionizing particles showing the larger deviations. For comparison, Fig. 7.8 shows the response of NE102A plastic scintillator to a number of different particles.

The first successful semi-empirical model for this behavior was put forward by *Birks* [7.5] in 1951. Assuming the response of organic scintillators to be ideally linear, Birks explained the deviations as being due to quenching interactions between the excited molecules created along the path of incident particle, i.e., interactions which drain energy which would otherwise go into luminescence. Since a higher ionizing power produces a higher density of excited molecules, more quenching interactions will take place for these particles. In this model, the light output per unit length, dL/dx, is related to the specific ionization [5] by

$$\frac{dL}{dx} = \frac{A \dfrac{dE}{dx}}{1 + kB \dfrac{dE}{dx}} \tag{7.6}$$

---

[5] The specific ionization is defined as the average number of ion pairs created by the passing particle per unit length. If ε is the mean energy lost for each ion pair created, then the specific ionization is $(dE/dx)/\varepsilon$.

Table 7.4. Measured values of kB for NE102 plastic scintillator (from *Badhwar* et al. [7.8])

Particle	Energy [MeV/nucl.]	dE/dx [MeV/g cm^2]	kB [mg/(cm^2 MeV)]
Compton electrons and recoil protons	<4	>97	9.1 ± 0.6
Compton electrons and alpha particles	<1.3	>272	9.8 ± 0.8
Compton electrons and protons	1.2 – 14	>34	10 ± 1
Recoil protons	<2.3	>150	10
Recoil protons	<8.4	>50	2
			3 ± 1
Protons	<100	>7	3.7 – 7.5
Protons	28 – 148	5.5 – 20	13.2 ± 2.5
Deuterons	23 – 60	10 – 23	
Nitrogen ions	3 – 9.5	$(1-2) \times 10^3$	<10
Protons to oxygen ions	Rigidity 1.5 – 1.6 GV	2.0 – 120	10
Oxygen-iron nuclei	Rigidity 1.5 – 1.6 GV	120 – 1300	10 $C = -5 \times 10^{-6}$
Protons	36 – 220	4.2 – 12.3	12.6 ± 2.0
^{4}He	38 – 220	17 – 49	7.2 ± 1.0
Carbon nuclei	95	265	7.8
Oxygen nuclei	105	550	$C = -7 \times 10^{-6}$

with A: absolute scintillation efficiency; kB: parameter relating the density of ionization centers to dE/dx.

In practice, kB is obtained by fitting Birk's formula to experimental data. Some values of kB for different particles in NE102A plastic are given in Table 7.4.

While Birk's formula has been relatively successful, deviations have made it necessary to turn to "*higher order*" formulae in order to better fit the data. Thus, expressions [7.9] such as

$$\frac{dL}{dx} = \frac{A \dfrac{dE}{dx}}{1 + B \dfrac{dE}{dx} + C \left(\dfrac{dE}{dx} \right)^2} \tag{7.7}$$

or [7.10]

$$\frac{dL}{dx} = \frac{A}{2B} \ln \left(1 + 2B \frac{dE}{dx} \right) \tag{7.8}$$

have been suggested. In all these cases the formulae reduce to a linear relationship for small dE/dx,

$$\frac{dL}{dx} \simeq \frac{dE}{dx} \tag{7.9}$$

as is observed experimentally. However, for large dE/dx, the formulae differ in their predictions. Birk's formula, for example, implies saturation, i.e.,

$$\frac{dL}{dx} \simeq \frac{A}{kB} \qquad (7.10)$$

which, if integrated, yields a fluorescent output proportional to the range, $R(E)$ of the particle in the scintillator,

$$L \simeq \frac{A}{kB} R(E) \ . \qquad (7.11)$$

The higher order formulae predict either a continuing increase of dL/dx with dE/dx (7.8) or a passage through a maximum [(7.7), assuming C is positive]. Experimentally, however, all these formulae have been shown to be incomplete as some further dependence of dL/dx on the specific particle type in addition to dE/dx is observed.

In the most detailed analysis so far, *Voltz* et al. [7.11 – 13] have considered this dependence and have taken into account the kinetics of the fast and slow scintillation components. They obtain some relatively complicated expressions for these components, which seem to be in agreement with the measured data. However, the situation still remains inconclusive because of large variations in the experimental results. More information, of course, would be desirable.

In inorganics, the differential light output, dL/dx, also varies with energy, however the dependence is generally weaker so that deviations are small. In NaI, linearity is maintained down to an energy of about 400 keV where a distinct deviation occurs. Figure 7.9 illustrates this nonlinearity. For accurate work in this energy region, therefore,

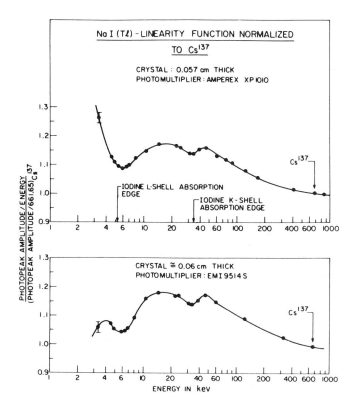

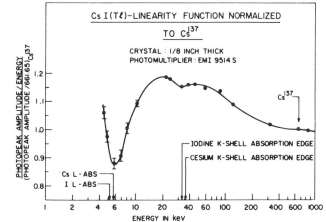

Fig. 7.9. Response of NaI(Tl) and CsI(Tl). The pulse height of the 661 keV gamma ray line from ^{137}Cs is defined as 1.0 (from *Aitken* et al. [7.14]). Note the nonlinearities that appear particularly at energies corresponding to the K- and L-edges in Iodine. This is because photoelectrons ejected by incident gamma rays just above the K energy have very little kinetic energy so that the response drops. Just below this energy, however, K-shell ionization is not possible and L-shell ionization takes place. Since the binding energy is lower, the photoelectrons ejected at this point are more energetic which causes a rise in the response. A similar argument applies to the L shell, etc. (Picture © 1967 IEEE)

this behavior should be taken into account. A similar behavior is also observed in CsI. While Birk's formula is only applicable to organic scintillators, some portions of the NaI response seem also to be well described by this expression.

7.6.2 Temperature Dependence

The light output of most scintillators is also a function of the temperature. This dependence is generally weak at room temperatures, but should be considered if operation at temperatures very different from normal is desired.

In organic scintillators, the light output is practically independent of temperature between $-60\,°C$ and $+20\,°C$ and only drops to 95% of this value at $+60\,°C$. Inorganic crystals, on the hand, are more sensitive as shown in Fig. 7.10. Both CsI(Tl) and CsI(Na), for example, show relatively strong variations in the normal range of temperatures, while NaI appears much less sensitive. BGO light output has also been found to exhibit a strong temperature dependence, increasing by about 1% per degree Celsius as the temperature is decreased. As the temperature decreases, the decay time for BGO also increases however.

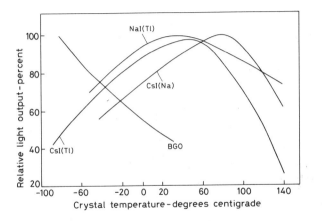

Fig. 7.10. Temperature dependence of light output from inorganic crystals (from *Harshaw Catalog* [7.3])

7.6.3 Pulse Shape Discrimination (PSD)

While the light emission of most scintillators is dominated by a single fast decay component, some materials, as we have mentioned, exhibit a substantial slow component. In general, both of these components depend on dE/dx to some degree or another. In scintillators where this dependence is strong, the overall decay time of the emitted light pulse will, therefore, vary with the type of exciting radiation. Such scintillators are thus capable of *pulse shape discrimination,* i.e., they are capable of distinguishing between different types of incident particles by the shape of the emitted light pulse. Figure 7.11 illustrates the different decay times, and hence different pulse shapes, exhibited by stilbene when excited by different particles. Similar differences are also observed in other organics, particularly liquid scintillators, and inorganic crystals. In CsI(Tl), for example, overall decay times of 0.425 µs for α-particles, 0.519 µs for protons and 0.695 µs for electrons are found [7.15].

The explanation for this effect lies in the fact that the fast and slow components arise from the deexcitation of different states of the scintillator. Depending on the spe-

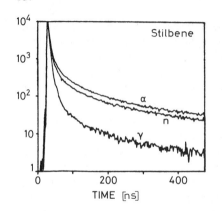

Fig. 7.11. Pulse shape of stilbene light for alpha particles, neutrons and gamma rays (from *Lynch* [7.71]; picture © 1975 IEEE)

cific energy loss of the particle (*dE/dx*), these states are populated in different proportions, so that the relative intensities of the two components are different for different *dE/dx*. In alkali halides such as CsI, for example, a high ionization loss produces a higher density of free electrons and holes which favors their recombination into loosely bound systems known as *excitons*. These excitons then wander through the crystal lattice until they are captured as a whole by impurity centers, exciting the latter to certain radiative states (fast component). The singly free electrons and holes, on the other hand, are captured successively resulting in the excitation of certain metastable states (slow component) not accessible to excitons. At low ionization density, exciton formation is less likely so that the proportion of excitons relative to free electrons and holes is lower. The proportion of radiative to metastable excited states will be different, therefore, and hence the pulse shape.

In organic scintillators, a high *dE/dx* produces a high density of excited molecules which results in increased intermolecular interactions. These reactions hinder the normal singlet internal degradation process leading to the radiative S^* state by draining their energy through other channels. The proportion of the fast component emitted relative to the slow component is thus reduced. In stilbene, for example, the slow component was found to account for $\simeq 35\%$, 54% and 66% of the total light output for electrons, protons and α-particles respectively [7.17]. A strong difference is also observed

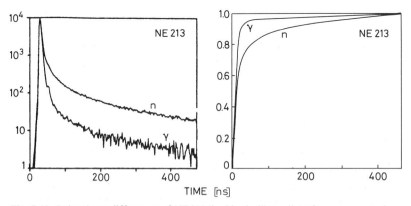

Fig. 7.12. Pulse shape differences of NE213 liquid scintillator light for neutrons and gamma rays. The time integral of the light pulses is also shown. A discrimination between these radiations may be obtained by measuring the time it takes for the integrated pulse to reach a certain fixed level (from *Lynch* [7.17]; picture © 1975 IEEE)

in liquid scintillators. Figure 7.12 shows the pulse shape difference between neutrons and γ-rays in NE213 liquid scintillator and the integrals of these curves. This latter graph illustrates a widely used method for picking out the different pulse shapes. Here the time difference between the start of the pulse and the point at which the integral of the pulse reaches a certain fixed value is measured. From the graph, it is clear that the different shapes will give different time measurements.

It should be noted that the differences in decay time are also sensitive to external factors such as impurities in the scintillator, temperature, etc. – so that variations in the two components can be expected from counter to counter. Nevertheless, pulse shape discrimination is used extensively today, particularly in neutron counting with liquid scintillators. (Here the neutron is detected by scattering from protons while the γ-ray interacts through the usual photoelectric, Compton and pair production processes – thereby accounting for the difference.) This technique allows a discrimination of γ-rays and thus an effective means for suppressing background from these sources.

7.7 Intrinsic Detection Efficiency for Various Radiations

In principle, scintillation phosphors will respond to any radiation which can directly or indirectly excite the molecules or atoms of the phosphor. However, for a given type of radiation with a given scintillator, one will not always find that a usable signal is efficiently produced. Indeed, one must consider the mechanisms by which the radiation interacts with the molecules of that particular scintillator material, the probability of these interactions occurring in the scintillator volume and the response in light output. The latter quantity is governed by the luminescence mechanisms in the scintillator and is discussed in the preceding sections. The second quantity is given by the mean free path of the radiation in the scintillator material. For a not too high energy charged particle in normal matter, this distance is generally on the microscopic level, so that the probability of it losing some energy in any scintillator of normal dimensions is almost 100%. For neutral particles, however, the mean free path in some materials can be quite large, so that a prohibitively large detector might be required in order to assure a reasonable efficiency. In addition, one must also consider if energy information is desired. In such a case, the requirements become even stronger since the particle must now deposit all of its energy rather than just part of it. In this section, we consider the several types of radiation most commonly encountered in nuclear physics, discuss the problems involved and the most suitable types of scintillator detector for each radiation.

7.7.1 Heavy Ions

While the heavy ions are capable of ionizing any of the phosphors discussed above, scintillators are generally not very suitable for these particles due to a reduced light output. This is due to the very high ionizing power of these particles which, as we have seen, induces quenching effects. For α's in organic scintillators, for example, the light output is only about 1/10 of that for electrons of the same energy. Strong non-linearities in the pulse response are also found. In the inorganic materials, the output intensity is also reduced, but remains higher than the organics, varying from 50 to 70% of that for electrons. Better linearity is found as well.

Consequently, where scintillators are desired, the inorganic crystals such as NaI have been traditionally used for the detection of these particles, their higher light output and high stopping power providing better energy resolution. For α's, ZnS is still employed also, although it has a poorer energy resolution and is only suitable for low count rates because of a long decay time.

7.7.2 Electrons

The efficiency of most scintillators for electrons is almost 100% in the sense that very few entering electrons will fail to produce a detectable signal. However, because of its small mass, the electron is susceptible to large angle scatterings in matter. This can cause an incident electron to be backscattered (or sidescattered) out of the detector before its full energy has been deposited. Obtaining a satisfactory energy measurement under these conditions, then, can be quite difficult.

The backscattering effect depends strongly on the atomic number of the material, however, and increases rapidly with increasing Z. Since the organic scintillators have the lowest effective Z, they in fact have proved to be the more advantageous. Indeed, in a set-up with a β-source held at a small distance from a flat NaI (high Z!) crystal, for example, between 80 to 90% of the incident electrons will be backscattered, whereas only about 8% are reflected with a plastic scintillator [7.18]. These percentages can be reduced somewhat by collimating the source so that the electrons are incident at angles closer to the perpendicular, for example, or eliminated altogether by completely enveloping the source with the detector in a 4π geometry. With the source external to the detector, however, the organic scintillators are clearly superior.

At very high electron energies the use of inorganics no longer becomes disadvantageous. Here electron energy loss is mainly through the production of bremsstrahlung and the subsequent electron showers which it produces, so that, a high-Z material is needed in order to facilitate shower production (see Sect. 2.4.2). Having the highest density and atomic number, the inorganics in fact become the more preferable.

7.7.3 Gamma Rays

In contrast to electrons, γ-rays are more efficiently detected by high Z materials. This difference can be understood by recalling the three basic interactions by which photons react with matter: the photoelectric effect, Compton scattering and pair production. In the photoelectric effect and pair production processes, the γ-ray is completely absorbed, being transformed into a charged particle or particles. In the Compton process, however, the γ retains its identity transferring only a part of its energy to a recoil electron. If the scattered γ-ray does not suffer another interaction in the scintillator material, it is possible for it to escape so that only a part of its energy is deposited. To make an efficient γ-ray detector, therefore, one must use a material in which the photoelectric and pair production cross sections are large compared to the Compton scattering cross section.

Fortunately, the former two cross sections are much more strongly dependent on Z, going roughly as Z^5 and Z^2 respectively, while the Compton process only varies linearly with Z. The high Z inorganic phosphors are thus the more favored for γ detection. Figure 7.13 illustrates the difference in the three cross sections for γ-rays in NaI and NE102A plastic. While the Compton cross sections in both materials are comparable, the photoelectric and pair cross sections are several orders of magnitude higher in the NaI. The probability of absorption versus scattering is thus much greater.

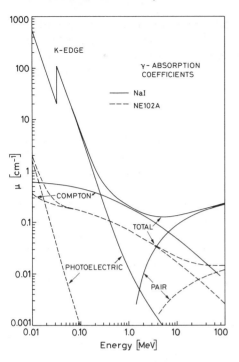

Fig. 7.13. Gamma-ray absorption coefficients for NaI and NE102A plastic scintillator. Note the difference in the relative magnitudes of the photoelectric and Compton cross sections

7.7.4 Neutrons

Like γ-rays, the detection of neutrons requires a transfer of all or part of its energy to a particle capable of ionizing and exciting the scintillator material.

For *fast* and higher energy neutrons, detection relies mainly on detecting the recoil proton in (n, p) scattering processes. Plastic and other organics are particularly convenient here, since they contain large amounts of hydrogenous material. The standard scintillator in neutron spectroscopy is liquid organic scintillator (e.g. NE213). This material offers excellent pulse shape discrimination properties and allows a rejection of the γ-ray background which usually accompanies neutron reactions. As well it is easily handled and can be adapted to a wide variety of geometries.

For thermal neutrons, detection is most efficiently done using the (n, γ) or (n, α) nuclear reactions. Scintillators which contain elements with high cross sections for these reactions, e.g. ^{6}Li or ^{10}B, or are capable of being loaded with these elements are therefore the most convenient. LiI(Eu), for example, is a particularly good thermal neutron detector. Indeed, a 2 cm thick crystal is almost 90% efficient for thermal neutrons. It, unfortunately, is also sensitive to γ radiation, which is a major source of background. The advantages of LiI(Eu) are therefore somewhat diminished. More effective are the glass scintillators which are particularly well suited since they can be made with either enriched ^{6}Li or ^{10}B. They are also sensitive to β and γ radiation, although some glasses offer the possibility of pulse shape discrimination. Liquid scintillator is again, however, the most effective here since it can also be loaded with elements such as ^{6}Li or ^{10}B, in addition to offering pulse shape discrimination.

8. Photomultipliers

Photomultipliers (PM's) are electron tube devices which convert light into a measurable electric current. They are extremely sensitive and, in nuclear and high-energy physics, are most often associated with scintillation detectors, although their uses are quite varied. It is nevertheless in this context that we will discuss the basic design and properties of photomultipliers, their characteristics under operation and some special techniques.

8.1 Basic Construction and Operation

Figure 8.1 shows a schematic diagram of a typical photomultiplier. It consists of a cathode made of photosensitive material followed by an electron collection system, an electron multiplier section (or dynode string as it is usually called) and finally an anode from which the final signal can be taken.[1] All parts are usually housed in an evacuated glass tube so that the whole photomultiplier has the appearance of an old-fashion electron tube.

During operation a high voltage is applied to the cathode, dynodes and anode such that a potential "*ladder*" is set up along the length of the cathode – dynode – anode structure. When an incident photon (from a scintillator for example) impinges upon the photocathode, an electron is emitted via the photoelectric effect. Because of the applied voltage, the electron is then directed and accelerated toward the first dynode, where upon striking, it transfers some of its energy to the electrons in the dynode. This causes secondary electrons to be emitted, which in turn, are accelerated towards the next dynode where more electrons are released and further accelerated. An electron cascade down the dynode string is thus created. At the anode, this cascade is collected to give a current which can be amplified and analyzed.

Photomultipliers may be operated in continuous mode, i.e., under a constant illumination, or in pulsed mode as is the case in scintillation counting. In either mode, if the cathode and dynode systems are assumed to be linear, the current at the output of the PM will be directly proportional to the number of incident photons. A radiation detector produced by coupling a scintillator to a PM (the scintillator produces photons in proportion to the energy deposited in the scintillator) would thus be capable of providing not only information on the particle's presence but also the energy it has left in the scintillator.

Let us now turn to a more detailed look at the various parts of the photomultiplier.

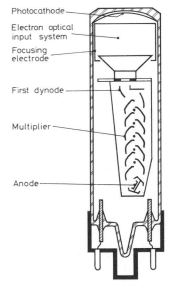

Fig. 8.1. Schematic diagram of a photomultiplier tube (from *Schonkeren* [9.1])

---

[1] An alternative structure rarely used with scintillation counters is the *side-on* PM. Here the photocathode is oriented so as to face the side of the tube rather than the end window. The dynode chain is then usually arranged in a circular fashion around the axis of the tube rather than linearly along it. The basic operating principle remains exactly the same, however.

8.2 The Photocathode

As we have seen, the photocathode converts incident light into a current of electrons by the photoelectric effect. To facilitate the passage of this light, the photosensitive material is deposited in a thin layer on the inside of the PM window which is usually made of glass or quartz. From Einstein's well-known formula,

$$E = h\nu - \phi \ , \tag{8.1}$$

where E is the kinetic energy of emitted electron, ν is the frequency of incident light and ϕ is the work function, it is clear that a certain minimum frequency is required before the photoelectric effect may take place. Above this threshold, however, the probability for this effect is far from being unity. Indeed, the efficiency for photoelectric conversion varies strongly with the frequency of the incident light and the structure of the material. This overall spectral response is expressed by the *quantum efficiency*, $\eta(\lambda)$,

$$\eta(\lambda) = \frac{\text{number of photoelectrons released}}{\text{number of incident photons on cathode } (\lambda)} \ , \tag{8.2}$$

where λ is the wavelength of the incident light. An equivalent quantity is the *radiant cathode sensitivity* which is defined as

$$E(\lambda) = \frac{I_k}{P(\lambda)} \ , \tag{8.3}$$

where I_k is the photoelectric emission current from the cathode and $P(\lambda)$ is the incident radiant power. The radiant cathode sensitivity is usually given in units of ampere/watts and is related to the quantum efficiency by

$$E(\lambda) = \lambda \eta(\lambda) \frac{e}{hc} \ . \tag{8.4}$$

For E in [A/W] and λ in nanometers

$$E(\lambda) = \frac{\lambda \eta(\lambda)}{1240} \ [A/W] \ . \tag{8.5}$$

A third unit is the *luminous cathode sensitivity* which is defined as the current per lumen of incident light flux. Since the lumen is essentially a physiological unit defined relative to the response of the human eye, the luminous sensitivity is not a unit to be recommended.

Figure 8.2 shows a graph of quantum efficiency vs λ for some of the more common photoelectric materials used in photomultipliers today. In general, the spectral response of these materials is such that only a certain band of wavelengths is efficiently converted. When choosing a PM, therefore, the primary consideration should be its sensitivity to the wavelength of the incident light. For the photocathodes shown in Fig. 8.2, the efficiency peaks near $\simeq 400$ nm, thereby making them quite suitable for use with scintillators. More than 50 other types of materials are in use, however, with spectral sensitivities varying from the infra-red to the ultraviolet. A brief list of the most

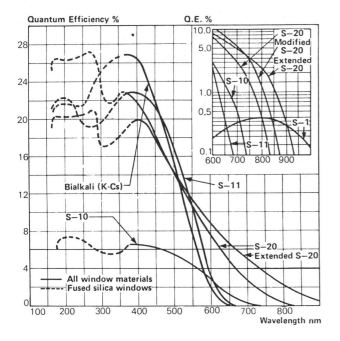

Fig. 8.2. Quantum efficiency of various photocathode materials (from *EMI Catalog* [8.2])

Table 8.1. Photocathode characteristics (from RTC catalog [8.3])

Cathode type	Composition	λ at peak response [nm]	Quantum efficiency at peak
S1 (C)	Ag – O – Cs	800	0.36
S4	SbCs	400	16
S11 (A)	SbCs	440	17
Super A	SbCs	440	22
S13 (U)	SbCs	440	17
S20 (T)	SbNa – KCs	420	20
S20R	SbNa – KCs	550	8
TU	SbNa – KCs	420	20
Bialkali	SbRb – Cs	420	26
Bialkali D	Sb – K – Cs	400	26
Bialkali DU	Sb – K – Cs	400	26
SB	Cs – Te	235	10

common types of photocathode is given in Table 8.1 along with their characteristics. Note that the different materials have been given standard type and code designations, an indication of their high frequency of use today.

Most of the photocathodes employed today are made of semiconductor materials formed from antimony plus one or more alkali metals. The choice of semiconductors rather than metals or other photoelectric substances lies in their much greater quantum efficiency for converting a photon to a *usable* electron. Indeed, in most metals, the quantum efficiency is not greater than 0.1% which means that an average of 1000 photons is needed to release one photoelectron. In contrast semiconductors have quantum efficiencies of the order of 10 to 30%, some two orders of magnitude higher! This difference is explained by their different intrinsic structures. Suppose, for example, an electron absorbs a photon at some depth x in the material. In traveling to the surface, this electron will suffer an energy loss, $\Delta E \simeq x(dE/dx)$, due to collisions with the atomic electrons along its path. In metals, these atomic electrons are essentially free so that large energy transfers result, i.e., the dE/dx is high. The probability of it reaching the surface with enough energy to overcome the potential barrier is therefore greatly reduced. This essentially restricts the usuable volume of the material to a very thin layer near the surface. The thickness of this layer is known as the *escape depth*. In contrast, semiconductors have an energy band structure with only a few electrons, those in the conduction and valence bands, being approximately free. The rest are tightly bound to the atoms. A photoelectron ejected from the conduction or valence band thus encounters less free electrons before reaching the surface. The only other possible collisions are with electrons bound to the lattice atoms. But due to the larger mass of the latter, little energy is transferred in these collisions. The photoelectron, therefore, is much more likely to reach the surface with a sufficient amount of energy to escape. The escape depth is thus much greater and the conversion efficiency higher.

A recent development in the construction photocathodes has been the use of negative electron affinity materials such as gallium phosphide (GaP) heavily doped with zinc and a small quantity of cesium. In these materials the band structure near the surface is bent so that the bottom energy level of the conduction band is actually above the potential of the vacuum. The work function is thus negative. Without a potential barrier, then, an electron need only have enough energy to reach the surface in order to escape. Such materials, therefore, have greatly improved quantum efficiencies reaching as high as 80%! Unfortunately a number of problems still remain in constructing them as photocathodes so that only limited use has been made of them so far.

8.3 The Electron-Optical Input System

After emission from the photocathode, the electrons in the PM must be collected and focused onto the first stage of the electron multiplier section. This task is performed by the electron-optical input system. In most PM's, collection and focusing is accomplished through the application of an electric field in a suitable configuration. Magnetic fields or a combination of electric and magnetic fields may also be employed in principle, but their use is extremely rare. Figure 8.3 gives a schematic diagram of a typical electron-optical input system. Here an accelerating electrode at the same potential as the first dynode of the electron multiplier is used in conjunction with a focusing electrode placed on the side of the glass housing. Some lines of equipotential are shown along with some possible electron paths.

Regardless of the design, two important requirements must be met:

1) Collection must be as efficient as possible, i.e. as many emitted electrons as possible must reach the electron-multiplier section regardless of their point of origin on the cathode.
2) The time it takes for an emitted electron to travel from the cathode to the first dynode must be as independent as possible of the point of emission.

The second requirement is particularly important for *fast* photomultipliers which are used in timing experiments since it determines the time resolution of the detector. This is discussed further in Sect. 8.6.

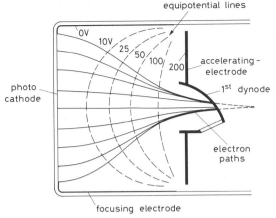

equipotential lines

photo cathode

accelerating-electrode

1ˢᵗ dynode

electron paths

focusing electrode

Fig. 8.3. Electron-optical input system of a typical *PM* (from *Schonkeren* [8.1])

8.4 The Electron-Multiplier Section

The electron-multiplier system amplifies the weak primary photocurrent by using a series of secondary emission electrodes or *dynodes* to produce a measurable current at the anode of the photomultiplier. The gain of each electrode is known as the *secondary emission factor, δ.* The theory of secondary electron emission is very similar to that described for photoelectric emission except that the photon is now replaced by an electron. On impact, energy is transferred directly to the electrons in the dynode material allowing a number of secondary electrons to escape. Since the conducting electrons in metals hinder this escape, as we have seen, it is not surprising that insulators and semiconductors are also used here as well.

One difference exists, however, in that an electric field must be maintained between the dynodes to accelerate and guide the electrons along the multiplier. Thus the secondary emission material must be deposited on a conducting material. A common procedure used today is to form an alloy of an alkali or alkaline earth metal with a more noble metal. During the mixing process, only the alkaline metal oxidizes, so that a thin insulating coating is formed on a conducting support. Materials in common use today are $Ag - Mg$, $Cu - Be$ and $Cs - Sb$. These have varying advantages but all meet the requirements of a good dynode material:

1) high secondary emission factor, δ, i.e., the average number of secondary electrons emitted per primary electron;
2) stability of secondary emission effect under high currents;
3) low thermionic emission, i.e., low noise.

Most conventional PM's contain 10 to 14 stages, with total overall gains of up to 10^7 being obtained.

Like the photocathode, use has also been made of negative affinity materials as dynodes, in particular GaP. With this material the individual gain of each dynode is greatly increased so that the number of stages in a PM can be reduced. A 5-stage PM made of GaP dynodes, for example, would provide the same overall gain as a 14-stage conventional PM. This reduction, as well, would diminish fluctuations in time as the path through which the cascade electrons must travel would also be much shorter.

8.4.1 Dynode Configurations

Dynode strings can be constructed in many ways and depending on the configuration, affect the response time and range of linearity of a photomultiplier. At present, five types of configurations are in use:

1) Venetian blind
2) Box and Grid
3) Linear focused
4) Circular focused (used in side-on PM's)
5) Microchannel plate.

  The first four types are the more conventional structures and are illustrated in Fig. 8.4. In the Venetian blind configuration, the dynodes are wide strips of material placed

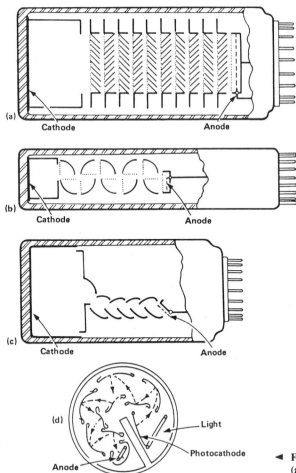

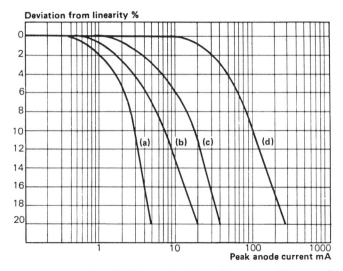

Fig. 8.5. Linearity of different dynode configurations: (*a*) box and grid, (*b*) venetian blind with standard voltage divider, (*c*) venetian blind with high current voltage divider, (*d*) linear focused with very high current divider (from *EMI Catalog* [8.2])

◄ **Fig. 8.4a – d.** Various dynode configurations for PM's (from *EMI Catalog* [8.2]): (**a**) venetian blind, (**b**) box and grid, (**c**) linear focused, (**d**) side-on configuration

at an angle of 45 degrees with respect to the electron cascade axis. This is a simple system which offers a large input area to the incident primary electrons. The dynodes are easily placed in line and the dimensions are not critical. The disadvantage, however, is that it is impossible to prevent a fraction of the primary electrons from passing straight through. This results in a low gain and large variations in transit time. This is avoided in the box and grid, linear focused and circular types which reflect the electrons from one dynode to the next. Other inherent advantages of these latter types are that: (1) space is efficiently used, so that many dynodes can be used, and (2) the cathode and anode are well isolated, so that there is no risk of feedback.

The response linearity of the various types are compared in Fig. 8.5. From the point of view of overall performance, it is clear that the linear focused type is the more favorable of the four. However, the specific application must be considered. If linearity over a modest current range is desired, for example, a Venetian blind configuration would do just as well as a linear focused type and for a lower cost.

In recent years a new design has appeared which uses *microchannel plate multipliers*. This device, originally invented for use in image intensifiers, consists of a lead glass plate perforated by an array of microscopic channels (typically $10 - 100$ μm in diameter) oriented parallel to each other (see Fig. 8.6). The inner surfaces of the channels are treated with a semiconductor material so as to act as secondary electron emitters while the flat end surfaces of the plate are coated with a metallic alloy so as to allow a potential difference to be applied along the length of the holes. Electrons entering a channel are thus accelerated along the hole until they eventually strike the wall to release further electrons which, in turn, are accelerated and so on. Each channel thus acts as a continuous dynode.

Typical microchannel plates have from $10^4 - 10^7$ holes and can provide multiplication factors of $10^3 - 10^4$. Two or three plates may also be cascaded to provide a higher overall gain. A common geometry is the *chevron* configuration shown in Fig. 8.7. Here

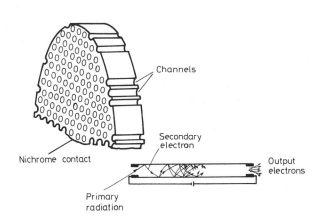

Fig. 8.6. Schematic diagram of a microchannel plate. The many channels act as continuous dynodes (from *Dhawan* [8.4]; picture © 1975 IEEE)

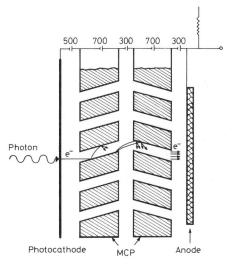

Fig. 8.7. Chevron configuration in a microchannel plate photomultiplier (after *Dhawan* [8.4]). A further increase in gain may be obtained by adding a third plate to form a "Z" configuration (picture © 1975 IEEE)

the channels are oriented at an angle with respect to the plate surfaces and to each other so as to avoid troublesome feedback from positive ions which occasionally form in the channels and drift backwards along the hole. The gain of the chevron configuration ranges from 10^5 to 10^7, which is comparable to the more conventional dynode structures. The advantage, however, is in the much improved timing properties due to the smaller dimensions of the microchannel plates. Transit times are only a few nanoseconds compared to a few tens of nanoseconds for the conventional types. This results in timing resolutions of ≤ 100 ps [8.4]. As well this smaller size makes microchannel plate PM's much less sensitive to magnetic fields. Tests, in fact, have shown, the immunity of these PM's to fields as high as 2 kG [8.4]. The disadvantages are high cost and their yet to be proven reliability. One particular problem which arises is that a cascade in one channel *drains* the neighboring channels for several μseconds [8.5] leading to nonlinearities for count rates above a few thousand per second.

8.4.2 Multiplier Response: The Single-Electron Spectrum

Ideally, the electron-multiplier system should provide a constant gain for all fixed energy electrons which enter the dynode system. In practice, this is not possible because of the statistical nature of the secondary emission process. Single electrons of the same energy entering the system will thus produce different numbers of secondary electrons, resulting in fluctuations in gain. This may be further amplified by variations in the secondary emission factor over the surface of the dynodes, differences in transit time, etc. A good measure of the extent of the fluctuations in a given multiplier chain is the *single electron spectrum*. This is the spectrum of PM output pulses resulting from the entry of single electrons only into the multiplier system. This distribution essentially gives the response of the electron-multiplier and can be measured by illuminating the PM with a very weak light source such that the probability of more than one single electron entering the multiplier at the same time is small. A more detailed description of the technique is given in the paper by *Hyman* et al. [8.6]. Because of the previously mentioned effects, the output pulse shapes will generally be different for each single-electron event. By integrating each current pulse, however, a new pulse whose amplitude is pro-

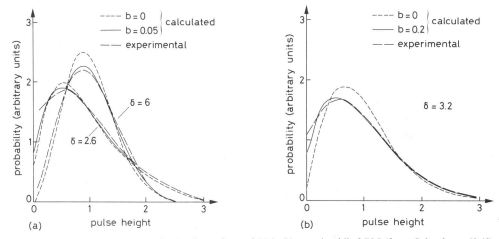

Fig. 8.8. Single-electron spectra for **(a)** linear-focused PM, **(b)** venetian blind PM (from *Schonkeren* [8.1])

portional to the total charge is obtained and thereby the gain for each event. Plotting each event versus gain then gives the response to the multiplier and thus the inherent gain fluctuations.

Figure 8.8 illustrates some measured single-electron distributions for a linear-focused PM and a Venetian blind PM. Analytically, these spectra are best described by a *Polya* distribution (also called a *negative-binomial* distribution or *compound-Poisson* distribution). These are also shown in Fig. 8.8. The parameter b shown in Fig. 8.8 is the rms deviation from perfect uniformity of the secondary emission factor over the surface of the dynode. As can be seen, the Venetian blind configuration is generally subject to more gain fluctuations than is the linear-focused type. This is due to better focused electrons in the latter which minimizes the effect of dynode nonuniformities.

8.5 Operating Parameters

8.5.1 Gain and Voltage Supply

The overall amplication factor or gain of a PM depends on the number of dynodes in the multiplier section and the secondary emission factor δ, which is a function of the energy of the primary electron. Figure 8.9 shows this dependence for several materials. In the multiplier chain, the energy of the electrons incident on each dynode is clearly a function of the potential difference between the dynodes so that we can write

$$\delta = KV_d \ . \tag{8.6}$$

Assuming the applied voltage is equally divided among the dynodes, the overall gain of the PM is then

$$G = \delta^n = (KV_d)^n \ . \tag{8.7}$$

From (8.7) it is interesting to calculate the number of stages n, required for a fixed gain G with a minimum supply voltage V_b. Thus

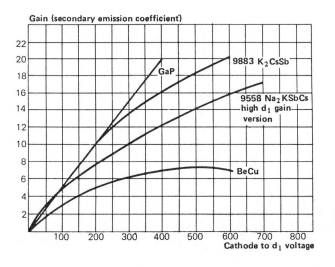

Fig. 8.9. Secondary emission factor for several dynode materials (from *EMI Catalog* [8.2])

$$V_b = nV_d = \frac{n}{K} G^{1/n} \ .$$

Minimizing, we find,

$$\frac{dV_b}{dn} = \frac{1}{K} G^{1/n} - \frac{n}{K} \frac{G^{1/n}}{n^2} \ln G = 0 \ , \qquad n = \ln G \qquad (8.8)$$

for operation at a minimum V_b. Apart from practical reasons, operating at the minimum voltage is desirable from the point of view of noise, etc. However, this comes into conflict with transit time spread and other factors (the number of pins, for example) which often results in the use of higher voltages.

An important relation which should be noted is the variation in gain with respect to supply voltage. From (8.7), we calculate

$$\frac{dG}{G} = n \frac{dV_d}{V_d} = n \frac{dV_b}{V_b} \qquad (8.9)$$

which for $n = 10$ implies a 10% variation in gain for a 1% change in V_b! Thus, to maintain a gain stability of 1%, the voltage supply must be regulated to within 0.1%! Modern supply voltages are regulated to better than 0.05%.

8.5.2 Voltage Dividers

In the previous section we saw how crucial it is to have a well regulated voltage applied to the dynodes. Ideally, batteries would be the best stabilized voltage source, however the required number makes such a scheme impractical. The most common method is to use a stabilized high voltage supply in conjunction with a *voltage divider* (see Fig. 8.10). This system consists of a chain of resistances chosen such as to provide the desired voltage to each of the dynodes. Variable resistances can also be placed, for example, between the cathode and accelerating electrode, so as to allow a fine adjustment.

In designing such a divider, however, it is important to prevent the occurrence of large potential variations between the dynodes due to the changing currents in the tube. Such variations would cause changes in the overall gain and linearity of the PM. For this reason, it is important that the current in the resistance chain, known as the *bleeder* current, be large compared to the tube current. A calculation shows, in fact, that the variation in gain with anode current is

$$\frac{\Delta G}{G} = \frac{I_{an}}{I_{bl}} \frac{n(1-\delta)+1}{(n+1)(1-\delta)} \ , \qquad (8.10)$$

with I_{an}: average anode current; I_{bl}: bleeder current; n: number of stages; δ: secondary emission factor.

To maintain a 1% linearity then, a bleeder current of about 100 times the average anode current would be necessary. In pulse mode operation, however, peak currents much larger than this current can still occur, particularly in the last few stages. To avoid momentary potential drops due to these peaks, the last stages can be maintained at a fixed potential by the addition of decoupling capacitors which provide the necessary charge during the peak period (see Fig. 8.10). These capacitors are then recharged

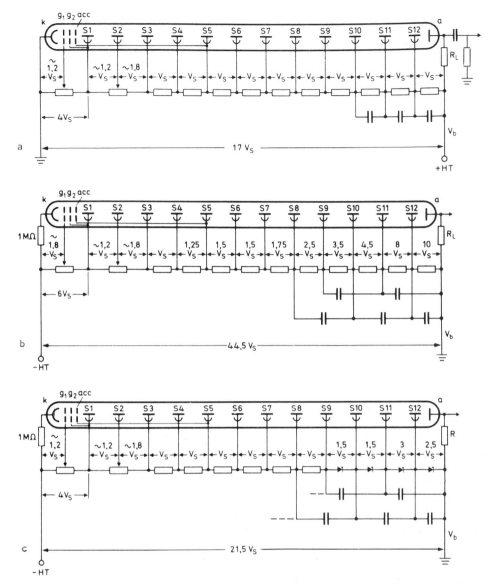

Fig. 8.10a–c. Examples of PM voltage divider networks (after examples from *Philips Catalog* [8.7]: (**a**) divider network using positive high voltage; note the AC coupling capacitor at the anode, (**b**) a network using negative high voltage and decoupling capacitors for maintaining the voltages between the last few dynodes, (**c**) example of the use of zener diodes to maintain voltages on the last few dynodes

during the non-peak periods. An alternate solution is to use Zener diodes in place of the resistances. These elements maintain a constant voltage for currents above a minimum threshold. In very high current applications, it may even be necessary to use a second external voltage supply to maintain the voltages on these last stages. Examples of the use of decoupling capacitors are shown in Fig. 8.10 (b) and (c).

The photomultiplier may be operated with either a positive or negative high voltage, as long as the potential of the dynodes is negative relative to the photocathode. If a positive high voltage is used, the photocathode should be kept at ground potential in order to avoid spurious discharges which might occur between the photocathode and the scintillator or outer envelope of the detector. Grounding the photocathode will also minimize noise from this component as well, an asset when doing pulse height spectroscopy. This advantage is somewhat offset, however, by the fact that the anode will be at constant positive potential. This requires that it be ac coupled through a capacitor in order to allow the pulse signal to pass at 0 dc level [see Fig. 8.10 (a)]. This is avoided if negative high voltage is used. The anode can then be kept at ground which allows it to be directly coupled to the detector electronics. For timing applications, this is particularly advantageous as the signal can be taken directly from the PM without suffering a reshaping (and thus loss of timing information) due to a coupling capacitor. The disadvantage, however, is that the cathode is now at a high negative potential. It thus becomes important to keep the glass well insulated so as to avoid leakage currents from the PM to the grounded material surrounding it.

8.5.3 Electrode Current. Linearity

The linearity of a PM depends strongly on the type of dynode configuration and the current in the tube. In general, photomultiplier linearity requires that the current at each stage be entirely collected by the following stage, so that a strict proportionality with the initial cathode current is maintained. Current collection, of course, depends on the voltage difference applied between stages. Figure 8.11, for example, shows the functional dependence of the cathode and anode currents on the applied voltages for various illuminations at the photocathode. These might be recognized as the Child-Langmuir characteristics for thermionic values. At a given initial current, the current increases with applied voltage until a saturation level is reached where all the current is collected. The initial dependence on voltage is due to the formation of a space charge around the emitting electrode. This *cloud* of electrons tends to nullify the electric field in this region and prevents the acceleration of subsequently emitted electrons towards the receiving electrode.[2] As the voltage is increased, however, this space charge is swept

---

[2] This phenomena is referred to by some authors as *space charge saturation* which is often abbreviated to plain *saturation*. This should not be confused with the saturation of the Child-Langmuir characteristics!

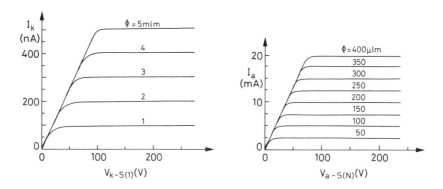

Fig. 8.11. Current-voltage characteristics of the PM cathode and anode under different illuminating light intensities (from *Schonkeren* [8.1])

away and all of the emitted current is collected. As a general rule, therefore, the cathode, dynode and anode currents should always be in the flat, saturated portion of the characteristic curve.

Maintaining these voltages during operation, however, requires more attention because of their dependence on the tube current. The resistivity of the photocathode, for example, is an important factor. This resistance is normally quite high, on the order of a few tenths of a MΩ or so. The emission of relatively small currents of photoelectrons can thus cause large changes in the potential of the photoemissive layer and a drop in its potential relative to the first dynode. This will, in turn, alter the collection efficiency. It is important, therefore, to work at a sufficiently high voltage so as to ensure staying on the flat part of the characteristic. This is also true for the dynodes, particularly the later stages where the current is high and the probability of space charge build-up is greater. To ensure linear operation, the distribution of voltage to these latter stages is generally increased over that of the earlier stages in most voltage dividers.

In the case of the anode, a similar effect occurs. Because it is connected in series to a load resistance (see Fig. 8.10), the anode voltage will fall as the anode current increases. Thus a change in the potential difference between the anode and the last dynode will occur. Since, the entire chain is kept at a constant voltage V_0, the potential differences between the earlier dynode stages will increase causing a change in gain and a loss of linearity. In order to stay on the saturated part of the characteristic, therefore, the current must be kept to within certain limiting values.

8.5.4 Pulse Shape

As we have seen, the output signal at the anode is a current or charge pulse whose total charge is proportional to the initial number of electrons emitted by the photocathode. In fact, more than any other device, the photomultiplier satisfies the requirements of an ideal current generator. As a circuit element, therefore, the PM may be equivalently represented [8.8] as a current generator in parallel with a resistance and capacitance (see Fig. 8.12). The resistance, R and the capacitance, C, here, represent the intrinsic resistance and capacitance of the anode plus those of any other elements which may be in the output circuit, e.g., the anode load resistor, cables, etc.

Let us examine the behavior of the signal at the circuit output. Assuming that the input is scintillator light described by an exponential decay, the current at the anode will be given by

$$I(t) = \frac{GNe}{\tau_s} \exp\left(\frac{-t}{\tau_s}\right), \qquad (8.11)$$

with G: gain of PM; N: the number of photoelectrons emitted by the cathode; e: charge of the electron; τ_s: decay constant of scintillator.

We then have an equation of the form,

$$I(t) = \frac{V}{R} + C\frac{dV}{dt} \qquad (8.12)$$

which has the solution

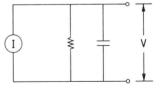

Fig. 8.12. Equivalent circuit for a photomultiplier. The PMT may be considered as an ideal current generator in parallel with a certain resistance and capacitance

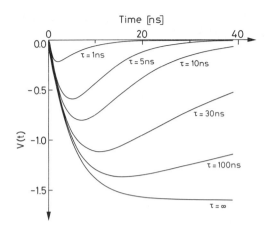

$$V(t) = \begin{cases} -\dfrac{GNeR}{\tau - \tau_s}\left[\exp\left(-\dfrac{t}{\tau_s}\right) - \exp\left(-\dfrac{t}{\tau}\right)\right] & \tau \neq \tau_s \\[3ex] \left(\dfrac{GNeR}{\tau_s^2}\right) t \exp\left(-\dfrac{t}{\tau_s}\right) & \tau = \tau_s \ , \end{cases} \qquad (8.13)$$

where $\tau = RC$. Taking some typical values: $G = 10^6$, $N = 100$, $C = 10$ pF and $\tau_s = 5$ ns, Fig. 8.13 shows this expression evaluated for different values of τ.

For $\tau \ll \tau_s$, the signal is small but faithfully reproduces the decay time of the initial signal. The rise time is rapid and is essentially given by the τ of the output circuit. This is known as the *current* mode of operation, since $V(t)$ is essentially given by the current through the resistance R. For $\tau \gg \tau_s$, the signal amplitude becomes larger but so does the decay time which is now essentially determined by the τ of the output circuit. In return, however, the rise time is approximately given by τ_s. This is known as the *voltage* mode of operation as $V(t)$ is now given by the voltage across the capacitance C. In this mode the current is essentially integrated by C.

As a general rule, the voltage mode is preferred since it gives a large signal which is free from fluctuations due to the integration by C. However, the longer decay time of the signal limits counting rates to $\simeq 1/\tau$, after which signal *pile-up* occurs. Operating in current mode would allow higher counting rates but the output signal would be small and much more sensitive to small fluctuations which occur at the photocathode. For optimum performance, PM output circuits must be tailored to the scintillator which is to be used. This usually involves altering the anode resistance so as to obtain a suitable τ. The capacitance C is usually kept as small as possible in order to maximize the amplitude.

8.6 Time Response and Resolution

Two principal factors affect the time resolution of photomultipliers:

1) Variations in the transit time of the electrons through the PM,
2) Fluctuations due to statistical noise.

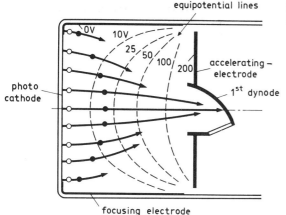

Fig. 8.14. Transit time difference (from *Schonkeren* [8.1])

Transit time variations may arise because of differences in the path length traveled by the electrons and in the energy with which they are emitted by the photocathode. Figure 8.14 illustrates the first effect for a conventional PM by showing the differences in distance traveled in a given period of time for electrons emitted from various points of the cathode. During the time it takes for the electrons on axis to travel to the first dynode, electrons emitted near the edge have only travelled $\simeq 1/3$ of the way. This difference is even further increased by the asymmetry of the dynode. Clearly the electrons emitted at the lower edge have a longer distance to travel than those from the upper edge of the figure. This effect is known as *transit time difference* and is associated with the geometry of the system. An obvious change is to use a spherical cathode, so as to better equalize the distances. A more effective way, however, is to grade the electric field in such a way that the electrons at the edge are accelerated more than those on the axis.

Apart from geometrical effects, there will also be variations which depend on the energy and direction of the emitted electrons. Clearly electrons emitted with a high energy will reach the dynode faster than those at lower energies. Similarly, electrons emitted in a direction closer to the normal of the cathode will arrive before those emitted in a direction more parallel to the surface. This effect is called *transit time spread* and is independent of the point at which the electron leaves the cathode. If we express the initial velocity of a photoelectron as a sum of its components in the direction perpendicular and parallel to the photocathode, i.e.,

$$v = v_\perp + v_\parallel$$

the transit time spread can be approximated by the formula

$$\Delta t = -\sqrt{\frac{2\,m_e\,W}{e^2 E^2}} \tag{8.14}$$

with m_e: electron mass, 9.1×10^{-28} g; e: charge of electron, 1.6×10^{-19} C; E: electric field strength [V/m]; W: energy component normal to cathode, i.e., $v_\perp^2 / 2\,m_e$.

For some typical values, $E = 4$ kV/m, $W = 0.4$ eV, Δt is about 0.5 ns. In modern *fast* PM's, the transit time is on the order of 0.2 to 0.5 ns which underlines the impor-

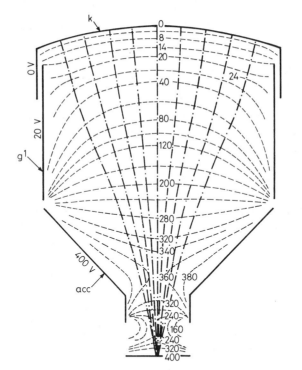

Fig. 8.15. Equipotential lines in the electron-optical input system of a fast photomultiplier (from *Hull* [8.9])

tance of this effect. To reduce the spread, the electric field must be increased as seen in (8.14).

The input system of a fast PM is shown in Fig. 8.15 and should be compared to the classic configuration in Fig. 8.3. In most modern fast PM's, the voltages on the focusing and accelerating electrodes are adjustable. Usually, the best values are found by trial and error.

The second source of timing *jitter* in PM's is due to natural fluctuations in the PM current because of the statistical nature of the photoelectric effect and secondary emission processes. This is known as *statistical noise* and constitutes a fundamental limitation to the time resolution of the PM. This phenomenon is discussed in more detail below.

8.7 Noise

8.7.1 Dark Current and Afterpulsing

Even when a photomultiplier is not illuminated, a small current still flows. This current is called the *dark current* and arises from several sources:

1) thermionic emission from the cathode and dynodes
2) leakage currents
3) radioactive contamination
4) ionization phenomena
5) light phenomena

with thermal noise being the principal component. This contribution is described by Richardson's equation

$$I = AT^2 \exp\left(\frac{-e\phi}{kT}\right) , \tag{8.15}$$

where A is a constant, ϕ the work function, T the temperature [K] and k Boltzmann's constant. Clearly, a lowering of temperature reduces this component of noise.

Leakage currents going through the electrode supports and the pins at the base also contribute a large component to the dark current. Reduction of this noise by insulation of the supports is difficult because of the small magnitude of the currents involved. Operation of the PM in a reduced atmosphere will, however, reduce leakage through the pins by lowering the breakdown voltage.

Radioactive materials in the glass housing or support materials can also cause electron emission from the photocathode or dynodes. The radiation from these contaminants can either directly strike the electrodes or cause fluorescence in the glass housing itself. In each case, a small current results.

In a similar manner, residual gases left or formed in the PM can also cause a detectable current. These gas atoms can be ionized by the electrons and since they are of the opposite charge, will accelerate back towards the cathode or dynodes where they can release further electrons. This often results in afterpulses occurring in a time equal to the time needed for the ions to transit the tube. This may be from a few hundred nanoseconds to microseconds. Under high current, afterpulses may also be caused by *electrode glow*, i.e., light emitted by the last few dynodes which travels to the photocathode. In such cases, afterpulses occurring between 30 to 60 ns after the true pulse are often seen. Since these are usually one-electron events, their amplitudes are generally small. A good discussion of afterpulsing is given by *Candy* [8.11].

In general, dark currents should be very small and in most PM's not more than a few nanoamperes.

8.7.2 Statistical Noise

Statistical noise is a direct result of the statistical nature of the photoemission and secondary emission processes. For a constant intensity of light, the number of photoelectrons emitted as well as the number of secondary electrons emitted will fluctuate with time. The current at the anode will thus fluctuate about some mean in the manner shown in Fig. 8.16. This noise is usually referred to as *shot* noise or the *Schottky Effect* and is measured by the variance of the fluctuations about the mean anode current.

Statistical fluctuations in a PM have two origins: (1) the photocathode, and (2) electron multiplier system. The first source arises from the statistical nature of the photoelectric effect and is the result of a fundamental physical limit. For a PM under constant illumination, these fluctuations can be calculated by assuming a Poisson distribution for the number of photons incident on the photocathode in a time period τ and a

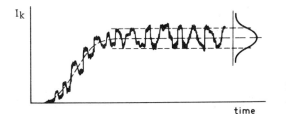

Fig. 8.16. Statistical noise from a PM (from *Philips Catalog* [8.7])

binomial distributed probability for the number of photoelectrons released. One then finds the rms deviation given by

$$\langle \Delta I^2 \rangle = Ie/\tau \tag{8.16}$$

with I: cathode current; e: electric charge of electron.

Added to this noise are fluctuations from the electron-multiplier system. These arise not only from the statistical nature of secondary emission, but from differences in electron transit times, nonuniformities in the secondary emission factor over the dynodes, and other factors. The extent of these fluctuations are probably best judged from the single-electron spectrum, discussed in the previous section. In general, however, multiplier noise accounts for not more then about 10% of the total statistical noise.

8.8 Environmental Factors

8.8.1 Exposure to Ambient Light

Since photomultipliers are extremely photosensitive, it is clear that care must be taken *not* to expose the PM to ambient light while it is under voltage. In such a case, the resulting high currents in the tube can give rise to instability (*fatigue*) effects or even destroy the PM entirely. In some cases, a tube can be recovered after a long period in darkness; however, there will most likely be a marked increase in the dark current. Even when not under voltage, it is best not to expose the photomultiplier to excessive illumination. The result is a higher dark current which, however, decays after a certain time. This recovery time depends on the intensity of illumination.

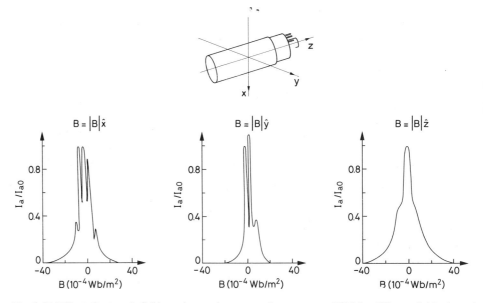

Fig. 8.17. Effect of magnetic fields on the anode current of an unscreened PM for different field orientations (from *Schonkeren* [8.1])

8.8.2 Magnetic Fields

Magnetic fields are one of the more important influences on the operation of PM's. It is easy to see, in fact, that a small magnetic field is enough to deviate the electron cascade from its optimum trajectory in a PM and thereby affect its efficiency. By far the most sensitive part of the PM to magnetic fields is the electron collection system. Here electrons may be so deviated that they may never reach the first dynode at all. As well, the orientation of the tube with respect to the field is clearly a determining factor as well as its symmetry with respect to its axis. Figure 8.17 illustrates the effect of a magnetic field on a PM oriented along three orthogonal axes. In general, the following conclusions can be made:

1) the anode current decreases as magnetic flux increases,
2) the influence of the field is least when oriented along the axis of the PM.

It is common practice to shield PM's with a mu-metal screen which fits around the PM tube. These are available commercially or can be made easily. Generally it is sufficient to shield only the area around the tube; however tests have shown that better results are obtained if the screen is extended pass the tube somewhat. Figure 8.18 shows this difference by comparing the effects of a magnetic field versus the positions and lengths of the mu-metal screen. For strong magnetic fields, it may also be necessary to use a further soft iron shield about the mu-metal. In such cases, care should be taken that parts of the PM do not become magnetized as well. More recently, new designs using a *close proximity focusing* scheme [8.12] have made their appearance on the com-

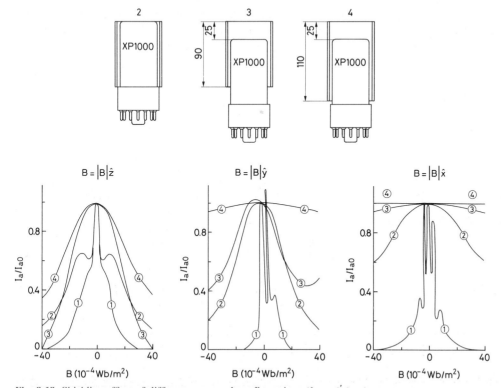

Fig. 8.18. Shielding effect of different mu-metal configurations (from *Schonkeren* [8.1])

mercial market. The distance between the photocathode and the first dynode is greatly reduced in these PM's making them much less sensitive to magnetic fields. Other types of phototubes resistant to magnetic fields are discussed in [8.13].

8.8.3 Temperature Effects

For most normal PM's, temperature effects are generally small compared to other factors. They can play a role, however, depending on the application. The dark noise, as given by Richardson's equation, for example, is obviously a function of T and should therefore be expected to vary. Figure 8.19 gives the variation measured for several photocathode materials.

The spectral sensitivity of the cathode also displays a dependence on temperature, although this effect varies with the type of cathode. Physically, it can be easily seen that the band structure and thus the Fermi level and resistance of the cathode will change; however, the exact effect of these changes is difficult to predict. Generally speaking, in the range between 25° to 50°C, a variation of $\simeq -0.5\%/°C$ with rising temperature is noted.

The variation of the gain of the PM with temperature has also been studied although the results are less conclusive due to large variations from one experiment to the other. In principle, the secondary emission factor does not depend directly on the temperature, however, it may be indirectly affected by temperature-related changes in the

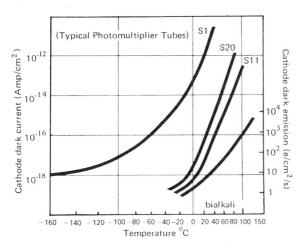

Fig. 8.19. Dark noise vs. temperature for various photocathodes (from *Wardle* [8.10])

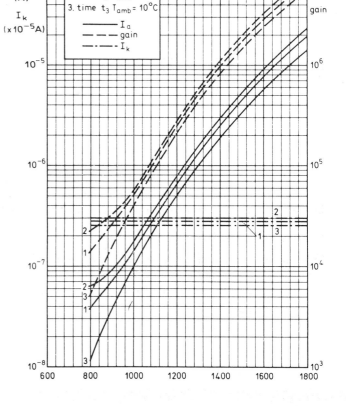

Fig. 8.20. Variation of PM gain, cathode current and anode current for three different temperatures (from *Schonkeren* [8.1])

surface properties of the dynodes, etc. Figure 8.20 shows plots of the measured gain variation with respect to temperature for a typical photomultiplier. Here, the variation is on the order of a few tenths of a percent per deg-K, however, it should be pointed out that this ratio not only seems to vary from one PM type to the other, but also among tubes of the same type, as well. Thus, the values shown should only be taken as order of magnitude estimates.

8.9 Gain Stability, Count Rate Shift

As we have seen, the gain stability of a PM is one of its most crucial characteristics and can be influenced by different factors. In general, variations in gain (*fatigue* effects) are most likely due to changes somewhere in the multiplier system. Two types of change can be distinguished:

1) *drift* − which is a variation in time under a constant level of illumination,
2) *shift* − a sudden change in the gain after the current has changed. In nuclear counting applications, this is sometimes known as *count-rate shift*, i.e., a shift in gain after the average counting rate is suddenly changed.

Figure 8.21 illustrates these two effects. While the causes are varied, their effects can be extremely important. A further discussion of these effects is given by *Yamashita* [8.15].

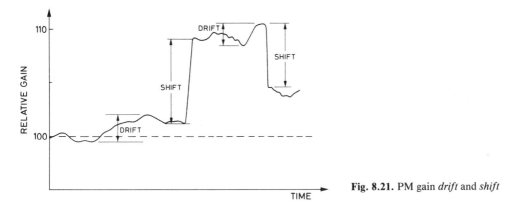

Fig. 8.21. PM gain *drift* and *shift*

To test a PM for stability, the following procedure can be used. Mount the PM in a scintillation counter and using a multichannel analyzer, observe the pulse height distribution from a ^{137}Cs source. In particular, note the position of the 662 keV peak.

Drift. To measure drift:

1) Adjust the source distance so that the count rate is $\simeq 1000\ \mathrm{s}^{-1}$;
2) let the PM operate for 3 hours at this count rate;
3) then once every hour for about 20 hours, determine the position of the peak.

The drift is then given by

$$\mathrm{DRIFT} = \frac{\sum_i |\bar{P} - P_i|}{nP} \qquad (8.17)$$

with P_i: ith measurement of the peak; n: number of measurements; $\bar{P}$: average of P_i over all n measurements. For an acceptable PM, the drift should not be more than 1%.

Shift. To measure shift:

1) Immediately after the measurement of drift, reduce the distance between PM and source so as to obtain a count rate of $\simeq 10000\ \mathrm{s}^{-1}$;
2) record the peak position every 10 minutes for 4 or 5 measurements.
The shift is then

$$\mathrm{SHIFT} = \sum_i \frac{|P_i - P_n|}{m P_n} \tag{8.18}$$

with P_n: the last measurement made for drift at 1000 counts/s; P_i: ith measurement made for shift; m: number of measurements. The value should not exceed 1% for an acceptable tube.

After long periods of inactivity PM's also show gain changes when anode currents of a few tens nA or more are induced. However, after a settling down period of a few hours, this drift stabilizes [8.16]. A long term gradual drop in gain is also noticed [8.17] − which appears related to the total charge collected by the anode.

9. Scintillation Detector Mounting and Operation

In the previous chapters, we considered the two basic components of the scintillation detector: the scintillator material and the photomultiplier, describing their intrinsic capabilities and limitations. In this present chapter, we will discuss the problem of coupling these two components to make an efficient detector and of putting it into operation.

The two most crucial points to consider when mounting a detector are those of light collection and transport. Indeed even with the highest quality scintillators and PM's, a detector is of no use, if none or only a small fraction of the scintillation photons emitted are transmitted to the PM. It is very important therefore to collect as many of the emitted photons as possible and to efficiently transport them to the PM photocathode.

9.1 Light Collection

The loss of light from a scintillator can occur in two basic ways: one is escape through the scintillator boundaries and the other is through absorption by the scintillator material. For small detectors, the latter effect is negligible. Only when the dimensions of the counter are such that the total path lengths traveled by the photons are comparable to the *attenuation length* will absorption begin to play a role. This parameter is defined as that length after which the light intensity is reduced by a factor e^{-1}. The light intensity as a function of length is then

$$L(x) = L_0 \exp\left(\frac{-x}{l}\right) , \qquad (9.1)$$

where l is the attenuation length, x the path length travelled by the light and L_0, the initial light intensity. Since a typical attenuation length is on the order of $\simeq 1$ m or more, it is clear that only very large detectors are affected.

By far, the most important loss is by transmission through the scintillator boundaries. This effect is perhaps best illustrated by referring to Fig. 9.1 which shows a simple detector arrangement. Light emitted at any given point in the scintillator travels in all directions and only a fraction of it directly reaches the PM, the remainder travels toward the scintillator boundaries where, depending on the angle of incidence, two things can happen. For light impinging at an angle greater than the Brewster angle, θ_B where

$$\theta_B = \sin^{-1}\left(\frac{n_{\text{out}}}{n_{\text{scint}}}\right) \qquad (9.2)$$

with n_{scint} being the index of refraction of the scintillator and n_{out} that of the surrounding medium, total internal reflection occurs so that this light is turned back into the

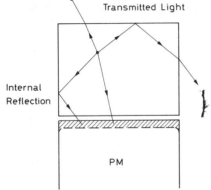

Transmitted Light   **Fig. 9.1.** Light collection in a typical scintillator

scintillator. At angles less than θ_B, partial reflection occurs and the remainder transmitted.

This loss, of course, reduces the efficiency and energy resolution of the detector. However, a second problem also arises: a non-uniformity in pulse height response over the volume of the detector − for depending on the point of emission, different fractions of the total light output will reach the photocathode. In practice, this non-uniformity is usually negligible for the type of small counters used in the nuclear physics lab. However, with larger detectors and with certain scintillator geometries, this non-uniformity can pose some serious problems.

To increase the efficiency of light collection, several methods can be employed.

9.1.1 Reflection

The simplest and most common practice is to redirect escaping light by external and/or internal reflection. This is illustrated in Fig. 9.2 which shows the counter from Fig. 9.1 now surrounded by an external reflector.

The light which was previously transmitted is now also directed back towards the PM by making one or more reflections. Of course, with each reflection some degradation occurs so that this method would not be satisfactory where large numbers of reflections occur.

The reflecting surface here may be *specular* (as shown in Fig. 9.2) or *diffuse*. With a specular surface, the reflections are *mirror-like* in the sense that the angle of reflection equals the angle of incidence. With a diffuse reflector, on the other hand, the reflections are essentially independent of the angle of incidence and follow Lambert's cosine law instead,

$$dI/d\theta \propto \cos\theta \tag{9.3}$$

with I: intensity of reflected light; θ: angle of reflection with respect to normal.

As a specular reflector simple aluminium foil has been found to be very satisfactory and is the most widely used. Of the diffuse reflectors, the most common are MgO, TiO_2 and aluminium oxide. These are usually found in the form of a powder or as a white paint. An important point to note is the reflectivity of these materials versus wavelength. This is shown in Fig. 9.3. While aluminium foil and MgO retain a relatively high reflectivity down to low wavelengths, TiO_2 drops off sharply at $\simeq 400$ nm where it

External Reflector

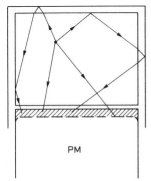

Fig. 9.2. Scintillator with an external reflector for improved light collection

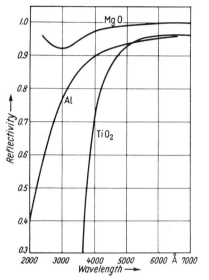

Fig. 9.3. Reflectivity of various materials (from *Mott* and *Sutton* [9.1])

becomes a poor reflector. Diffuse reflectors are generally considered to be slightly more efficient; however, the difference is small and varies according to the geometry of the detector.

While a good efficiency is generally obtained with the external reflector alone, studies have shown that the best results are obtained by also maximizing internal reflection at the same time. The medium surrounding the scintillator should therefore have an index of refraction which is as small as possible in order to minimize θ_B (9.2). Air, obviously, is the best and most convenient medium. Assuming n_{scint} is typically $\simeq 1.5$, this implies a $\theta_B \simeq 42°$. Therefore, a layer of air should be left between the reflector and the scintillator.

With plastic scintillators, internal reflection is facilitated by polishing the surfaces of the plastic. A common procedure for small detectors is to wrap the scintillator in aluminium foil and to follow this with a light-tight layer of black tape. To maximize internal reflection, the foil should be *loosely* applied so as to ensure a layer of air in contact with the scintillator. This type of mounting is illustrated in the next section.

With other scintillators different mountings are generally necessary because of their particular physical characteristics. NaI, for example, is hygroscopic and therefore requires special protection from air. Commercially, these crystals are usually found hermetically sealed in a metallic container with a lining of dry MgO powder along the walls. One end is sealed by a glass or quartz window to allow coupling to the PM and the other by a thin metal window to allow passage of radiations. Similar complications also arise for liquid and gas scintillators which must necessarily be sealed in containers.

9.2 Coupling to the PM

In contrast to the internal reflection requirement discussed above, the coupling between the scintillator and the PM must be made so as to allow a maximum of light transmission. Leaving air here would result in the total trapping of portions of the light in the scintillator. Optical contact between the two media should therefore be made with a material whose index of refraction is as close as possible to that of the scintillator and the PM window. The most common agent is silicone grease or oil. For organic scintillators, the optical coupling with silicone is almost perfect since the refractive indices of the scintillator, grease and PM window are almost identical. For inorganics, however, the match is not as good so that some trapping does occur.

9.3 Multiple Photomultipliers

Instead of the reflecting method described above, light collection could also have been increased by placing a second PM onto the other face of the scintillator (assuming our application allows the placement of another PM at this point!). Light escaping from this end would then be detected by this PM and its signal could be summed with the other to give the total pulse height. Such an arrangement, of course, involves added electronics and thus increased complexity and cost. For small detectors, such a solution would be somewhat extravagant as the simpler reflection technique provides excellent results. However, for large counters where many reflections would occur and/or where the attenuation of light becomes a factor, the use of multiple PM's may be the only way to obtain efficient light recovery and uniform response (Fig. 9.4).

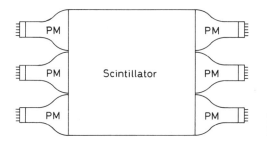

Fig. 9.4. A large scintillator viewed by multiple PM's for better light collection efficiency

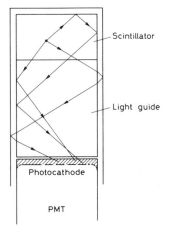

Fig. 9.5. Example of scintillator-PM coupling with a light guide

9.4 Light Guides

Often in many experiments, it is impossible or not desirable to couple the PM directly to the scintillator. This may be because of a lack of space, the presence of magnetic fields, an inconvenient scintillator shape or any number of other reasons. In such a situation, the scintillator light may be conducted to the PM via a *light guide* (or *light pipe*) as illustrated in Fig. 9.5. Such guides are usually made of optical quality plexiglass, lucite or perspex and work on the principle of internal reflection, that is, light entering from one end is "*guided*" along the pipe by internally reflecting it back and forth between the interior walls. Like plastic scintillators, the walls are usually polished for this purpose. Of course, only that fraction of the light incident at angles greater than the Brewster angle can be transferred in this way.

The light guide may be made in any variety of shapes and lengths to conform to the geometry desired. Thus, a counter which "*turns arounds a corner*" may be constructed, for example, or one which bends at a certain angle relative to the PM, etc. It is also par-

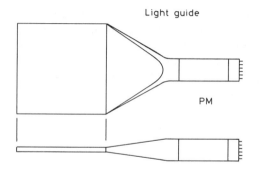

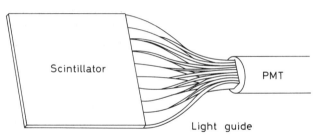

Fig. 9.6. Adapting a flat scintillator sheet to the circular face of a PM with a light guide

Fig. 9.7. The *twisted* light guide. Many strips of light guide material are glued on to the edge of the scintillator and then twisted 90° so as to fit onto the PM face

ticularly useful for adapting inconvenient scintillator shapes to the circular face of the PM. A common example is the case of a flat scintillator plate or sheet which must be viewed on edge. This situation is illustrated in Fig. 9.6.

One common solution is the "*fish tail*" light guide shown in Fig. 9.6, which slowly changes form from the long rectangular geometry of the scintillator edge to the round circular form of the PM. An alternate solution is the *twisted* light guide (Fig. 9.7), which consists of separate strips of lucite attached to the scintillator edge and twisted around so that they converge at the PM photocathode. This geometry is somewhat more complicated, but results in a better light collection.

Many studies have been made as to the most efficient light collecting design for a light guide. *Garwin* [9.2], in particular, has used phase space arguments to show that the flux density of photons in a light guide is "*incompressible*", that is, a given flux of light at the input can never be concentrated into a smaller cross-sectional area at the output. In fact, if the cross-sectional area at the input of the light pipe is A_i and the area at the output is A_o, where $A_o < A_i$, then at *best*, only a fraction, A_o/A_i, of the light can be transferred. This best case is for a so-called *adiabatic* light guide which changes its

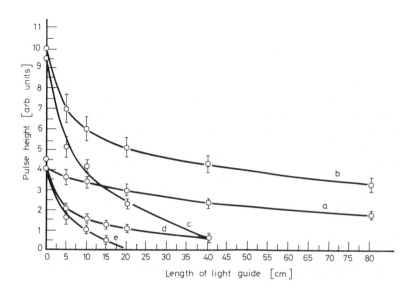

Fig. 9.8. Performance of a scintillation detector using light guides of varying lengths and different reflection schemes (from *Kilvington* et al. [9.5]): (*a*) total internal reflection only, no reflector, (*b*) total internal reflection with external reflector, (*c*) light guide painted with diffuse reflector, (*d*) specular reflector, no light guide, (*e*) diffuse reflector, no light guide

shape gradually without sharp bends or kinks. The quantity of light transferred then depends only on the cross-sectional area and not on the shape. An adiabatic light guide which keeps the same cross-sectional area throughout will therefore conduct all the light down the guide. *Keil* [9.3] has also examined some of the geometrical aspects of internal reflection. Other studies concerning light pipe design are given in [9.4].

The use of enveloping external reflectors with light guides has also been investigated and the conclusions are much the same as with scintillators: the best performance seems to be with loosely wrapped aluminium reflector allowing a maximum of internal reflection. Results from *Kilvington* et al. [9.5] are shown in Fig. 9.8 above.

In addition to plexiglass and lucite, a relatively new development for light guides is the use of optical fibers. These fibers, of course, work on the same principle, however, their use allows a flexible connection between the scintillator and the PM. They are generally used for small counters.

9.5 Fluorescent Radiation Converters

While light guides provide a good deal of flexibility, they can become impractical when applied to very large detectors such as high energy physics calorimeters or hodoscopes. Indeed, since the photons in a light pipe cannot be *"compressed"*, one must use many PM's to cover the available area. This becomes space-consuming (assuming that space is available) and expensive. On the other hand, reducing the number of PM's so that only a fraction of the light is collected engenders a loss in efficiency and resolution. An alternative method of light collection in such cases is the use of *fluorescent radiation converters* [9.3, 6]. These are essentially light guides doped with a highly efficient, fluorescent wavelength-shifting material. The scintillator is coupled to this converter either through an air gap or through direct contact with a layer of oil. Scintillator light entering the converter is absorbed by the fluorescent material, reemitted at another wavelength, and that fraction of light which is trapped transmitted down the guide by internal reflection to the PM. Figure 9.9 illustrates an example of such a set-up. The role of the wavelength shifter is very important here, since it prevents the light from being reabsorbed. It must absorb light in the frequency range emitted by the scintillator and reemit it at a frequency compatible with the spectral sensitivity of the PM photocathode. At present, the most widely used converter is BBQ which absorbs in the blue and emits in the green. While this is not quite optimum, there is sufficient overlap with the spectral sensitivity of most PM's to give a usable signal.

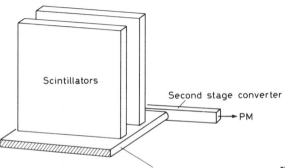

Fig. 9.9. Fluorescent converter set-up

Per unit area, the light collection efficiency of this method is greatly reduced when compared to light pipes. However, a number of advantages are gained. In particular, the fundamental constraint of *"incompressibility"* is no longer applicable so that light from the entire area of the detector may be collected and concentrated onto one or two PM's. The total light collected, therefore, is greater than that which would be collected by the same number of light pipes. In addition, since the light transferred originates in the converter itself, the converter may be made in highly symmetric forms, such as cylinders or rectangular bars, so as to maximize internal reflection. Moreover, several scintillators may be coupled to the same converter to form a compact, inexpensive system.

9.6 Mounting a Scintillation Detector: An Example

To illustrate the principles and ideas outlined in the previous sections, we present here a practical example of how to mount a simple plastic scintillation counter for use in the laboratory. For a more durable construction there are, of course, more elaborate ways than shown here, however, the basic technique is the same.

Step 1

Assemble the necessary materials. These include, silicone grease, aluminium foil, black tape, the scintillator, the PM, its base and shield (Fig. 9.10).

Fig. 9.10. Materials for mounting the scintillation detector

Step 2

Clean the scintillator and PM faces of any old grease or dirt. Alcohol may be used on the PM but *not* on the plastic scintillator. Dab a bit of silicone grease onto the center of the PM and press on the scintillator so that the grease spreads out radially to form a smooth, thin layer coupling the entire surface of the scintillator to the PM (Fig. 9.11). If some excess grease spills out over the edges, this is acceptable. *Do not* spread on the

Fig. 9.11. Couple the scintillator to the PM with optical grease

grease with your fingers and then press on the scintillator. Such a method risks trapping air bubbles between the scintillator and the PM and thus producing an inefficient optical coupling. Handling the plastic scintillator with bare hands should also be avoided as explained in Chap. 7.

Step 3

Wrap the scintillator and PM loosely in aluminium foil so as to assure a layer of air in contact with the scintillator (Fig. 9.12). While aluminium foil is the most convenient re-

Fig. 9.12. Wrap scintillator loosely in aluminium foil

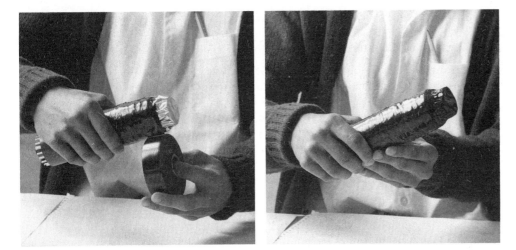

Fig. 9.13. Wrap in black electrical tape

flector, other materials may, of course, be used. In either case, however, it is important to consider the energy loss and absorption effects of this covering on the scintillator. If high energy particles or gamma rays are to be detected, this is usually negligible. However, if low energy β-particles are to be detected, energy loss may play an important role. In such a case, a thinner covering may be required on the detecting surface. This, of course, is more difficult as light tightness must also be maintained.

Step 4

Wrap the PM and detector *neatly* in black tape along its entire length so as to ensure light tightness (Fig. 9.13). Special attention should be paid to corners and sharp bends where light leaks will most likely occur. As above, the absorption effects of the tape on the detecting surface must also be considered.

Step 5

Mount the PM in its base along with its magnetic shielding (not visible) and outer protective shell (Fig. 9.14). The counter is now ready to be tested (see next section).

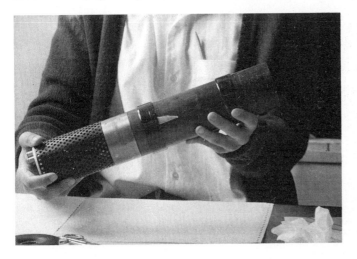

Fig. 9.14. Assembled scintillation detector

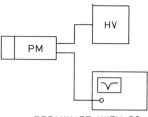

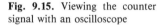

TERMINATE WITH 50Ω

Fig. 9.15. Viewing the counter signal with an oscilloscope

9.7 Scintillation Counter Operation

9.7.1 Testing the Counter

Newly mounted counters or counters whose working status is unsure may be tested by placing a suitable radioactive source in front of the counter and viewing the resulting signal directly with an oscilloscope (Fig. 9.15).

Applying the recommended PM voltage, a good, well formed signal should be observed. Figure 9.16 shows the signal from the plastic scintillation detector that was mounted in the previous section. For comparison, the signal from a *slow* NaI detector is also shown. A ^{207}Bi internal conversion source was used in the former case and a ^{137}Cs γ-source in the latter. For the plastic scintillator, *ringing*, i.e., small secondary pulses, appears on the tail of the signal. These may be tolerated as long as their amplitude remains smaller than the threshold value of the discriminator which will be used with the counter. Otherwise, they may cause multiple triggering of the discriminator. In contrast, the NaI signal shows a perfectly smooth exponential decay. As explained in Chap. 14, such long *tail* pulses should, in general, be reshaped by an amplifier before further processing.

Depending on the PM base, one or more potentiometers may be present to allow an adjustment of the electron optical input stage of the PM. This should be adjusted to give the maximum signal height, if this is not already the case. If no signal or a weak or distorted signal is obtained, this could be due to any number of other problems, including a malfunctioning base or PM, a bad optical coupling to the scintillator, etc.

By removing the source, the noise from the counter can be seen. This should generally be weak and much smaller than the signal. In a good tube, the noise should not be more than $\simeq 50$ mV in the recommended PM voltage range.

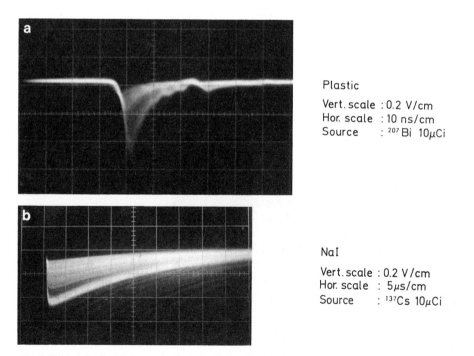

Plastic

Vert. scale : 0.2 V/cm
Hor. scale : 10 ns/cm
Source   : ^{207}Bi 10μCi

NaI

Vert. scale : 0.2 V/cm
Hor. scale : 5μs/cm
Source   : ^{137}Cs 10μCi

Fig. 9.16. Anode signals from plastic (**a**) and NaI (**b**) scintillation counters

If the signal appears unusually intense and noisy, check for light leaks in the wrappings of the counter. This can be done by covering the counter with a black cloth or darkening the room. If there is sharp change in the signal, this is most likely the problem. Check especially corners and edges where wrapping is difficult. The leak may sometimes be spotted by simply running the hands along the counter, while observing the signal for changes.

From the oscilloscope signal, a rough idea of the resolution can also be obtained. In Fig. 9.16, for example, two intensified lines can be readily distinguished in the plastic scintillator signal. These correspond to the 0.48 MeV and 0.97 MeV internal conversion electrons from ^{207}Bi. The fact that they are distinct indicates that the detector is producing well-defined pulse heights in response to these electrons. Similarly, the NaI signal shows a well-defined line corresponding to the 0.662 MeV γ-ray from ^{137}Cs. Moreover, the valley between the peak and the Compton plateau can easily be made out as well. For NaI or other γ-ray detectors, a good simple test is to use a ^{60}Co source which emits γ's of 1.17 MeV and 1.33 MeV. If these two lines can be distinguished visually on the oscilloscope, then clearly the resolution is $\simeq 12\%$ or better at these energies. More detailed measurements of detector resolution, however, require a multichannel analyzer. Spectrum measurements using known calibration sources can then be effectuated and the resolution determined precisely.

9.7.2 Adjusting the PM Voltage

As we saw in the last chapter, the voltage applied to the PM determines its overall gain and thus the pulse height of the output signals. In general, however, PM's have a wide range of possible working voltages extending over a 1000 V or more before the maximum rating is reached. For most applications, it is generally best to stay near the voltage recommended by the manufacturer. This may not always be possible, however, because the gain is too high or too low for the application desired. In such cases, a lower or higher voltage must be used which requires paying some attention to possible saturation or high current effects.

9.7.3 The Scintillation Counter Plateau

In counting applications, where the PM signal is analyzed by a discriminator, a simple procedure for finding the working voltage is to make a so-called *plateau* measurement. This is similar to the plateau curve for Geiger counters and involves measuring the total count rate from the counter-discriminator as a function of the applied PM voltage. The set-up for such a measurement is diagrammed in Fig. 9.17 and an example of a plateau curve, found for the plastic counter in Fig. 9.16, shown in Fig. 9.18.

Starting at the low voltage end, we see that very few counts are registered as the pulse heights are too small to pass through the discriminator. As the voltage increases, however, the curve rises sharply but then becomes flatter over a certain range of voltages. Above this range, however, the curve rises sharply once again. This latter rise marks the onset of *regeneration* effects in the PM, i.e., afterpulsing, discharges, etc., in analogy with the Geiger counter. Similarly, the flat part is known as the *plateau* and represents a region in which the counting rate is the least sensitive to changes in the applied voltage. Fixing the voltage on this plateau (usually in the middle), then, ensures a minimum of counting variations due to drifts in the PM gain or voltage supply. (With modern day regulated voltage supplies, the latter is not usually a problem, however).

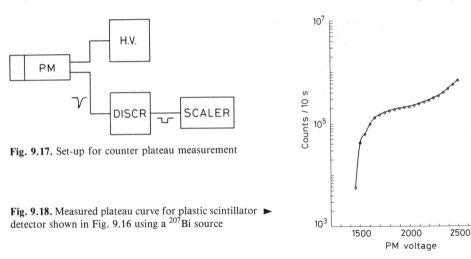

Fig. 9.17. Set-up for counter plateau measurement

Fig. 9.18. Measured plateau curve for plastic scintillator ▶ detector shown in Fig. 9.16 using a ^{207}Bi source

Despite its very strong similarity to the Geiger counter plateau, the origin and significance of the scintillator plateau are completely different. Indeed, whereas the Geiger plateau is the result of signal saturation, which depends entirely on the intrinsic characteristics of the Geiger counter, the scintillator plateau depends on both intrinsic and extrinsic factors, in particular,

1) the spectrum of the incident radiation and the response of the counter,
2) the gain-voltage relationship of the PM,
3) the threshold value of the discriminator.

Indeed, depending on the type of source used or scintillator crystal used, the same PM may give plateaux of differing flatness, width, etc.

Let us look more closely at how the plateau arises [9.7]. Figure 9.19 shows the spectrum of pulse heights from the scintillator in our example with the ^{207}Bi source. As we have seen, at very low gain this spectrum is entirely contained in a region below the threshold value of the discriminator so that there are no counts recorded. As the voltage is raised, however, this spectrum gets stretched proportionately to the right so that more and more of the spectrum passes over the threshold and is counted. (This, of course, is unlike the Geiger counter, where all the pulse heights progressively approach a fixed saturated value as voltage is increased.) This is illustrated in Fig. 9.20. At each voltage on the plateau curve, therefore, the count recorded is the integral of the spectrum from where it crosses the threshold to its uppermost value, the amount of the spectrum above the threshold being determined by the voltage.

Let us derive the plateau curve mathematically by first calculating the *integral counting curve* from the differential spectrum in Fig. 9.19, i.e.,

$$I(p_0) = \int_{p_0} \mathrm{Sp}(x)\,dx \ , \tag{9.4}$$

where p_0 is the lower limit and $\mathrm{Sp}(x)$ is the pulse height spectrum in Fig. 9.19. The result is shown in Fig. 9.21 where p_0 is given in arbitrary units. As can be seen, two plateaux arise as the lower limit of integration passes over each peak in the differential spectrum and into the following valley. This is to be expected (consider, for example,

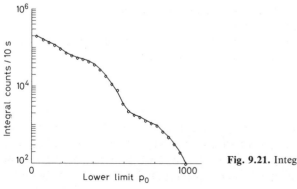

Fig. 9.19. Pulse height spectrum of ^{207}Bi observed with the detector in Fig. 9.16

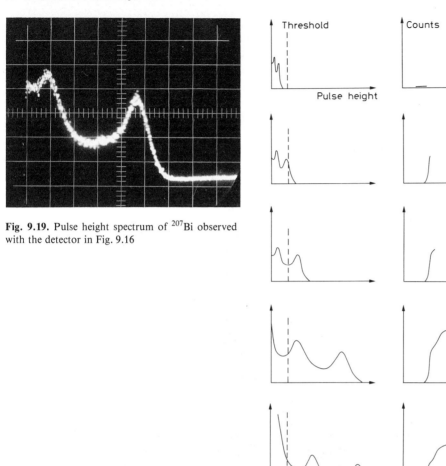

Fig. 9.20. Steps leading to the scintillator counting plateau

the integral of a spectrum containing one single peak), but is *not* responsible for our counting-voltage plateau.

Now for a given electronic threshold, what is the gain required to bring a point p_0 on the spectrum to the threshold level? Clearly

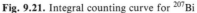

Fig. 9.21. Integral counting curve for ^{207}Bi

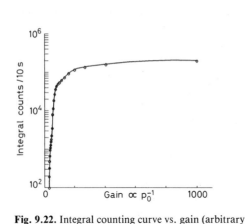

Fig. 9.22. Integral counting curve vs. gain (arbitrary units)

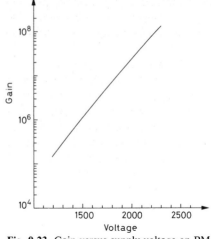

Fig. 9.23. Gain versus supply voltage on PM

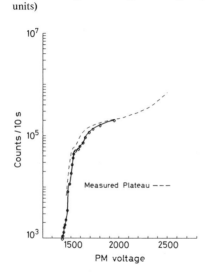

Fig. 9.24. Calculated plateau curve vs. measured plateau

$$K \cdot \text{gain} \cdot p_0 = \text{threshold} \ , \tag{9.5}$$

where K is a simple conversion constant. The required gain is then proportional to the inverse of p_0

$$\text{gain} = \text{const} \ \frac{1}{p_0} \ . \tag{9.6}$$

If we now transform our integral curve into units of gain, we find the form in Fig. 9.22.

To obtain the voltage plateau now, we must further convert the horizontal scale to applied voltage. Figure 9.23 shows the relationship between gain and supply voltage for this PM. With the appropriate conversion factors, we thus find the derived plateau curve in Fig. 9.24. The measured plateau curve is also included for comparison. As can be seen, much of the plateau is reasonably well reproduced. This could have been extended further, of course, had we gone lower in our measurement of the integral counting curve. However, the high voltage rise due to regeneration effects would not have been reproduced as the integral spectrum was measured at a PM voltage where such effects did not occur.

From this analysis, we can discern several important points. Indeed, as can already be seen in Fig. 9.22, the plateau corresponds to the very low end of the integral counting curve, i.e., small pulse heights and not the peak-valley plateaux (these are compressed into the two knee-like structures in Fig. 9.22). This is due to the change in scale coming from the p_0-gain-voltage relationship which stretches out the lower part of the integral curve. The *flatness* of the plateau, then, depends on the slope of the integral curve in this region and, therefore, the differential energy spectrum. Had we used a different radioactive source with a different spectrum, for example, a β-source, a different plateau shape would have been found, perhaps less flat. Similarly changing the crystal geometry or surrounding materials may also alter the spectrum and thus the plateau. And quite obviously, varying the threshold will displace the plateau.

It becomes obvious, then, that plateau measurements for a counter should be made with the same source of radiation as will be used during the experiment, if possible. As well, the discriminator thresholds should also be the same as will be used during the actual experiment. It is clear, also, that comparisons of different PM's through their plateaux should be made with care, because of differences in the gain-voltage relation. A more reliable way is to make pulse height resolution measurements with a multichannel analyzer.

9.7.4 Maintaining PM Gain

Before using the counter for measurements, a certain settling time should be allowed in order to stabilize the PM. Once the correct gain is set, of course, it must be maintained for all measurements; because of drifts and changes which occur in the PM with time, use and other factors, noting the voltage is *not* sufficient. It is not rare, in fact, to find that the same gain is not reproduced by returning to the same voltage, although a very large difference should not be observed. Thus, it is important to fix a standard to which the gain can always be referred and not simply note down the PM voltage. The best method is to choose a radioactive source with a distinguishable feature (e.g. a peak or eventually a Compton edge) in or near the energy region of interest and to require it to fall into a certain channel in a multichannel analyzer. A fixed, calibrated light source, such as an light-emitting diode (LED) arranged so as to shine on the PM or scintillator, may also be used. In this manner, the same gain (within the resolution of the multichannel analyzer) will always be reproduced. If it does not disturb the measurement, the source or light can be left permanently in front of the counter to allow a continuous check of the gain. Otherwise, periodic checks may be performed.

10. Semiconductor Detectors

Semiconductor detectors, as their name implies, are based on crystalline semiconductor materials, most notably silicon and germanium. These detectors are also referred to as *solid-state* detectors, which is a somewhat older term recalling the era when *solid-state* devices first began appearing in electronic circuits. While work on crystal detectors was performed as early as the 1930's [10.1], real development of these instruments first began in the late 1950's. The first prototypes quickly progressed to working status and commercial availability in the 1960's. These devices provided the first high-resolution detectors for energy measurement and were quickly adopted in nuclear physics research for charged particle detection and gamma spectroscopy. In more recent years, however, semiconductor devices have also gained a good deal of attention in the high energy physics domain as possible high-resolution particle track detectors. Much work is being put into developing these instruments [10.2, 3], along with gas detectors, as *the* detectors for the next generation of high energy experiments.

The basic operating principle of semiconductor detectors is analogous to gas ionization devices. Instead of a gas, however, the medium is now a solid semiconductor material. The passage of ionizing radiation creates electron-hole pairs (instead of electron-ion pairs) which are then collected by an electric field. The advantage of the semiconductor, however, is that the average energy required to create an electron-hole pair is some 10 times smaller than that required for gas ionization. Thus the amount of ionization produced for a given energy is an order of magnitude greater resulting in increased energy resolution. Moreover, because of their greater density, they have a greater stopping power than gas detectors. They are compact in size and can have very fast response times. Except for silicon, however, semiconductors generally require cooling to low temperatures before they can be operated. This, of course, implies an additional cryogenic system which adds to detector overhead. One of the problems in current semiconductor detector research, in fact, is to find and develop new materials which can be operated at room temperature. Being crystalline materials, they also have a greater sensitivity to radiation damage which limits their long term use.

We will survey, in this chapter, the different types of semiconductor detectors and their operation. Since an understanding of these detectors requires some knowledge of solid-state physics, however, we shall also review some of the basic physics aspects of semiconductors and in particular semiconductor junctions. A more thorough discussion can be found in some of the standard texts on solid state physics given in the bibliography.

10.1 Basic Semiconductor Properties

In this section, we will briefly review the basic properties of semiconductor materials and those electrical characteristics which are important for their use as radiation detectors. Our discussion here will be concerned with pure semiconductors which are also known as *intrinsic* semiconductors. The term *"pure"* here is relative, since in reality, no semiconductor is ever completely free of impurities in the lattice. For discussion purposes, however, we will assume that this is the case. Impurities, nevertheless, play an important role and may limit or enhance the characteristics of the material. Their effects will be discussed in a later section.

10.1.1 Energy Band Structure

Semiconductors are crystalline materials whose outer shell atomic levels exhibit an *energy band* structure. Figure 10.1 schematically illustrates this basic structure consisting of a *valence* band, a *"forbidden"* energy gap and a *conduction* band. The band configuration for conductors and insulators are also shown for comparison.

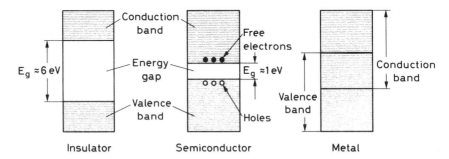

Fig. 10.1. Energy band structure of conductors, insulators and semiconductors

The *energy bands* are actually regions of many discrete levels which are so closely spaced that they may be considered as a continuum, while the *"forbidden"* energy gap is a region in which there are no available energy levels at all. This band structure arises because of the close, periodic arrangement of the atoms in the crystal which causes an overlapping of the electron wavefunctions. Since the Pauli principle forbids more than one electron in the same state, the degeneracy in the outer atomic shell energy levels breaks to form many discrete levels only slightly separated from each other. As two electrons of opposite spin may reside in the same level, there are as many levels as there are pairs of electrons in the crystal. This degeneracy breaking does not affect the inner atomic levels, however, which are more tightly bound.

The highest energy band is the conduction band. Electrons in this region are detached from their parent atoms and are free to roam about the entire crystal. The electrons in the valence band levels, however, are more tightly bound and remain associated to their respective lattice atoms.

The width of the gap and bands is determined by the lattice spacing between the atoms. These parameters are thus dependent on the temperature and the pressure. In conductors, the energy gap is nonexistent, while in insulators the gap is large. At normal temperatures, the electrons in an insulator are normally all in the valence band, thermal energy being insufficient to excite electrons across this gap. When an external electric field is applied, therefore, there is no movement of electrons through the crystal and thus no current. For a conductor, on the other hand, the absence of a gap makes it very easy for thermally excited electrons to jump into the conduction band where they are free to move about the crystal. A current will then flow when an electric field is applied. In a semiconductor, the energy gap is intermediate in size such that only a few electrons are excited into the conduction band by thermal energy. When an electric field is applied, therefore, a small current is observed. If the semiconductor is cooled, however, almost all the electrons will fall into the valence band and the conductivity of the semiconductor will decrease.

10.1.2 Charge Carriers in Semiconductors

At 0 K, in the lowest energy state of the semiconductor, the electrons in the valence band all participate in covalent bonding between the lattice atoms. This is illustrated in Fig. 10.2 for silicon and germanium. Both silicon and germanium have four valence electrons so that four covalent bonds are formed. At normal temperatures, however, the action of thermal energy can excite a valence electron into the conduction band leaving a *hole* in its original position. In this state, it is easy for a neighboring valence electron to jump from its bond to fill the hole. This now leaves a hole in the neighboring position. If now the next neighboring electron repeats the sequence and so on, the hole appears to move through the crystal. Since the hole is positive relative to the sea of negative electrons in the valence band, the hole acts like a positive charge carrier and its movement through the crystal also constitutes an electric current. In a semiconductor, the electric current thus arises from two sources: the movement of free electrons in the conduction band and the movement of holes in the valence band. This is to be contrasted with a metal where the current is carried by electrons only.

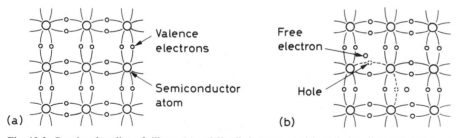

Fig. 10.2. Covalent bonding of silicon: (**a**) at 0 K, all electrons participate in bonding, (**b**) at higher temperatures some bonds are broken by thermal energy leaving a *hole* in the valence band

10.1.3 Intrinsic Charge Carrier Concentration

In a semiconductor, electron-hole pairs are constantly being generated by thermal energy. At the same time, there are also a certain number of electrons and holes which recombine. Under stable conditions, an equilibrium concentration of electron-hole pairs is established. If n_i is the concentration of electrons (or equally holes) and T the temperature, then

$$n_i = \sqrt{N_c N_v} \exp\left(\frac{-E_g}{2kT}\right) = AT^{3/2} \exp\left(\frac{-E_g}{2kT}\right) , \tag{10.1}$$

where N_c is the number of states in the conduction band, N_v the number of states in the valence band, E_g the energy gap at 0 K and k the Boltzmann constant. N_c and N_v can be calculated from Fermi-Dirac statistics and each can be shown to vary as $T^{3/2}$. Making this dependence explicit then gives the right-hand side of (10.1) where the constant A is independent of temperature.

Typical values for n_i are on the order of 2.5×10^{13} cm^{-3} for Ge and 1.5×10^{10} cm^{-3} for Si at $T = 300$ K. This should be put into perspective, however, by noting that there are on the order of 10^{22} atoms/cm^3 in these materials. This means that only 1 in 10^9 germanium atoms is ionized and 1 in 10^{12} in silicon! Despite the large exponents, therefore, the concentrations are very low.

10.1.4 Mobility

Under the action of an externally applied electric field, the drift velocity of the electrons and holes through a semiconductor can be written as

$$v_e = \mu_e E$$
$$v_h = \mu_h E \ ,$$

$$(10.2)$$

where E is the magnitude of the electric field and μ_e and μ_h are the *mobilities* of the electrons and holes respectively. For a given material, the mobilities are functions of E and the temperature T. For silicon at normal temperatures [10.4], μ_e and μ_h are constant (Table 10.1) for $E < 10^3$ V/cm, so that the relation between velocity and E is linear. For E between $10^3 - 10^4$ V/cm, μ varies approximately as $E^{-1/2}$, while above 10^4 V/cm, μ varies as $1/E$. At this point the velocity saturates approaching a constant value of about 10^7 cm/s. Physically, saturation occurs because a proportional fraction of the kinetic energy acquired by the electrons and holes is drained by collisions with the lattice atoms.

Table 10.1. Some physical properties of silicon and germanium

	Si	Ge
Atomic number Z	14	32
Atomic weight A	28.1	72.6
Density [g/cm^2]	2.33	5.32
Dielectric constant (relative)	12	16
Intrinsic resistivity (300 K) [Ωcm]	230000	45
Energy gap (300 K) [eV]	1.1	0.7
Energy gap (0 K) [eV]	1.21	0.785
Electron mobility (300 K) [cm^2/Vs]	1350	3900
Hole mobility (300 K) [cm^2/Vs]	480	1900

At temperatures between 100 and 400 K, μ also varies approximately as T^{-m}, where m depends on the type of material and on the charge carrier. Values of m in silicon are $m = 2.5$ for electrons and $m = 2.7$ for holes, while in germanium, $m = 1.66$ for electrons and $m = 2.33$ for holes [10.4]. Measured values of the drift velocity as a function of T and E are given in [10.5].

The mobilities, of course, determine the current in a semiconductor. Since the current density $J = \rho v$, where ρ is the charge density and v the velocity, J in a pure semiconductor is given by

$$J = e n_i (\mu_e + \mu_h) E \ ,$$

$$(10.3)$$

where we have substituted (10.2) for v and used the fact that current is carried by both electrons and holes. Moreover, $J = \sigma E$, where σ is the conductivity; comparison with (10.3), therefore, gives us the relation

$$\sigma = e n_i (\mu_e + \mu_h) \ . \tag{10.4}$$

This also gives us the resistivity which is just the inverse of σ.

10.1.5 Recombination and Trapping

An electron may recombine with a hole by dropping from the conduction band into an open level in the valence band with the emission of a photon. This process is known as direct *recombination* and is the exact opposite of electron-hole generation. Since both momentum and energy are conserved, however, the electron and the hole must have exactly the right values in order for this to occur. Such processes are therefore very rare and indeed theoretical calculations [10.6] show that electrons and holes should have lifetimes as long as a second if this were the only process. Experimental measurements, however, show carrier lifetimes to range from nanoseconds to hundreds of microseconds which clearly implies that other mechanisms are involved.

The most important mechanism is through *recombination centers* resulting from impurities in the crystal. These elements perturb the energy band structure by adding additional levels to the middle of the forbidden energy gap as shown in Fig. 10.3. These states may capture an electron from the conduction band and then do one of two things: (1) after a certain holding time, the electron is released back into the conduction band, or, (2) during the holding time, it may also capture a hole which then annihilates with the trapped electron. Such centers are particularly efficient as the impurity is left in its original state so that each center may participate in many recombinations.

For radiation detection, the existence of recombination impurities plays a detrimental role since they will reduce the mean time charge carriers remain free. This time, of course, should be longer than the time it takes to collect the charges otherwise charge loss will occur with a subsequent reduction in resolution. Semiconductor detectors therefore require relatively pure crystals. For large volume detectors, in particular, the impurity concentration cannot be more than 10^{10} impurities per cm^3 [10.7].

A second effect which arises from impurities is *trapping*. Some impurities, in fact, are only capable of capturing one kind of charge carrier, that is electrons or holes, but not both. Such centers simply hold the electron or hole and then release it after a certain characteristic time. If the trapping time is on the order of the charge collection time, then quite obviously charges will be lost and incomplete charge collection will result. If the trapping time is very much smaller, then little or no effect occurs. Recombination centers are also trapping centers as we have seen.

While impurities are the principal source of recombination and trapping, structural defects in the lattice may also give rise to similar states in the forbidden band. Such de-

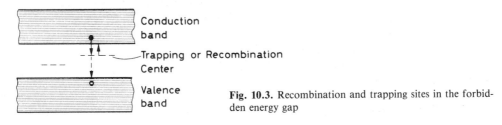

Fig. 10.3. Recombination and trapping sites in the forbidden energy gap

fects include simple *point* defects such as vacancies in the lattice or atoms which occupy positions in between lattice points, and *dislocations,* in which an entire line of atoms is displaced. These structural defects may arise during growth of the crystal or may be caused by thermal shock, plastic deformation, stress and bombardment by radiation. The latter source, of course, is of particular concern for detectors and is discussed in a later section.

While we have thus far discussed the detrimental effects of impurities, the addition of certain elements to pure semiconductors may also enhance the characteristics of the material. This is discussed in the next section on doped semiconductors. The difference between these impurities and recombination and trapping impurities is in the depth of the energy levels created in the forbidden band. As we will see, doping impurities create *shallow* levels very close to the conduction or valence bands, whereas recombination and trapping impurities produce *deep* levels near the center. Electrons and holes in shallow levels are easily excited out into the conduction and valence bands and thus are not trapped for very long periods.

10.2 Doped Semiconductors

In a pure semiconductor crystal, the number of holes equals the number of electrons in the conduction band. This balance can be changed by introducing a small amount of impurity atoms having one more or one less valence electron in their outer atomic shell. For silicon and germanium which are tetravalent, this means either pentavalent atoms or trivalent atoms. These impurities integrate themselves into the crystal lattice to create what are called *doped* or *extrinsic* semiconductors.

If the dopant is pentavalent, the situation in Fig. 10.4a arises. In the ground state, the electrons fill up the valence band which contains just enough room for four valence electrons per atom. Since the impurity atom has five valence electrons, an extra elec-

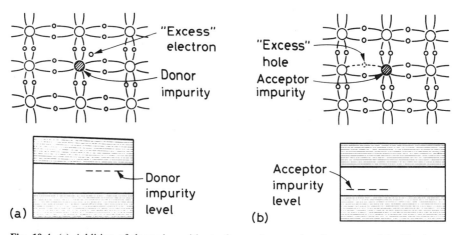

Fig. 10.4. (a) Addition of donor impurities to form n-type semiconductor materials. The impurities add *excess* electrons to the crystal and create donor impurity levels in the energy gap. **(b)** Addition of acceptor impurities to create p-type material. Acceptor impurities create an *excess* of holes and impurity levels close to the valence band

tron is left which does not fit into this band[1]. This electron resides in a discrete energy level created in the energy gap by the presence of the impurity atoms. Unlike recombination and trapping states, this level is extremely close to the conduction band being separated by only 0.01 eV in germanium and 0.05 eV in silicon. At normal temperatures, therefore, the *extra* electron is easily excited into the conduction band where it will enhance the conductivity of the semiconductor. In addition, the extra electrons will also fill up holes which normally form, thereby decreasing the normal hole concentration. In such materials, then, the current is mainly due to the movement of electrons. Holes, of course, still contribute to the current but only as *minority carriers*. Doped semiconductors in which electrons are the majority charge carriers are called *n-type* semiconductors.

If the impurity is now trivalent with one less valence electron, there will not be enough electrons to fill the valence band. There is thus an *excess* of holes in the crystal (Fig. 10.4b). The trivalent impurities also perturb the band structure by creating an additional state in the energy gap, but this time, close to the valence band as shown in Fig. 10.4b. Electrons in the valence band are then easily excited into this extra level, leaving extra holes behind. This excess of holes also decreases the normal concentration of free electrons, so that the holes become the majority charge carriers and the electrons minority carriers. Such materials are referred to as *p-type* semiconductors.

In practice, *donor* elements such as arsenic, phosphorous and antimony are used to make n-type semiconductors, while gallium, boron and indium are most often employed as *acceptor* impurities for p-type materials. The amount of dopant used is generally very small with typical concentrations being on the order of a few times 10^{13} atoms/cm^3. Since the densities of germanium and silicon are on the order of 10^{22} atoms/cm^3, this implies impurity concentrations of only a few parts per billion!

Great use is also made of heavily doped semiconductors particularly as the electrical contacts for semiconductors. Impurity concentrations in these materials can be as high as 10^{20} atoms/cm^3, so that they are highly conductive. To distinguish these semiconductors from normally doped materials, a "+" sign after the material type is used. A heavily doped p-type semiconductor is therefore written as p$^+$ and a heavily doped n semiconductor as n$^+$.

Regardless of the type of dopant, the concentration of electrons and holes obey a simple law of mass action when in thermal equilibrium. If n is the concentration of electrons and p is the concentration of holes, then their product is

$$np = n_i^2 = AT^3 \exp\left(\frac{-E_g}{kT}\right) , \qquad (10.5)$$

where n_i is the intrinsic concentration given in (10.1). Since the semiconductor is neutral, the positive and negative charge densities must be equal, so that

$$N_D + p = N_A + n , \qquad (10.6)$$

where N_D and N_A are the donor and acceptor concentrations. In an n-type material, where $N_A = 0$ and $n \gg p$, the electron density is therefore

$$n \simeq N_D , \qquad (10.7)$$

---

[1] Note this extra electron does not imply that the crystal is now charged. Electrical neutrality is assured since the nucleus of the impurity atom also contains an *extra* positive proton.

i.e., the electron concentration is approximately the same as the dopant concentration. Using (10.5), then, the minority carrier concentration is

$$p \simeq \frac{n_i^2}{N_D} \; .$$

(10.8)

From (10.4), the conductivity or resistivity of an n-type material thus becomes

$$\frac{1}{\rho} = \sigma \simeq e N_D \mu_e \; .$$

(10.9)

An analogous result is found for p-type materials.

10.2.1 Compensation

A question one can ask now is what happens if both n- and p-type impurities are added to a semiconductor. In reality, all semiconductors do, in fact, contain impurities of both types. Since the extra electrons from donor atoms will be captured by extra holes from acceptor materials, it is easy to see that a cancellation effect occurs. What counts, therefore, is the net concentration, $|N_D - N_A|$, where N_D and N_A are the concentrations of donor and acceptor atoms. If $N_D > N_A$, then the semiconductor is an n-type and vice-versa. Similarly, semiconductors with equal amounts of donor and acceptor impurities remain intrinsic or at least retain most of the characteristics of intrinsic materials. Such materials are known as *compensated* materials and are designated with the letter *"i"*. (This should *not* be confused with *intrinsic*).

It is quite obviously difficult to make compensated materials because of the need to have *exactly* the same amounts of donor and acceptor impurities. The first breakthrough came in the 1960's when Pell developed the lithium drifting method for compensating p-type silicon and germanium. While we will not go into detail, the process consists essentially of diffusing lithium, which acts as a donor, onto the surface of a p-type material. This changes that end of the semiconductor into an n-type material. The concentration of lithium drops off in a Gaussian manner so that the junction between the n-type and p-type regions lies at some depth under the surface. A bias voltage is then applied across the junction such that the positive lithium ions are pulled further into the bulk of the p-region. To increase lithium mobility, the temperature of the material is raised. As the lithium donor concentration increases in the p-region, its concentration decreases in the n-region. Because of the dynamics of the local electric field, however, the donor concentration of the p-side cannot exceed the acceptor concentration and vice versa on the n-side. Indeed, if this happens the local field essentially reverses direction and sweeps the ions back in the opposite direction. An equilibrium point is thus reached in which the lithium donor ions spread themselves over a region such that the number of donors is exactly equal to the number of acceptors. A compensated region is thus formed. To obtain thick compensated regions from $10-15$ mm, this drift process can take several days. The interested reader may find a more detailed description of the process in [10.8].

Because of the high mobility of the lithium ions at room temperature, particularly in germanium, it is necessary to cool the material to liquid nitrogen temperatures once the desired compensation is obtained. This temperature must be maintained at all times in order to preserve the compensation. In the case of silicon, however, lithium mobility is lower so that short periods at room temperature usually do no harm.

The most important property of compensated material is its high resistivity. Rough measurements on silicon have yielded values as high as $100\,000\ \Omega\,cm$. This is still much lower than the intrinsic maximum value of $\sim 230\,000\ \Omega\,cm$, however, so that although the drifting process is effective, it is still not perfect. We will see how these materials are used for radiation detection in Sects. 10.5.4 and 10.7.1.

10.3 The np Semiconductor Junction. Depletion Depth

The functioning of all present-day semiconductor detectors depends on the formation of a semiconductor *junction*. Such junctions are better known in electronics as rectifying diodes, although that is not how they are used as detectors. Semiconductor diodes can be formed in a number of ways. A simple configuration which we will use for illustration purposes is the pn junction formed by the juxtaposition of a p-type semiconductor with an n-type material. These junctions, of course, cannot be obtained by simply pressing n- and p-type materials together. Special techniques must be used instead to achieve the intimate contact necessary for junction formation. One method, for example, is to diffuse sufficient p-type impurities into one end of a homogeneous bar of n-type material so as to change that end into a p-type semiconductor. Others methods are also available and will be discussed in later sections.

The formation of a pn-junction creates a special zone about the interface between the two materials. This illustrated in Fig. 10.5. Because of the difference in the concentration of electrons and holes between the two materials, there is an initial diffusion of holes towards the n-region and a similar diffusion of electrons towards the p region. As a consequence, the diffusing electrons fill up holes in the p-region while the diffusing holes capture electrons on the n-side. Recalling that the n and p structures are initially neutral, this recombination of electrons and holes also causes a charge build-up to occur on either side of the junction. Since the p-region is injected with extra electrons it thus becomes negative while the n-region becomes positive. This creates an electric field gradient across the junction which eventually halts the diffusion process leaving a region of immobile space charge. The charge density and the corresponding electric field profile are schematically diagrammed in Fig. 10.5. Because of the electric field, there is a potential difference across the junction. This is known as the *contact* potential. The

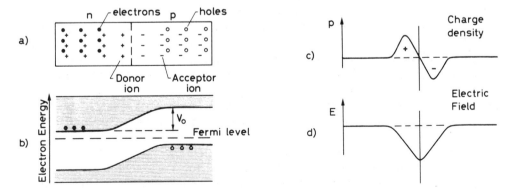

Fig. 10.5. (a) Schematic diagram of an np junction, (b) diagram of *electron* energy levels showing creation of a contact potential V_0, (c) charge density, (d) electric field intensity

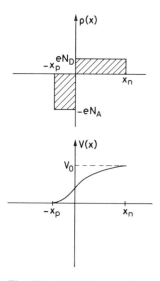

Fig. 10.6. Model for calculating the depletion depth of an np junction. For simplicity, assume a uniform charge distribution in the depletion zone

energy band structure is thus deformed as shown in Fig. 10.5, with the contact potential generally being on the order of 1 V.

The region of changing potential is known as the *depletion* zone or *space charge region* and has the special property of being devoid of all mobile charge carriers. And, in fact, any electron or hole created or entering into this zone will be swept out by the electric field. This characteristic of the depletion zone is particularly attractive for radiation detection. Ionizing radiation entering this zone will liberate electron-hole pairs which are then swept out by the electric field. If electrical contacts are placed on either end of the junction device, a current signal proportional to the ionization will then be detected. The analogy to an ionization chamber thus becomes apparent.

10.3.1 The Depletion Depth

The width of the depletion zone is generally small and depends on the concentration of n and p impurities. If the charge density distribution in the zone, $\rho(x)$ is known, this can be determined from Poisson's equation,

$$\frac{d^2 V}{dx^2} = -\frac{\rho(x)}{\varepsilon} \ , \tag{10.10}$$

where ε is the dielectric constant.

As an illustration, let us take the simple example of a uniform charge distribution about the junction [10.9]. This is illustrated in Fig. 10.6. Letting x_n denote the extent of the depletion zone on the n-side and x_p the depth on the p-side, we then have,

$$\rho(x) = \begin{cases} eN_D & 0 < x < x_n \\ -eN_A & -x_p < x < 0 \ , \end{cases} \tag{10.11}$$

where e is the charge of the electron and N_D and N_A are the donor and acceptor impurity concentrations. Since the total charge is conserved, we also have the relation

$$N_A x_p = N_D x_n \ . \tag{10.12}$$

Now integrating (10.10) once, we find

$$\frac{dV}{dx} = \begin{cases} -\dfrac{eN_D}{\varepsilon} x + C_n & 0 < x < x_n \\[2mm] \dfrac{eN_A}{\varepsilon} x + C_p & -x_p < x < 0 \ , \end{cases} \tag{10.13}$$

where C_n and C_p are integration constants. Since $dV/dx = 0$ at $x = x_n$ and $-x_p$, (10.13) becomes

$$\frac{dV}{dx} = \begin{cases} -\dfrac{eN_D}{\varepsilon} (x - x_n) & 0 < x < x_n \\[2mm] \dfrac{eN_A}{\varepsilon} (x + x_p) & -x_p < x < 0 \ . \end{cases} \tag{10.14}$$

Equation (10.14), of course, represents the electric field in the space charge region. One more integration now yields,

$$V(x) = \begin{cases} -\dfrac{eN_D}{\varepsilon}\left(\dfrac{x^2}{2}-x_n x\right)+C & 0 < x < x_n \\[3mm] \dfrac{eN_A}{\varepsilon}\left(\dfrac{x^2}{2}+x_p x\right)+C' & -x_p < x < 0 \ . \end{cases} \tag{10.15}$$

Since the two solutions must join at $x = 0$, it is clear that $C = C'$. Now at $x = x_n$, $V(x) = V_0$ which is the contact potential; thus

$$V_0 = \frac{eN_D}{2\varepsilon}x_n^2 + C \ . \tag{10.16}$$

Similarly on the p-side $V = 0$ at $x = -x_p$, so that

$$0 = -\frac{eN_A}{2\varepsilon}x_p^2 + C \ . \tag{10.17}$$

Eliminating C, we then obtain

$$V_0 = \frac{e}{2\varepsilon}(N_D x_n^2 + N_A x_p^2) \ . \tag{10.18}$$

Using (10.12) then yields,

$$x_n = \left(\frac{2\varepsilon V_0}{eN_D(1+N_D/N_A)}\right)^{1/2} \ , \qquad x_p = \left(\frac{2\varepsilon V_0}{eN_A(1+N_A/N_D)}\right)^{1/2} \ . \tag{10.19}$$

From (10.19), we can see that if one side is more heavily doped than the other, which is usually the case, then the depletion zone will extend farther into the lighter-doped side. For example, if $N_A \gg N_D$, then $x_n \gg x_p$ which means that the depletion region is almost entirely on the n-side of the junction.

The total width of the depletion zone can now be easily found

$$d = x_n + x_p = \left(\frac{2\varepsilon V_0}{e}\frac{(N_A+N_D)}{N_A N_D}\right)^{1/2} \ . \tag{10.20}$$

If $N_A \gg N_D$, as in our example above, then (10.20) is approximately,

$$d \simeq x_n \simeq \left(\frac{2\varepsilon V_0}{eN_D}\right)^{1/2} \tag{10.21}$$

Using the expression in (10.9) for the resistivity ρ, (10.21) can be expressed as

$$d \simeq (2\varepsilon\rho_n\mu_e V_0)^{1/2} \ , \tag{10.22}$$

where ρ_n is the resistivity of the n-region. For the case of a heavily doped n-side ($N_D \gg N_A$), the depletion region will be entirely on the p-side, so that the factor $\rho_n \mu_e$ in (10.22) should be replaced by $\rho_p \mu_h$. Evaluating some of the constants now, we find the formulae

Silicon
$$d \simeq \begin{cases} 0.53 \, (\rho_n V_0)^{1/2} \, \mu m & \text{n-type} \\ 0.32 \, (\rho_p V_0)^{1/2} \, \mu m & \text{p-type} \end{cases}$$

Germanium   (10.23)
$$d \simeq \begin{cases} (\rho_n V_0)^{1/2} \, \mu m & \text{n-type} \\ 0.65 \, (\rho_p V_0)^{1/2} \, \mu m & \text{p-type} \end{cases}$$

where ρ is in Ω cm and V_0 in Volts. If we take typical values of $\rho \sim 20\,000 \, \Omega$ cm for high-resistivity n-type silicon and $V_0 = 1$ V, this yields $d \simeq 75 \, \mu m$, which is a rather small sensitive depth.

10.3.2 Junction Capacitance

Because of its electrical configuration, the depletion layer also has a certain capacitance which, as we will see later, affects the noise characteristics when the junction is used as a detector. For a planar geometry, the capacitance is

$$C = \varepsilon \frac{A}{d} \, , \tag{10.24}$$

where A is the area of the depletion zone and d its width. Substituting in the formulae of (10.23) then yields

Silicon
$$C = \begin{cases} 2.2 \, (\rho_n V_0)^{-1/2} \, pF/mm^2 & \text{n-type} \\ 3.7 \, (\rho_p V_0)^{-1/2} \, pF/mm^2 & \text{p-type} \end{cases}$$

Germanium   (10.25)
$$C = \begin{cases} 1.37 \, (\rho_n V_0)^{-1/2} \, pF/mm^2 & \text{n-type} \\ 2.12 \, (\rho_p V_0)^{-1/2} \, pF/mm^2 & \text{p-type} \end{cases}$$

10.3.3 Reversed Bias Junctions

While the pn-junction described above will work as a detector, it does not present the best operating characteristics. In general, the intrinsic electric field will not be intense enough to provide efficient charge collection and the thickness of the depletion zone will be sufficient for stopping only the lowest energy particles. As we will see later, this small thickness also presents a large capacitance to the electronics and increases noise in the signal output. Better results can be obtained by applying a reverse-bias voltage to the junction, i.e., a negative voltage to the p-side, as shown in Fig. 10.7. This voltage will have the effect of attracting the holes in the p-region away from the junction and towards the p contact and similarly for the electrons in the n-region. The net effect is to

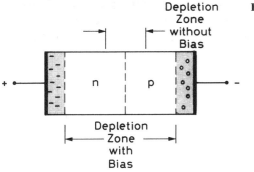

Fig. 10.7. Reversed-bias junction

enlarge the depletion zone and thus the sensitive volume for radiation detection – the higher the external voltage, the wider the depletion zone. Moreover, the higher external voltage will also provide a more efficient charge collection. The maximum voltage which can be applied, however, is limited by the resistance of the semiconductor. At some point, the junction will breakdown and begin conducting.

Under a reverse bias, the width of the depletion layer can be calculated from (10.20) by replacing V_0 with $V_0 + V_B$, where V_B is the bias voltage. In general $V_0 \ll V_B$, so that V_0 may usually be neglected and equations (10.20) to (10.23) used directly with the substitution $V_0 \Rightarrow V_B$. This is also true for the capacitance in (10.25). An interesting point to note is that because of the difference in mobility between electrons and holes, the same bias voltage V_B will yield a somewhat larger depletion depth if the material is n-type rather than p-type.

If we now calculate the depth for n-type silicon, we can see that a depletion layer greater than 1 mm can be obtained if a reverse bias of about $V_B = 300$ V is applied. This, of course, is a great improvement over the few tens of microns in an unbiased junction. With current high resistivity silicon, depletion depths up to 5 mm can be obtained before breakdown occurs. In order to obtain greater depletion widths, even higher resistivity material is necessary which means using higher purity semiconductors or compensated material. These will be discussed in a later section.

10.4 Detector Characteristics of Semiconductors

Having reviewed some of the basic properties of semiconductor materials and junctions, we now turn to some of the characteristics of semiconductors as detectors of radiation.

Figure 10.8 shows the basic configuration used for operating a junction diode as a radiation detector. In order to be able to collect the charges produced by radiation, electrodes must be fitted onto the two sides of the junction. With semiconductors, however, an *ohmic* metal contact cannot, in general, be formed by directly depositing the metal onto the semiconductor material. Indeed, as we shall see in Sect. 10.5.2, contact between many metals and semiconductors results in the creation of a rectifying junction with a depletion zone extending into the semiconductor. To prevent this formation, a heavily doped layer of n^+ and p^+ material is used between the semiconductor and the metal leads. Because of the high dopant concentrations, the depletion depth is

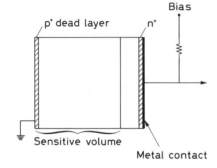

Fig. 10.8. Basic layout of a junction diode detector

then essentially zero as can be seen from (10.19). This then forms the desired *ohmic* contact.

For signal isolation purposes, the bias voltage to the detector is supplied through a series resistor rather than directly. To collect the charge signal from the detector, a preamplifier of the charge-sensitive type is generally used. Because of the low-level of the signal, this preamplifier must have low-noise characteristics. Signal processing after the preamplifier also requires pulse shaping in order to obtain the best signal-to-noise characteristics as explained in Chap. 14.

10.4.1 Average Energy per Electron-Hole Pair

The primary advantage of semiconductors over other detectors is the very small average energy needed to create an electron-hole pair. Like gases, the average energy at a given temperature is found to be independent of the type and energy of the radiation and only dependent on the type of material. Table 10.2 summarizes these values for various semiconductors at normal and liquid nitrogen temperatures.

Table 10.2. Average energy for electron-hole creation in silicon and germanium

	Si	Ge
300 K	3.62 eV	–
77 K	3.81 eV	2.96 eV

For the same radiation energy, the number of charge carriers created will therefore be almost an order of magnitude greater in these materials than in gases. Compared to the number of photoelectrons created in a scintillation counter, this difference approaches two orders of magnitude. Not surprisingly, semiconductors should provide a greatly improved energy resolution.

As small as the average energies are it is interesting to compare these values to the band gaps in Table 10.1. Since the gaps are only on the order 1 eV wide, it is clear that less than a third of the energy deposited by passing radiation is actually spent on the production of electron-hole pairs. The other two thirds, in fact, go into exciting lattice vibrations.

10.4.2 Linearity

Assuming that the depletion region is sufficiently thick to completely stop all particles, the response of semiconductors should be perfectly linear with energy. If E is the energy of the radiation then E/w electron-hole pairs should be created, where w is the average energy from Table 10.2. Assuming a collection efficiency n, then a charge $Q = nE/w$ is collected on the electrodes. Since the depletion region has a capacitance C, as we saw in Sect. 10.3.2, the observed voltage on the electrodes is then

$$V = \frac{Q}{C} = n\frac{E}{wC} \qquad (10.26)$$

which varies linearly with E. Moreover, since w is independent of particle type, the response is, in principle, also independent of the type of radiation. This turns out to be true only for lightly ionizing radiation such as electrons and protons. For heavier ions, *plasma* effects occur which affect the collection efficiency and lead to differences in the pulse height between different particles with the same energy. This is discussed further in Sect. 10.9.5.

If the depletion zone is smaller than the range of the radiation, it is clear then that a nonlinear response should be expected since the full energy is not totally deposited in the sensitive volume. What is measured instead is the energy loss ΔE which is a nonlinear function of energy. For a given depletion depth, the response is therefore linear up until the range of the particles exceeds this depth.

10.4.3 The Fano Factor and Intrinsic Energy Resolution

The intrinsic energy resolution, as we saw in Chap. 5, is dependent on the number of charge carriers and the Fano Factor. Despite numerous experiments, the Fano factor for both silicon and germanium is still not well determined. However, it is clear that F is small and on the order of 0.12. This, of course, contributes greatly to enhancing the resolution of semiconductors which already profit greatly from the small average energy required to create an electron hole pair.

From (5.6), the expected resolution is

$$R = 2.35\sqrt{\frac{F}{J}} = 2.35\sqrt{\frac{Fw}{E}} \ , \qquad (10.27)$$

where w is the average energy for electron-hole creation and $J = E/w$. For a 5 MeV alpha particle, the intrinsic resolution expected for silicon is therefore $R \simeq 0.07\%$ or 3.5 keV. Typical measured resolutions are about 18 keV, however, which indicates that contributions from other sources, e.g., electronics, play important roles.

10.4.4 Leakage Current

Although a reversed biased diode is ideally nonconducting, a small fluctuating current nevertheless flows through semiconductor junctions when voltage is applied. This current appears as noise at the detector output and sets a limit on the smallest signal pulse height which can be observed.

The leakage current has several sources. One is the movement of minority carriers, i.e., holes from the n-region which are attracted across the junction to the p-side and electrons from the p-region which are similarly attracted to the opposite side. This current is generally quite small and in the range of nanoamperes per cm^2. A second source is thermally generated electron-hole pairs originating from recombination and trapping enters in the depletion region. These centers cannot capture electrons or holes since they have all been swept out, but they can catalyze the creation of electrons and holes from the valence band by serving as intermediate states. The contribution from this source depends on the absolute number of traps in the depletion region and thus on their concentration and the volume of the zone. In general, current densities on the order of a few $\mu A/cm^2$ can be expected. The third and by far the largest source of leakage current is through surface channels. This component is complex and depends on very many factors including the surface chemistry, the existence of contaminants, the surrounding atmosphere, the type of mounting, etc. Clean encapsulation is generally required here to minimize this component.

10.4.5 Sensitivity and Intrinsic Efficiency

For charged particles, the intrinsic detection efficiency of semiconductors is close to 100% as very few particles will fail to create some ionization in the sensitive volume. The limiting factor on sensitivity is the noise from leakage currents in the detector and the associated electronics which set a lower limit on the pulse amplitude which can be detected. To ensure an adequate signal, the depletion depth must therefore be chosen sufficiently thick such that enough ionization will be produced to form a signal larger than the noise level. These events are then detected with almost 100% efficiency. If energy measurements are being made, however, the depletion depth must also be larger than the range of the particles. Figure 10.9 shows the range of various particles in silicon.

For gamma-ray detection, germanium is preferred over silicon because of its higher atomic number. However, because of the smaller band gap, the leakage current in germanium at normal temperatures is too high to be acceptable and must be cooled to liq-

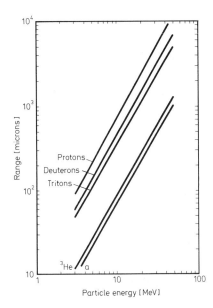

Fig. 10.9. Range of various particles in silicon (from *Skyrme* [10.10])

uid nitrogen temperatures. For low energy x-rays below ~30 keV, silicon detectors are preferred because of the K-edge in germanium which is located at ~11 keV. Photon absorption at these energies, of course, would yield almost zero energy photoelectrons.

10.4.6 Pulse Shape. Rise Time

Because the collection time for electrons and holes depends on the location of the charges with respect to the electrodes, pulse shapes from semiconductors vary in form and rise time.

As in gas detectors, the electrical pulse on the electrodes arises from induction caused by the movement of the charges rather than the actual collection of the charge itself. Assuming two parallel electrodes, (6.29) gives the change in potential energy for a charge q which moves a distance dx. Since we are interested in the charge collected we can reexpress this as

$$dQ = \frac{q\,dx}{d} \; , \tag{10.28}$$

where d is the distance between electrodes. Although (10.28) is derived for the case of empty space between the electrodes, it can be shown that this is also valid [10.11] in the presence of a space-charge as well.

Let us take the example of a pn-junction detector. These are generally formed from a p-type material which is heavily doped on one side with n-type donors. As we saw in Sect. 10.3.1, the depletion zone then extends almost entirely into the p-side. This is illustrated in Fig. 10.10 which also shows the electric field in the zone as given by (10.13). Using the coordinate system shown in the figure (10.13) is rewritten as

$$E = -\frac{eN_A}{\varepsilon} x \; . \tag{10.29}$$

From (10.9), we have the result that the conductivity, $\sigma \simeq eN_A \mu_h$. Substituting into (10.29) then gives us

$$E = -\frac{x}{\mu_h \tau} \; , \tag{10.30}$$

where we have defined $\tau = \varepsilon/\sigma = \rho\varepsilon$ and ρ is the resistivity.

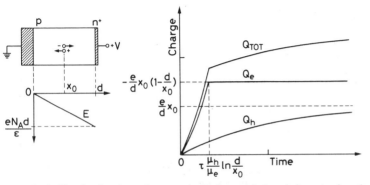

Fig. 10.10. Signal pulse shape due to a single electron-hole pair in an np junction

Assume now that an electron-hole pair is created at a point x in the depletion zone. The electron will thus begin to drift towards the n^+ layer and the hole towards the p electrode. From the definition of mobility, we have for electrons

$$v = \frac{dx}{dt} = -\mu_e E = \frac{\mu_e}{\mu_h} \frac{x}{\tau} . \tag{10.31}$$

Assuming that the mobilities are independent of E, this yields the solution

$$x(t) = x_0 \exp \frac{\mu_e t}{\mu_h \tau} . \tag{10.32}$$

The time it takes for the electron to reach the electrode at $x = d$ is then

$$t = \tau \frac{\mu_h}{\mu_e} \ln \frac{d}{x_0} . \tag{10.33}$$

The charged induced as a function of t during this period is thus

$$Q_e(t) = -\frac{e}{d} \int \frac{dx}{dt} dt = \frac{e}{d} x_0 \left(1 - \exp \frac{\mu_e t}{\mu_h} \right) . \tag{10.34}$$

Similarly for the hole, we have the equation,

$$v_h = \frac{dx}{dt} = \mu_h E = -\frac{x}{\tau} \tag{10.35}$$

which yields the solution,

$$x(t) = x_0 \exp \frac{-t}{\tau} . \tag{10.36}$$

The collection time in this case is infinite and the induced charge,

$$Q_h(t) = -\frac{e}{d} x_0 \int \exp \frac{-t}{\tau} \frac{dt}{\tau} = -\frac{e}{d} x_0 \left(1 - \exp \frac{t}{\tau} \right) . \tag{10.37}$$

The pulse shape is now given by the total charge induced and this is diagrammed in Fig. 10.10 along with the contributions Q_e and Q_h. The total charge collected is $Q_{tot} = -e$ as can be shown by taking the maximum limits on (10.34) and (10.37).

The parameter τ, as can be seen, determines the rise time of the signal. In silicon, this is given roughly by $\tau = \rho \cdot 10^{-12}$ s, where ρ is in Ω cm. For typical 1000 Ω cm material, τ is thus on the order of a nanosecond.

The above calculation, of course, was only for a single electron-hole pair. To calculate the pulse shape due to incident radiation, it is necessary to know the particle trajectory, the density of ionization along the track, the variation of mobilities, the electric field distribution, etc. and integrate all these factors − a complicated task!

10.5 Silicon Diode Detectors

For charged particle detection, silicon is the most widely used semiconductor material. As we have noted, it has the advantage of room temperature operation and wide availability. One of the disadvantages of silicon detectors is their relatively small size. Current devices are limited to surface areas of a few ten's of square centimeters although some development is being done to increase this limit.

Silicon diodes may be fabricated in a number of ways which result in a number of different types of detectors. These are described below.

10.5.1 Diffused Junction Diodes

Diffused junction diodes were among the first devices fabricated for radiation detection. These diodes are usually produced by diffusing n-type impurities, such as phosphorus, into one end of a homogeneous p-type semiconductor at high temperature ($\sim 1000\,^{\circ}$C). By adjusting the concentrations and diffusion time, junctions lying at depths of a few tenths of a micron to two microns in the semiconductor can be produced.

With the diffusion process, the surface layer becomes heavily doped, so that the depletion layer extends mainly into the p-side. This, unfortunately, leaves a relatively thick *dead* layer through which radiation must pass before reaching the sensitive volume. For energy measurement, this is particularly disadvantageous as the energy lost in this layer goes unrecorded. As well, it sets a lower limit on the particle energy which can be detected. In general, the dead layer is smaller than the depletion zone and does not vary very strongly with the bias. Methods for measuring the thickness have been developed and are given in [10.9, 12]. An additional disadvantage of these detectors is that the diffusion process must be performed at high temperatures. This tends to reduce charge carrier lifetime and increase noise in the detector.

The advantage of diffusion junctions is their ruggedness relative to other semiconductor detectors and their greater resistance to contamination on the detector surface. Nevertheless, because of the relatively thin windows which can be obtained with other detectors types, diffused junctions are not often used.

10.5.2 Surface Barrier Detectors (SSB)

By far the most widely used silicon detectors for charged particle measurements are the *surface barrier* type. These detectors rely on the junction formed between a semiconductor and certain metals, usually n-type silicon with gold or p-type silicon with aluminium. Because of the different Fermi levels in these materials, a contact *emf* arises when the two are put together. This causes a lowering of the band levels in the semiconductor as illustrated in Fig. 10.11. This situation, of course, is similar to the np junction (see Fig. 10.5) and a depletion zone extending entirely into the semiconductor is formed. Such junctions are also known as *Schottky barriers* and possess many of the characteristics of pn-junctions.

The depletion depth in a surface barrier detector can be calculated using (10.21) which is also valid for this configuration. With current high resistivity silicon, depths of ~ 5 mm can be attained.

The fabricating process for surface barrier detectors is actually simpler than that for diffused junctions and presents many advantages over the latter. They are made at

Fig. 10.11. Formation of a Schottky barrier junction

Fig. 10.12. Schematic diagram of a surface barrier detector (from *Ortec* [10.28])

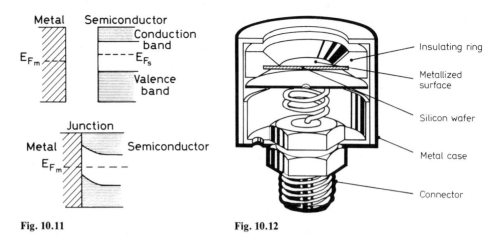

Fig. 10.11   Fig. 10.12

room temperature by first etching the silicon surface and then depositing a thin layer (~ 40 μg/cm^2) of gold by evaporation. In this process, it is also necessary to allow the surface to oxidize slightly before deposition. The junction is then mounted in a insulating ring with metallized surfaces for electrical contact. Figure 10.12 shows a typical SSB.

Surface barrier detectors can be made with varying thickness and depletion zone regions. If the detector is not too thick, a fully depleted detector is possible. Here the depletion zone extends through the entire thickness of the silicon wafer. Such detectors are useful as transmission detectors for measuring the energy deposition of passing charged particles, i.e., *dE/dx*. Moreover by increasing the bias on fully depleted detectors, a gain in the collection time of the charges can be obtained resulting in a faster signal risetime. This is not the case with partially depleted detectors where an increase in bias also extends the depletion zone and thus the distance over which charges must be collected.

One of the disadvantages of surface barriers is their sensitivity to light. The thin gold covering is insufficient to stop ambient light and since visible wavelengths have energies of 2 to 4 eV, (while the energy gap is only 1.1 eV wide) a signal will be obtained. A light tight enclosure of some sort must therefore be provided.

SSB's are also more sensitive to surface contamination and care must be taken to keep this surface clean. Touching the surface with a finger, of course, is to be avoided. If the detector is used in a vacuum chamber, attention must also be paid to the possibility of oil from the vacuum pump being deposited on the surface.

10.5.3 Ion-Implanted Diodes

Ion-implanted junctions are formed by bombarding the semiconductor crystal with a beam of impurity ions from an accelerator. The dopants are thus literally *shot* into the crystal. By adjusting the beam energy to have a certain range in the semiconductor, the impurity concentration and depth profile can be controlled.

Since some radiation damage is incurred in the process, the semiconductor must be annealed at temperatures of about 500 °C before use. This is still much less than the temperature used in the diffusion process, however, so that carrier lifetime is much less affected. A new method combining oxide passivation with ion-implantation has also been developed [10.3] which reduces surface leakage current and thus noise.

Ion-implanted detectors are generally more stable than surface barrier detectors and can have entrance windows as thin as 34 nm Si equivalent [10.3]. At present, they offer the best characteristics of all the silicon detector types and are those generally chosen for use in high-energy physics. They are, however, much more expensive.

10.5.4 Lithium-Drifted Silicon Diodes – Si(Li)

One of the original problems in fabricating semiconductor detectors was the insufficient thickness of the depletion zone. To obtain thicknesses greater than a few millimeters required very high resistivities which could only be obtained with intrinsic materials or eventually compensated semiconductor. Both, of course, were difficult or impossible to achieve.

As described in Sect. 10.2.1, this problem was solved with the development of the lithium-drifting process for forming compensated material. Junctions formed with compensated materials are known as p-i-n-junctions and possess properties different from regular np junctions. In particular, there is no space charge in the compensated zone. This implies an almost constant electric field. Figure 10.13 illustrates a typical planar p-i-n diode and along with the potential and electric field in the various region.

Detectors made from lithium-drifted silicon are known as Si(Li) (pronounced as *silly*) detectors. The thickness of the compensated zone is generally limited to about 10 to 15 mm which makes them suitable for beta particle and low-energy x-ray detection. Because of the greater sensitive region, the noise contribution from thermally generated electrons and holes is much greater than in normal silicon diodes so that cooling to low temperatures is necessary for high resolution operation. Moreover, in order to maintain the lithium-drifted compensation, Si(Li) must be stored at low temperatures, although short periods at room temperature generally do no harm. A review of typical applications of Si(Li) detectors is given in [10.12].

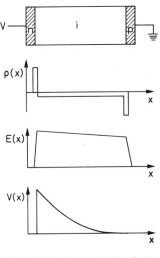

Fig. 10.13. Characteristics of p-i-n junctions. These is no space charge in the compensated zone so that the electric field is essentially constant

10.6 Position-Sensitive Detectors

As we have mentioned, semiconductor devices have recently gained considerable attention in the high-energy physics domain as possible high-resolution spatial detectors. Interest in position-sensing was already present early in the development of the semiconductor detector, however, because of the need for an all-electronic detector to replace the photographic emulsions being used in magnetic spectrographs [10.13]. Two types of detector using different methods of obtaining spatial information were developed. The first uses a *continuous* readout with a *resistive charge division* method, while the second employs a discrete array of readout elements. Detectors providing both one- and two-dimensional information have been developed using these methods.

10.6.1 Continuous and Discrete Detectors

Figure 10.14 shows a schematic diagram of a one-dimensional continuous detector and its equivalent circuit. Basically the detector is a rectangular diode with a uniform, resistive electrode on the front face and a low resistive back electrode. A typical length for such detectors is 5 cm. If now a charged particle passes through the diode, the charge

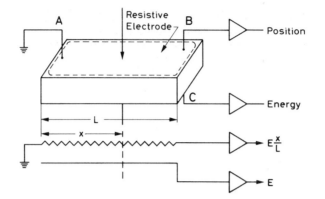

collected at contact B will be proportional to the energy E of the particle and the resistance of the electrode between the contact and the point of incidence, i.e.,

$$B = E \frac{x}{L} \, , \qquad\qquad (10.38)$$

where x is the distance between the point of incidence and the contact A, and L is the length of the resistive layer. The signal at electrode C, on the other hand, is proportional to the energy. Dividing B by C, thus yields a signal proportional to the position x,

$$x = L \frac{B}{C} \, . \qquad\qquad (10.39)$$

This is known as the method of resistive charge division.

One of the problems with this type of detector is to ensure linearity of the position signal. This requires the semiconductor and the resistive layer to be highly uniform and homogeneous. Correct shaping of the output signals is also required. If care is taken, nonlinearities can be less than 1% of the detector length. Spatial resolutions can then be on the order of a ~ 250 µm. This appears to be the inherent limit of these detectors, however [10.13].

In contrast to the continuous device, the discrete-type detector consists of a series of individual electrode strips placed on the same semiconductor base. Each electrode then acts as a separate detector. Figure 10.15 illustrates a simple two dimensional configuration. These detectors are also referred to as *matrix* detectors while the one-dimensional discrete devices are called *strip* detectors.

The obvious disadvantage of discrete detectors is the quantity of electronics required. Since each electrode is a separate detector, preamplifiers and other units must be provided for each strip. This, of course, becomes quite costly and cumbersome. In detectors with many electrodes, the elements may be connected to an external resistive divider network as shown in Fig. 10.15. In such a case, however, the discrete detector becomes a continuous detector with essentially the same characteristics. Nevertheless discrete detectors offer better timing and energy resolution with spatial resolutions only limited by the electrode width. Typical widths on these devices have been on the order of $0.2 - 0.4$ mm. A good review of these devices and their applications is given by *Laegsgaard* [10.13].

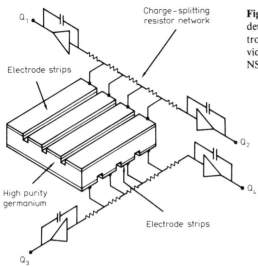

Fig. 10.15. Layout of a two-dimensional *matrix* detector. To reduce the readout electronics, the electrodes may be connected to an external resistive divider [from *Gerber* et al.: IEEE Trans. Nucl. Sci. NS-**24**, No. 1, 182 (1977)]

10.6.2 Micro-Strip Detectors

While the use of semiconductors in high energy physics was also investigated early in the history of these devices, the more favorable characteristics of gas detectors along with the rapid advances being made in this domain during the 1970's dominated the attention of most experimentalists. In the 1980's, however, interest was renewed by the demonstration of a 5 μm spatial resolution with a silicon *micro-strip* detector [10.14].

This device is composed of separate readout strips arranged at intervals of 20 μm. Figure 10.16 shows a schematic diagram of the detector along with the layout of the strips. High resistivity n-type silicon (2000 Ω cm) is used as the base material onto which p^+ diode strips with aluminium contacts are implanted. An n^+ electrode is similarly implanted on the opposite face. The thickness of the detector is on the order of 300 μm which implies an operating voltage of 160 V to obtain full depletion. For minimum ionizing particles, the average energy loss in silicon [10.15] is about 39 keV/100 μm so that about 100 electron-ion pairs/μm are created. A total of 30000 pairs is therefore expected in the detector.

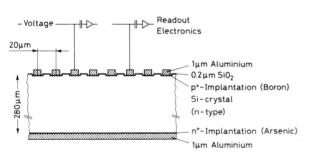

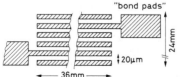

Fig. 10.16. Layout of a micro-strip detector and readout strips (from *Hyams* et al. [10.14])

To reduce the number of channels, only strips at every 60 µm are read out. A capacitive charge division method can then be used between these readout strips. By calculating the center of gravity of the charge collected, the position of the impinging particle can then be obtained to within 5 µm.

Because of their small size and full depletion, micro-strip detectors offer a fast response time and present many advantages as triggering devices for high energy particle physics experiments. Charge collection in such devices can be performed in less than 10 ns which should allow for high counting rates. Moreover, if every channel is read out, a resolution of 2 µm should be possible [10.14]. Such a large number of channels may be feasible using large scale integration techniques [10.16] to include the associated electronics on the same chip as the detector.

The relative sensitivity of semiconductors to radiation damage does pose somewhat of a problem, however. This generally results in a deterioration of resolution and an increase in the leakage current. Some compensation can be obtained by increasing the electric field. As well, different methods of fabricating junctions may improve their resistance to radiation damage.

10.6.3 Novel Position-Sensing Detectors

The success of the micro-strip detector has stimulated a number of new ideas for space localizing semiconductor detectors. Among some of the novel devices is the *silicon drift chamber* proposed by *Gatti* and *Rehak* [10.17]. Their idea is based on creating a drift channel along the center plane of a flat silicon wafer by fully depleting the wafer from the flat sides and one edge.

Figure 10.17 illustrates the basic principle. Starting with a flat n-type silicon wafer, p^+ contacts are deposited on the flat sides of the material. This creates two depleted regions sandwiching a central undepleted zone. A third n^+ contact is now implanted on one edge of the wafer. Applying a reverse bias then fully depletes the wafer. The potential inside the wafer is then parabolic in form (Fig. 10.17) with a minimum along the central dividing plane. Electrons created at some point in the wafer will then *fall* down the potential well to the central valley where they drift along the longitudinal component of the electric field towards the n^+ electrode. Measuring this drift time then pro-

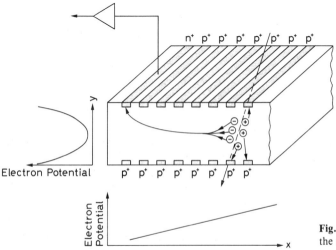

Fig. 10.17. Operating principle of the silicon drift chamber

vides spatial information in a manner analogous to gas drift chambers. Tests with a prototype chamber [10.18] have shown a spatial resolution of 5 μm with minimum ionizing particles which is equivalent to that obtained with micro-strip detectors. The great advantage of these devices, of course, is the small volume of electronics required.

Some of the other new detectors proposed rely on the use of *charged-coupled devices* (CCD). These are silicon devices consisting of a two-dimensional array of tiny potential wells each covering a surface area of a few square microns. One chip contains tens of thousands of elements. When struck by radiation, electrons are released which are then trapped in these wells. This charge information is then readout by successively shifting the charge from one well to the next until it reaches the output electronics. A more detailed description of these detectors is given in [10.2, 19]. CCD's have been mainly used for light imaging purposes and are extremely sensitive, low noise devices. As a detector, spatial resolutions for these instruments are expected to be better than 2 μm. However, they are very limited in counting rate. A review of these and other devices may also be found in the articles by *Charpak* and *Sauli* [10.2] and *Klanner* [10.20].

10.7 Germanium Detectors

For gamma-ray detection, germanium is preferred over silicon because of its much higher atomic number ($Z_{Si} = 14$, $Z_{Ge} = 32$). The photoelectric cross section is thus about 60 times greater in Ge than Si. Germanium, however, must be operated at low temperatures because of its smaller band gap. This inconvenience is offset, however, by its greater efficiency.

Germanium may also be used for charged-particle detection, however, apart from its greater stopping power, it offers no advantage over silicon and, in fact, becomes disadvantageous because of its need for cooling.

10.7.1 Lithium-Drifted Germanium – Ge(Li)

In order to obtain a sufficient sensitive thickness for the detection of gamma rays, the first detectors were made from lithium compensated germanium. These detectors are known as Ge(Li) (pronounced as *jelly*) detectors. Since maximum obtainable thicknesses for compensated germanium are about 15 or 20 mm, a coaxial geometry is generally use to maximize the sensitive volume. This is schematically diagramed in Fig. 10.18. In this configuration, lithium is drifted in from the outer surface of a cylindrical crystal of p-type germanium to form a cylindrical shell of compensated material. A central core of insensitive p material is then left. If this core extends along the entire length of the axis, the configuration is then known as a *true coaxial* or *open-ended coaxial* detector. To increase sensitive volume even further, lithium may also be drifted in from the front face of the cylinder. The extent of the insensitive core is then reduced. These are known as *closed-end* coaxial detectors. For high counting efficiency, the central core may also be removed to form a *well type* detector. The various configurations are illustrated in Fig. 10.18. For lower gamma energies, Ge(Li) detectors may also be fabricated with the conventional planar geometry.

Because of the high mobility of the lithium ions in germanium, even at room temperature, Ge(Li) detectors must be kept at liquid nitrogen temperatures at all times.

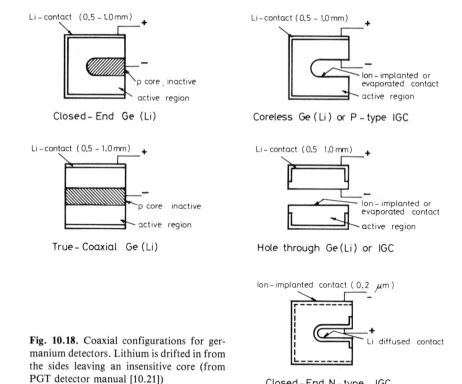

Fig. 10.18. Coaxial configurations for germanium detectors. Lithium is drifted in from the sides leaving an insensitive core (from PGT detector manual [10.21])

This requires mounting the crystal in a mechanically rigid cryostat with an accompanying dewar for liquid nitrogen. This, of course, puts severe constraints on the experimental geometries which can be used with germanium detectors. Nevertheless, detectors with various cryostat positions and dewars are available commercially.

The sensitivity of coaxial Ge(Li) detectors is generally limited by the thickness of the dead layer formed on the face of the crystal by lithium drifting and the cryostat window which absorb low-energy photons. Typical limits are on the order of 30 keV. For planar detectors, the window contact can be made from a thin layer of gold which results in a lower limit on the order of a few keV.

A more complete discussion of the operating characteristics of Ge(Li) detectors is given in the book by *Knoll* [10.12].

10.7.2 Intrinsic Germanium

In more recent years, advances in semiconductor growth technology have allowed the fabrication of very high purity germanium with impurity concentrations of less than 10^{10} atoms/cm^3. Detectors of this type have the advantage of not having to be kept at low temperatures at all times. Cooling is only necessary when a high voltage is applied.

Intrinsic germanium (also called HPGe for *High Purity Germanium*) detectors are constructed and operated in the same way as Ge(Li) detectors and are now gradually replacing the latter. One advantage with intrinsic detectors is the possibility of using n-type semiconductor rather than the p-type required for the lithium-drifting process. A very thin window can then be formed by ion-implantation to extend the sensitivity of

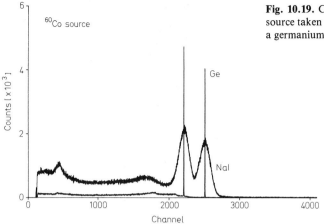

Fig. 10.19. Comparison of spectra from a ⁶⁰Co source taken with a NaI detector (*top curve*) and a germanium detector

the coaxial detector to below 10 keV. These detectors are also somewhat more resistant to radiation damage.

10.7.3 Gamma Spectroscopy with Germanium Detectors

The principal application of germanium detectors is gamma ray spectroscopy. At present, germanium detectors offer the highest resolution available for gamma-rays energies from a few keV up to a 10 MeV. This is illustrated in Fig. 10.19 which compares the spectrum from ⁶⁰Co taken with a NaI detector to the same spectrum as measured by an intrinsic Ge detector. The difference is quite dramatic: at 1.33 MeV the Ge resolution is about 0.15% while for NaI this value is ~8%! In addition, the peak to Compton ratio is much greater due to the higher photoelectric cross section of germanium.

For precision spectrum measurements, the energy resolution and signal to noise ratio are the most important parameters. It is important therefore to shield the detector with lead so as to minimize background. The signal-to-noise ratio can also be increased with the use of an optical feedback preamplifier. Attention should also be paid to the count rates. These should not be too high so as to avoid pile-up effects which can distort the spectrum.

In order to measure the absolute intensities, a calibration of the absolute detection efficiency is necessary. This must be performed with calibration sources which span the energy region of interest. Gamma sources whose outputs are calibrated to within 1 or 2% can be obtained commercially. In most cases, it is the *full peak efficiency*, i.e., the efficiency for photoelectric conversion, which is desired. This is given by the total count rate in the photopeak for each gamma-ray divided by the total output of the source. The Compton scattered part is ignored.

Attention must be paid to the source-detector geometry. The calibration must be performed at the source-to-detector distance to be used and this distance must be reproducible. It is a good idea to construct a rigid source holder which mounts onto the detector so as to insure this reproducibility. Another factor is the size of the radioactive sources to be measured. If these are distributed sources, this must also be taken into account in the calibration. Most commercial calibration sources can be considered as point-like and it may be possible to simulate the distributed source by displacing the calibration source off-axis and measuring the efficiency for several points. The total efficiency can then be estimated by integrating over these measured points.

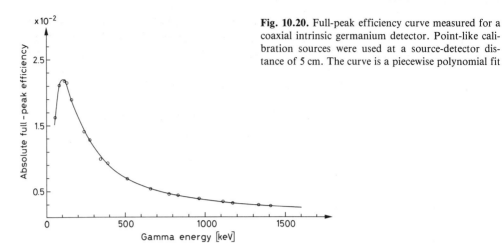

Fig. 10.20. Full-peak efficiency curve measured for a coaxial intrinsic germanium detector. Point-like calibration sources were used at a source-detector distance of 5 cm. The curve is a piecewise polynomial fit

As with spectrum measurements, it is important to keep the count rate at reasonable levels. If possible the calibration should be performed at these rates. At high rates, accidental coincidences between two gamma-rays from the same source can reduce the number of counts in their true gamma-ray peaks by pile-up. The two gamma rays are thus registered at an energy corresponding to their sum rather their individual energies. This *summing-effect* is particularly important for sources emitting many cascade photons. If the decay scheme and emission angles between all the gamma rays are known, the effect can, in principle, be calculated. However, for most sources this is generally not possible.

Figure 10.20 shows a measured full-peak efficiency curve for a coaxial intrinsic germanium detector at a source-to-detector distance of 5 cm. The points are fitted piecewise by polynomial functions. A number of empirical forms have also been proposed to describe the form of this curve [10.22].

A final factor to consider is the effect of dead time which can also lead to distortion as well as losses. This is discussed in Sect. 5.5.

10.8 Other Semiconductor Materials

Because germanium must be cooled during operation, a good deal of effort is being put into finding new high-Z semiconductor materials which can be operated at room temperature. Many compounds have been investigated, however, only two show some promise as of this moment. The first is cadmium telluride (CdTe) which is commercially available and the second mercuric iodide (HgI_2) which is still under development.

Cadmium telluride was the first material (other than silicon) to have been developed as a room temperature detector. It has an energy gap of 1.45 eV and atomic numbers of 48 and 52. This, of course, makes it highly efficient for gamma ray detection. Mercuric iodide has an even higher Z (80 and 53) and an energy gap of 2.14 eV. The average energy for electron-hole creation is somewhat higher than Si and Ge being on the order of 4.4 eV.

While these detectors present many favorable properties for gamma detection, they are still fraught with many problems. Mercuric iodide, in particular, is plagued by incomplete charge collection caused by hole trapping and polarization effects leading to space charge build-up which limit its efficiency and resolution. CdTe is on somewhat firmer ground and is now sold commercially. In both cases, however, it is still difficult to fabricate large volume detectors from these materials because of nonuniformities. The low yield of good crystals also makes them very expensive.

Research and development are still continuing in these domains and it is hoped that a better understanding of the physics of these materials will allow more reliable detectors to be constructed. For a review of the current status of these detectors, the reader is referred to [10.2, 26].

10.9 Operation of Semiconductor Detectors

10.9.1 Bias Voltage

The bias voltage, as we saw in (10.23) and (10.24), determines the thickness of the depletion layer and also the capacitance of the detector. Higher voltages thus reduce the noise by increasing the depletion thickness; however, a greater risk of breakdown is also incurred. For commercial detectors, the optimum bias values are supplied by the manufacturer and should not be surpassed. Typical values for SSB's, for example, range from $50 - 300$ V while for germanium detectors these voltages can be as high as $4000 - 4500$ V.

When voltage is applied, this should be done slowly, raising the voltage a few tens of volts at a time and allowing the detector to "*settle*" for a few seconds after each step. A good procedure is to observe the noise signal on the oscilloscope as the voltage is raised. Immediately after each increase in voltage, the signal may disappear from the oscilloscope display but should reappear after a second or two. If there are any sudden, intermittent increases in noise, this is a sign of incipient breakdown and one should proceed slowly allowing more settling time. If there is a sudden, very large increase in noise, breakdown has occurred and the voltage should be removed immediately to prevent irreversible damage. After the desired voltage has been reached, it is a good idea to allow the detector to stabilize for a few hours, especially if it has not been used for a long period. In some cases, the noise level will also diminish further.

Particular care should be taken with very thin detectors as small increases in voltage correspond to large increases in the electric field intensity. A 10 Volt increase on a 20 μm detector, for example, corresponds to an electric field increase of 5000 V/cm!

In the case where the operating voltage is unknown, the bias may be determined by observing the noise level on an oscilloscope while *slowly* raising the applied voltage. The noise amplitude should diminish to some minimum and then slowly rise again. Setting the voltage at the minimum point then should normally provide the best results. Needless to say, care should be taken not to apply too much voltage.

10.9.2 Signal Amplification

Because of the small signals obtained from semiconductor detectors, care must be taken to use low-noise electronics for signal processing. In particular, a preamplification is necessary before any further treatment can be made. Because the capacitance of semiconductors change with temperature, this is performed by a *charge-sensitive* preamplifier which, after the detector itself, is the second most important part of a semiconductor detector system. This type of preamplifier is preferred because of its insensitivity to changes in capacitance at its input as described in Sect. 14.1. To ensure stability, it is necessary for the preamplifier capacitance to be much larger than all other sources of capacitance at the input, i.e., the detector, cables, etc. Since typical detector capacitances are on the order of ten's of picofarads, preamplifier dynamic capacitances are generally on the order of a few 10's of nanofarads. With very thin SSB's, however, the detector capacitance can approach 1 or 2 nF, in which case, higher preamplifier capacitance would be required for optimum performance.

The noise of the preamplifier is particularly important as it affects the ultimate resolution of the detector. Since the signal from the detector appears as electric charge, electronics noise is usually quantified by giving its *equivalent noise charge* (ENC). If V_{rms} is the average voltage noise level appearing at the output, then

$$\text{ENC} = e\,\frac{V_{rms}}{w}\,C\;,\tag{10.40}$$

where C is total input capacitance of the detector and preamplifier and w the average energy required to create an electron-hole pair. The noise may also be expressed in terms of an equivalent energy corresponding to the ENC. This is usually given as a peak width [i.e., the full width at half maximum (FWHM)] using the relation

$$\frac{\text{FWHM}}{2.35\,w} = \text{ENC}\;.\tag{10.41}$$

From (10.40), it is clear that minimizing noise requires minimizing the input capacitance to the preamplifier. For this reason, the preamplifier is generally mounted as close as possible to the detector in order to reduce capacitance from cables, etc. Coupling to the detector can be performed either directly (dc coupling) or through an additional capacitor (ac coupling) [10.27]. These are diagrammed in Fig. 10.21. For high resolution operation at low temperatures, such as with germanium gamma-ray detectors, a direct coupling is made for the minimum in input capacitance. A direct coupling also allows monitoring of the leakage current which is an advantage. An inconvenience, however, is that neither electrode of the detector is at ground potential which implies some additional work in designing the detector mounts to ensure a good insulation. Typical noise values range from less than 1.5 keV at 0 input capacitance to 18 keV at 1000 pF [10.28].

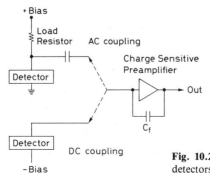

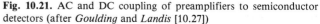

Fig. 10.21. AC and DC coupling of preamplifiers to semiconductor detectors (after *Goulding* and *Landis* [10.27])

For charged particle spectroscopy with room temperature detectors, the ac coupling method is generally quite adequate. In this configuration, one electrode is grounded which facilitates mounting, however, the additional load capacitor contributes to the noise. If the particles are of sufficiently high energy, however, the effect is usually minimal.

For fast timing applications, such as time-of-flight measurements, in which a timing signal, as well as an energy signal from the detector are desired, some additional considerations must be made. If the collection time of the detector is greater than the rise-time of the preamplifier, then both energy and timing signals may be derived from the same charge-sensitive preamplifier. Very often, however, the detector is a thin, high capacitance dE/dx detector in which the collection times are much shorter. In such cases, a hybrid system involving charge-sensitive and voltage sensitive or current sensitive preamplifiers might be necessary. These and other timing problems are reviewed by *Spieler* [10.29].

10.9.3 Temperature Effects

As we have seen, temperature has an important effect on the conductivity of the detector. Germanium detectors must *always* be operated at low temperatures otherwise the high leakage current will cause irreversible damage to the crystal.

For silicon detectors, increasing temperature will also result in higher leakage currents and greater noise. For each 10 °C rise in temperature, there is roughly a three-fold increase in the leakage current [10.28]. The maximum temperature limit for silicon is generally between 45° to 50°C at which point breakdown occurs.

Decreasing the temperature with silicon detectors, of course, greatly reduces noise. However, the expansion coefficients of the mounting should first be considered before cooling. In particular, the bonding epoxy could crack if it is not made to withstand low temperatures. Another point to be aware of, is that the band gap for silicon increases by about 0.1 eV (see Table 10.1) when going from room temperature to liquid nitrogen temperatures. For the same gain on the associated electronics, there will therefore be a slight shift between spectra recorded at these two temperatures.

10.9.4 Radiation Damage

As we have mentioned, semiconductor detectors are relatively sensitive to radiation damage. Incident particles colliding with lattice atoms cause point defects by "*knocking*" them out of their normal positions. These structural defects then give rise to discrete trapping levels in the forbidden band gap which reduce the number of charge carriers in the semiconductor. There also appears to be changes in the resistivity of the base material as well. A review of the radiation damage problem can be found in [10.30].

The main effects of radiation damage on detector performance are an increased leakage current and a degradation of energy resolution. In more heavily damaged detectors, double peaks in spectra have also been reported. If the damage is not too great, however, an increase in the bias voltage can compensate resolution loss to some extent by decreasing the collection time.

For a given fluence[2] of radiation, the increase in leakage current can be estimated through the relation

---

[2] The fluence is defined as the total accumulated number of particles incident per unit area. It is thus the integral of the particle flux over time.

Table 10.3. Damage constants in silicon for various radiations (from [10.25, 26])

Particle type	K [cm^2/s]	
	n-type	p-type
Electrons		
3 MeV	$2-10\times10^{-8}$	3×10^{-9}
Muons		
GeV	1.4×10^{-7}	
Neutrons		
Fission	0.5×10^{-5}	2.5×10^{-6}
1 MeV	1×10^{-5}	2.5×10^{-6}
14 MeV	2×10^{-6}	0.7×10^{-6}
Protons		
2 MeV	2×10^{-8}	
20 MeV	$2-10\times10^{-5}$	1.3×10^{-5}
207 MeV	5×10^{-6}	2×10^{-6}
590 MeV	1.2×10^{-6}	0.9×10^{-6}
3 GeV	10^{-6}	
24 GeV	3.8×10^{-8}	

$$J = q\,n_i\,dK\,\frac{\phi}{2} \qquad\qquad (10.42)$$

where ϕ is the radiation fluence, n_i the intrinsic carrier concentration from (10.1), d the depletion depth, and K the damage constant which depends on the radiation type and its energy. Table 10.3 summarizes some measured damage constants for various radiations at different energies. By choosing the maximum tolerable leakage current, the maximum allowable fluence for a given type and energy of incident radiation can be calculated by inverting (10.42).

10.9.5 Plasma Effects

For very heavily ionizing particles, i.e., heavy ions, fission fragments, etc., the high density of electron-hole pairs created in semiconductors leads to space-charge phenomena which affect the rise time and pulse height of the resulting signal. In effect, the high ionizing power of these particles produces a dense cloud of space charge along the particle trajectory which locally nullifies the external applied electric field. The charge in the cloud, therefore, is not immediately swept up. Gradually, of course, diffusion dissipates the cloud and the charge is collected after a characteristic delay time.

This delay affects the signal in a number of ways:

1) The risetime of the pulse is much slower since the effective collecting field is much lower. This change in risetime is known as the *plasma time*.
2) During the delay, electrons and holes in the cloud have time to recombine so that total collected charge is less than what is created. This leads to what is called *pulse height defect*.

The main consequence of pulse height defect is that the detector calibration is different for different particle types. In general, the energy-pulse-height relation can be given by a formula [10.31, 32] of the type

$$E(X, M) = (a + a_1 M) X + b + b_1 M \; , \tag{10.43}$$

where E is the energy of the ion, M, its mass and X the pulse height. The coefficients a, a_1, b and b_1 are experimentally determined by measuring the spectrum of fission fragments from a standard source such as ^{252}Cf. A new formula based on more recent measurements has also been proposed by *Ogihara* et al. [10.33].

11. Pulse Signals in Nuclear Electronics

As we have seen, modern detectors provide a variety of information on detected radiation in the form of electrical signals. In order to extract this information, however, the signal must be further processed by an electronics system. This system can be designed to perform an enormous variety of tasks. For example, it can be made to sort out the various signals from the detectors, extract energy information, determine the relative timing between two signals, etc., and based on this information, make a decision as to the acceptability of the event. It can then take appropriate action, for example, by activating a recording device. Indeed, many systems virtually run themselves!

In the following chapters, we will try to present an introduction to some of the logic and techniques of setting up an electronics system in nuclear and particle physics experiments and touch on some of the problems which will most likely be encountered. As will be seen, *nuclear electronics* today has been largely standardized into a modular form, i.e., the circuits for the basic processing functions (e.g., amplification, discrimination, etc.) have been built into separate electronic modules of standard mechanical and electrical specifications which are then interconnected as desired. These systems, (NIM, CAMAC), which we will discuss later, are extremely advantageous as they allow the design of many different systems using the same set of modules. Moreover, since the physical and electrical specifications have been standardized and accepted throughout the world, modules from different laboratories can be freely exchanged without any problem in compatibility. An analogous example, perhaps more familiar to the reader, is the modern Hi-Fi system. While the various components (e.g. amplifier, tuner, tape deck, etc.) are constructed separately by many different manufacturers, they are all compatible with each other so that a variety of Hi-Fi systems can be created with no problem.

In both cases, as well, a detailed knowledge of electronics at the level of circuit design is not necessary in order to set up the system, only an understanding of the logic. In the following chapters, therefore, we will assume only an elementary knowledge of circuits and give explanations where some more sophisticated knowledge is needed.

We begin in this chapter with a discussion of pulse signals and their characteristics.

11.1 Pulse Signal Terminology

The coding of information in nuclear electronics is generally done in the form of pulse signals. These are brief surges of current or voltage in which information may be contained in one or more of its characteristics, for example, its polarity, amplitude, shape, its occurrence in time relative to another pulse, or simply its mere presence. This mode of coding, as opposed to other systems, for example, amplitude or frequency modulation of a sinusoidal signal, is natural in nuclear and particle physics, for, as we have

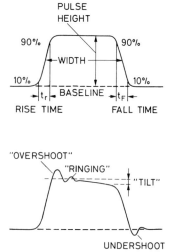

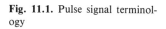

Fig. 11.1. Pulse signal terminology

seen most modern particle detectors are pulse devices. To begin our discussion, let us first identify some basic characteristics of pulse signals and define the terms which are associated with these features. Figure 11.1 (a) shows an ideal rectangular pulse, either in voltage or current, as a function of time. In nuclear electronics, this time scale may vary from micro-seconds to fractions of a nanosecond. We can define the following features:

1) *Baseline.* The baseline of the signal is the voltage or current level to which the pulse decays. While this is usually zero, it is possible for the baseline to be at some other level due to the superimposition of a constant dc voltage or current, or to fluctuations in the pulse shape, count rate etc.
2) *Pulse Height or Amplitude.* The amplitude is the height of the pulse as measured from its maximum value to the instantaneous baseline below this peak.
3) *Signal Width.* This is the full width of the signal usually taken at the half-maximum of the signal (FWHM).
4) *Leading Edge.* The leading edge is that flank of the signal which comes first in time.
5) *Falling Edge.* The falling edge or tail is that flank which is last in time.
6) *Rise Time.* This is the time it takes for the pulse to rise from 10 to 90% of its full amplitude. The rise time essentially determines the rapidity of the signal and is extremely important for timing applications.
7) *Fall Time.* In analogy with rise time, the fall time is the time it takes for the signal to fall from 90 to 10% of its full amplitude.
8) *Unipolar and Bipolar.* Signal pulses may also be unipolar or bipolar. A unipolar pulse is one which has one major lobe entirely (excepting a small possible undershoot) on one side of the baseline. In contrast, bipolar pulses cross the baseline and form a second major lobe of opposite polarity. Figure 11.2 illustrates these two types. Both are used in nuclear electronics.

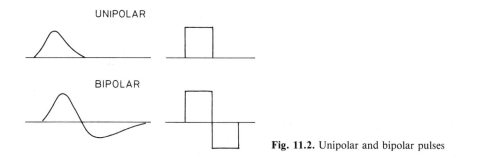

Fig. 11.2. Unipolar and bipolar pulses

   In practice, however, it will be found that pulses are very often distorted by various factors in the circuit. Figure 11.1 also illustrates some of various deviations in form which may be observed and the terminology which is used to describe the effects.

11.2 Analog and Digital Signals

Pulse signals, and signals in general, carry information in two forms: *analog* or *digital*. An analog signal codes continuously-valued information by varying one or more of its characteristics, (e.g. amplitude and/or shape), in some fixed relation to the information value. The voltage signal from a microphone, for example, *analogically* varies its amplitude continuously in proportion to the intensity of the sound it picks up. Similarly, a scintillation detector, as we have seen, generates pulses whose amplitudes are in proportion to the energy deposited in the detector. If a beam of particles with a continuous spectrum of energies is allowed to strike the detector, then a continuous analog spectrum of pulse heights will result. More generally, if we consider each possible amplitude or shape of a pulse as a *state*, then the analog signal can be said to have an infinite, noncountable number of states. In nuclear physics, most analog pulses are of the amplitude varying type, although in certain cases, such as pulse shape discrimination, information is also contained in the shape of the pulse. Since the proportionality between pulse height and energy is usually linear, these pulses are also referred to as *linear* pulses.

In contrast to the continuum of amplitudes or shapes which are possible for the analog pulse, the *digital* or *logic* signal may only take on a discrete number of states; the information represented is thus of a *quantized* nature. For example, the signal from a Geiger-Müller counter has essentially two states: present or not present.[1] The corresponding information is then simply: *yes*, radiation was detected, or *no*, radiation was not detected. No finer distinction is possible. Similarly, we might imagine a ten-state rectangular signal which can only take on the amplitudes 0, 1, 2, 3, ... 9 V, for example. Such a signal might then be used to represent the decimal integers from 0 to 9, only. Technically, however, it is difficult to find electrical devices which have more than two naturally quantized states. In practice, therefore, all logic signals are limited to two-states only. When reference is made to logic or digital signals, therefore, this generally implies the two-state variety.

Although in a certain sense, the logic signal carries *less* information than the analog signal, from a technical point of view it is the more reliable since the exact amplitude or form of the signal need not be perfectly preserved. Indeed, distortion or noise, which are always present in any circuit, will easily alter the information in an analog signal but would have much less effect on the determination of what state a logic signal is in. Moreover, the limited information-carrying ability of the logic signal may be overcome by using several logic signals to represent equivalent analog information in numerical form, that is, each logic signal would represent a digit in a number (hence the name *digital* signal) whose value corresponds to the analog information. Since only two states are available, this number must be in the binary system. One logic state, for example, *pulse-present* would represent the binary digit *1*, while the other state *pulse-absent* would represent *0*. The contrary, of course, would also be an equally valid system, as would a system involving positive and negative signals, etc. The important point is to keep the same convention throughout. An entire binary number would then consist of a string of logic signals, which could be transmitted serially, i.e., one after the other, or simultaneously along an equal number of parallel lines. This involves, of course, more signals and electronics, however, with the advent of miniaturization and large scale integrated electronics, this disadvantage is removed giving digital systems a clear superiority over analog systems.

In nuclear electronics, the two electrical states of the logic signal are standardized by the NIM convention. One logic state is usually taken as 0 V, i.e., no pulse at all, and the other at a fixed voltage level. Because of the obvious difficulty in generating a pulse

---

[1] Recall that the Geiger counter signal is saturated so that when it is present, it has the same amplitude and shape regardless of the energy of the radiation detected.

with exactly the right voltage level, a band of voltages into which the signal must fall is defined instead. These limits are given in the next chapter. As a general rule, the analog signals from radiation detectors are changed into logic signals at sometime or another in a nuclear electronics chain. Usually this is performed relatively early after analyzing the detector signal for certain conditions, for example, a certain minimum energy. Here, the detector signal is passed through a *discriminator* which tests the amplitude of the signal for the minimum height. If *yes*, a logic signal is issued, if *no*, no signal (i.e., logic state *no*) is issued. This signal may then be fed into other modules for processing, and depending on the outcome, may initiate further operations: for example, incrementing a counting device or starting a timer, etc. In spectroscopy measurements, where the analog signal is to be analyzed for the exact value of its amplitude, the signal is often *digitized* and the numerical result treated with a computer. Electronic devices which perform such a conversion are know as *analog-to-digital converters* (ADC's). In a similar manner, a digital signal may be converted into an analog signal by a *digital-to-analog converter* (DAC).

As will be seen in Chap. 14, a variety of electronic modules exist for signal processing, some treating analog signals only, others logic signals only, and some converting the two. We emphasize here the importance of distinguishing between the different types. Sending a logic signal into a module expecting an analog pulse or vice-versa does no harm generally, however, unless there is a specific reason for doing so, the result is meaningless.

11.3 Fast and Slow Signals

For technical reasons which we will consider later, it is important to distinguish between *fast* and *slow* pulses in an electronics system. *Fast* signals generally refer to pulses with rise times of a few nanoseconds or *less* while *slow* signals have rise times on the order of hundreds of nanosecond or *greater*. This definition includes both linear and logic signals.

Fast pulses are very important for timing applications and high count rates; in these applications it is very important to preserve their rapid rise times throughout the electronics system. Slow pulses, on the other hand, are generally less susceptible to noise and offer better pulse height information for spectroscopy work.

While it would certainly be more convenient to be able to work with both these types in an identical manner, fast signals, unfortunately, must be treated differently from slow pulses. This is because of their much greater susceptibility to distortion from small, stray capacitances, inductances and resistances in the circuits and interconnections. These elements can combine to form inadvertent, *parasitic* circuits; for example, equivalent RC or RL circuits which have fast transient responses due to the small values of R, C and L (recall that the time constant of the RC circuit is $\tau = RC$). Compared to slow signals these transients are negligibly short. Compared to fast signals, however, these transients are of the same order of magnitude in duration. A fast signal passing through one of these inadvertent circuits can thus be quickly deformed.

A second problem, which will be treated in Chap. 13, is distortion from reflections in the interconnecting cables. This arises because of the short duration of fast pulses relative to their time of transit in the interconnection. Processing fast signals, therefore, not only requires special attention in the circuit design but also in the interconnections between modules. For this reason, essentially two standardized NIM systems have arisen: one designed for fast nanosecond pulses and the other for slower pulses. Within each system the electronic modules are compatible with each other; however, mixing the two without proper adaptation will invariably lead to problems. And, as already mentioned, special attention must be paid to how the interconnections are made in fast systems. This is one of the most important points to keep in mind when setting up nuclear electronics.

11.4 The Frequency Domain. Bandwidth

Our description and visualization of pulses so far has been in terms of its variation in time. A complete understanding of pulse electronics, especially pulse distortion, however, also requires viewing the pulse in terms of its frequency components. From Fourier analysis, it is well known that a pulse form can be decomposed into a superposition of many pure sinusoidal frequencies. Indeed, if we have a pulse whose shape in time is represented by the function $f(t)$, where t is time, then it may be decomposed as

$$f(t) = \frac{1}{\sqrt{2\pi}} \int_{-\infty}^{\infty} g(\omega) \exp(i\omega t)\, d\omega \ , \tag{11.1}$$

where $g(\omega)$ is the *Fourier transform* or frequency spectrum of the pulse. Inverting (11.1) gives

$$g(\omega) = \frac{1}{\sqrt{2\pi}} \int_{-\infty}^{\infty} f(t) \exp(-i\omega t)\, dt \ . \tag{11.2}$$

As a representative example, consider an ideal rectangular pulse of width T shown in Fig. 11.3. For simplicity, we have taken the pulse to be centered at $t = 0$. Thus,

$$f(t) = \begin{cases} A & |t| < T/2 \\ 0 & |t| > T/2 \ . \end{cases} \tag{11.3}$$

If we Fourier analyze this function, we find the spectrum

$$g(\omega) = \frac{1}{\sqrt{2\pi}} \int A \exp(-i\omega t)\, dt = \frac{AT}{\sqrt{2\pi}} \frac{\sin(\omega T/2)}{(\omega T/2)} \ , \tag{11.4}$$

which is plotted in Fig. 11.3 as a function of frequency $f = \omega/2\pi$. As can be seen, $f(t)$ contains a continuous spectrum of frequency components from 0 to ∞, arranged in a band structure reminiscent of optical diffraction. The energy or power contained in each frequency component is the square of $g(\omega)$:

$$E(\omega) = |g(\omega)|^2 \ . \tag{11.5}$$

The negative frequencies are, of course, purely imaginary.

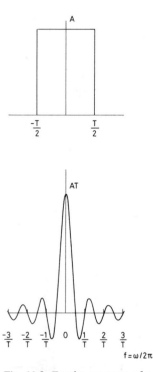

Fig. 11.3. Fourier spectrum of a rectangular pulse

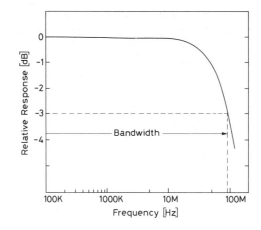

Fig. 11.4. Typical frequency response curve. The frequency range between the points at which the curve falls by 3 dB from its maximum value is defined as the response bandwidth. The lower end of the band is not seen on this graph

All frequencies play a role in the shaping of the function $f(t)$. Thus, in order for an electronic device to faithfully treat the information contained in this signal, the device must be capable of responding uniformly to an infinite range of frequencies. In any real circuit, of course, this is impossible. There will always be resistive and reactive components present, which will filter out some frequencies more than others, so that the response is limited to a finite range in ω. This is also true for the interconnecting cables, as will be seen later. Figure 11.4 shows a typical response curve. The range of frequencies delimited by the points at which the response falls by 3 dB is defined as the *bandwidth* and represents the range of accepted frequencies. Frequencies outside this range are attenuated or cutoff.

While complete and faithful signal reproduction is desirable, it is of course not absolutely necessary that this be ideally so. Indeed, what is important is that those parts of the signal carrying information be reproduced with good fidelity. For nuclear pulses, these parts are the amplitude and, more particularly, the fast rising edge. To investigate the effect of bandwidth on these characteristics, Fig. 11.5 shows the resulting pulse shapes obtained by integrating (11.4) from $f = 0$ to various cut-off frequencies.

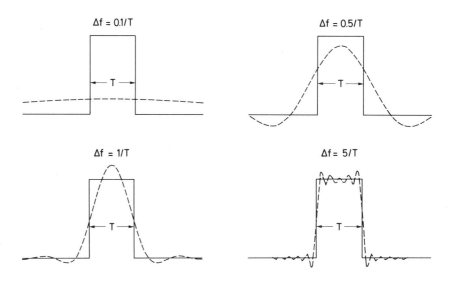

Fig. 11.5. Effect of limited bandwidth on a rectangular pulse

As can be seen, a minimum bandwidth of $\Delta f \geq 1/T$, is necessary to give a reasonable approximation of the pulse. This is not too surprising as most of the frequencies are contained in this region as seen in Fig. 11.3. Moreover, by comparing the various figures, it can be seen that the high frequency components allow the signal to rise sharply, while the lower frequencies account for the flat parts. For a typical fast pulse of say 5 ns width this would mean $\Delta f \geq 200$ MHz. For slow pulses, this limit, of course, is lower. Fast nuclear electronics, therefore, must be capable of accepting frequencies up to $\simeq 500$ MHz. In addition, we can also define a lower limit. Since we are only interested in the fast rising edge and not the flat parts, eliminating some of the lower frequencies should not affect the information in our signal too much either. For nanosecond pulses, it can be shown, in fact, that frequencies up to $\simeq 100$ kHz can be removed with no harm. As a practical rule, therefore, only the frequencies between $\simeq 100$ kHz and $\simeq$ several hundred MHz to 1 GHz are of importance in nuclear electronics. As we will see in a later chapter, achieving a few hundred MHz bandwidth is not a simple task.

12. The NIM Standard

The first (and simplest) standard established for nuclear and high energy physics is a modular system called *NIM (Nuclear Instrument Module)*. In this system, the basic electronic apparatus, for example, amplifiers, discriminators, etc., are constructed in the form of modules according to standard mechanical and electrical specifications. These modules, in turn, fit into standardized *bins* which supply the modules with standard power voltages. Any NIM module will fit into any NIM bin. A specific electronic system for a given application can easily be created, then, by simply collecting the necessary modules, (e.g. an amplifier, a discriminator and a scaler for a simple counting system), installing them in a NIM bin and cabling them accordingly. After the experiment, the modules can be transferred to another NIM system, for example, or rearranged and/or combined with other modules for another application, or stored for later use. The NIM system offers enormous advantages in flexibility, interchange of instruments, reduced design effort, ease in updating instruments, etc. – all of which leads to reduced costs and more efficient use of instruments. For this reason the NIM system is now adopted worldwide by research laboratories and commercial enterprises.

12.1 Modules

Mechanically, NIM modules must have a minimum standard width of 1.35 inches (3.43 cm) and a height of 8.75 in (22.225 cm). They can, however, also be built in multiples of this standard, that is, double-width, triple-width, etc. Figure 12.1 shows an

Fig. 12.1. NIM modules

example of single and double width modules. Power for these modules is supplied through a rear connector which fits into a corresponding connector in the bin. Apart from these mechanical restrictions and the power voltages to be described below, however, the individual is free to design his module in any way desired, thus allowing for new developments and improvements.

12.2 Power Bins

The standard NIM bin is constructed to accept up to 12 single-width modules or a lesser number of multiple-width modules. Figure 12.2 shows a typical example. The external bin dimensions are such as to allow mounting in 19-inch racks or cabinets. The rear power connectors must provide, at the very least, four standard dc voltages, -12 V, $+12$ V, -24 V and $+24$ V, as designated by the NIM convention. However, many bins also provide -6 V and $+6$ V. Prior to 1966, these voltages were not officially part of the NIM standard, but in the past decade their use has become so increasingly common that they are essentially standard now.

Fig. 12.2
A NIM bin

The pin assignments of the rear connectors are shown in Fig. 12.3. All pins are bussed to all of the 12 rear connectors, that is, all #10 pins, for example, are connected to a common conductor, as are all #11 pins, etc. However, since only the standard pins (marked with an asterisk) are used, manufacturers have tended to bus these pins only, so as to discourage use of the others. In any case, pins marked as *reserved* are not to be used since the NIM committee retains these as options for possible use at a later date. Pins marked as *spare* can be used as desired by the individual.

12.3 NIM Logic Signals

NIM modules include both analog and digital instruments. It should be recalled that in analog signals, information is carried in the amplitude or shape of the signal, thus they

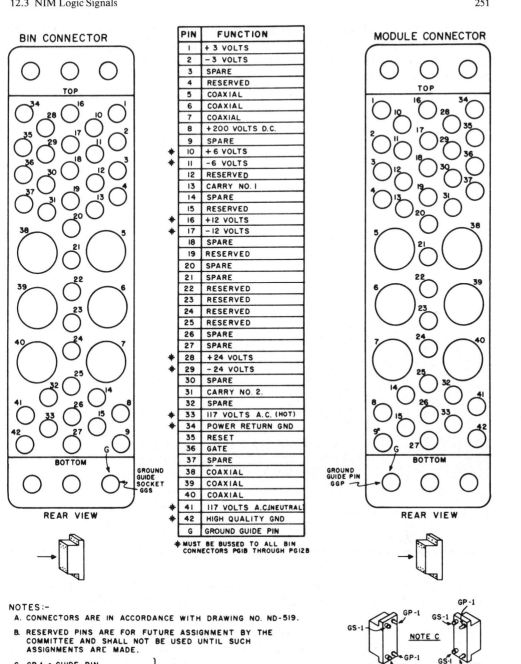

BIN CONNECTOR

TOP

REAR VIEW

PIN	FUNCTION
1	+ 3 VOLTS
2	− 3 VOLTS
3	SPARE
4	RESERVED
5	COAXIAL
6	COAXIAL
7	COAXIAL
8	+200 VOLTS D.C.
9	SPARE
* 10	+ 6 VOLTS
* 11	− 6 VOLTS
12	RESERVED
13	CARRY NO. 1
14	SPARE
15	RESERVED
* 16	+12 VOLTS
* 17	−12 VOLTS
18	SPARE
19	RESERVED
20	SPARE
21	SPARE
22	RESERVED
23	RESERVED
24	RESERVED
25	RESERVED
26	SPARE
27	SPARE
* 28	+24 VOLTS
* 29	− 24 VOLTS
30	SPARE
31	CARRY NO. 2.
32	SPARE
* 33	117 VOLTS A.C. (HOT)
* 34	POWER RETURN GND
35	RESET
36	GATE
37	SPARE
38	COAXIAL
39	COAXIAL
40	COAXIAL
* 41	117 VOLTS A.C. (NEUTRAL)
* 42	HIGH QUALITY GND
G	GROUND GUIDE PIN

*MUST BE BUSSED TO ALL BIN
CONNECTORS PG1B THROUGH PG12B

MODULE CONNECTOR

TOP

REAR VIEW

GROUND GUIDE SOCKET GGS

GROUND GUIDE PIN GGP

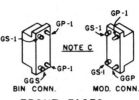

NOTES:−

A. CONNECTORS ARE IN ACCORDANCE WITH DRAWING NO. ND-519.

B. RESERVED PINS ARE FOR FUTURE ASSIGNMENT BY THE
COMMITTEE AND SHALL NOT BE USED UNTIL SUCH
ASSIGNMENTS ARE MADE.

C. GP-1 = GUIDE PIN
   GGP = GROUND GUIDE PIN   } PER ND-519
   GS-1 = GUIDE SOCKET
   GGS = GROUND GUIDE SOCKET

D. SEE ALSO SECTION F OF THIS REPORT.

NOTE C

BIN CONN.   MOD. CONN.

FRONT FACES
SHOWING GUIDE PINS & SOCKETS

Fig. 12.3. Pin assignments on a NIM rear connector (from *Costrell* [12.1])

are of continuously varying heights and form. Digital or logic signals, on the other hand, are of fixed shape and have only two possible states: *yes* or *no*. It is customary to refer to the two states as *logical 0* and *logical 1*; which signal is chosen as *1* or *0* is arbitrary however. For example, one might designate a + 5 V signal as *logical 1* and 0 V as *logical 0* or equally, a − 5 V signal as *1* and − 1 V as *0*. Such a signal might then be used to indicate the presence or absence of a particle in a detector, for example. Thus, it can be used to increment a scaler or to form a coincidence with another signal, etc. This, of course, requires that the scaler or coincidence unit recognize the signals. In practice, a voltage range into which the logic signal must fall is defined rather than a definite level. This allows for fluctuations in the signal due to noise or interference, etc.

While it is not an official part of the NIM convention, an essential standardization has also been set for the voltage levels of logic signals. These levels have been designated as *Preferred Practice* and are generally accepted by manufacturers and laboratories.

Two types of standards exist: *slow-positive* logic and *fast-negative* logic. The first refers to signals of relatively slow rise time, on the order of hundreds of nanoseconds or more. They are of positive polarity and are used with slow detector systems. Table 12.1 defines the voltage levels for this logic. Note that the definition is in terms of a voltage across a 1000 Ω impedance. This implies that the current carried by the signal is very small. The consequence of this is that slow-positive signals cannot be transmitted through long cables. As we will see in the next chapter, the characteristic impedance of most cables is not more than about 100 Ω. After a meter or two of cable, therefore, the signal becomes highly attenuated.

Table 12.1. Slow-positive NIM logic

	Output must deliver	Input must accept
Logic 1	+ 4 to + 12 V	+ 3 to + 12 V
Logic 0	+ 1 to − 2 V	+ 1.5 to − 2 V

Input impedance must be 1000 Ω or more
Source impedance 10 Ω or less

Fast-negative logic, often referred to as *NIM logic*, employs extremely fast signals with rise times on the order of 1 ns and comparable widths. This type is often used in experiments using fast plastic counters (in high-energy physics, for example) where high count rates or fast timing is desired. The NIM logic levels are defined in Table 12.2. Note that unlike slow-positive logic, the definition is current based rather than voltage based. As well, the input and output impedances of all fast NIM modules are

Table 12.2. Fast-negative NIM logic

	Output must deliver	Input must accept
Logic 1	− 14 mA to − 18 mA	− 12 mA to − 36 mA
Logic 0	− 1 mA to + 1 mA	− 4 mA to + 20 mA

Current into 50 Ω
Neither risetime nor width is defined

required to be 50 Ω, as are the characteristic impedances of the connecting cables. The corresponding voltage levels are thus 0 V and − 0.8 V for logic *0* and *1* respectively. In contrast to slow positive signals, the fast NIM signals can be transmitted through relatively long lengths of cable.

12.4 TTL and ECL Logic Signals

While not part of the NIM standard two other logic families are often found in nuclear and particle physics electronics. The first is the *TTL* (Transistor-Transistor Logic) logic family. This is a positive going logic which is very often found on NIM electronics modules. The levels are defined in Table 12.3.

The second is a logic family which is becoming increasingly popular in high-energy physics. This is the *emitter-coupled* logic (ECL) family which is currently the fastest form of digital logic available. These levels are also defined in Table 12.3.

Table 12.3. TTL and ECL signal levels

	TTL	ECL
Logic 1	2 − 5V	− 1.75 V
Logic 0	0 − 0.8 V	− 0.90 V

ECL is, in fact, used in most modern fast-logic NIM and CAMAC circuits. However, to make them compatible with the NIM standards the levels at the input and output stages are generally adapted. This adds additional cost and power consumption to the module.

ECL modules do away with the adapter stages and enter directly with the ECL levels. This allows a better noise immunity with no loss of bandwidth and minimizes ground loops. Since the input impedance is high ($\simeq 100 \, \Omega$), cheaper twisted pair cables may be used allowing higher density modules. Flat ribbon cable may also be used but only for short lengths.

NIM signals are not directly compatible with ECL, however, if frequency is not a problem, a very simple translator can be constructed as shown below (Fig. 12.4).

12.5 Analog Signals

Three ranges of analog signals: 0 to 1 V, 0 to 10 V and 0 to 100 V are specified in the NIM convention, however, only 0 to 10 V has found any appreciable usage and indeed almost all commercially available modules work with this definition.

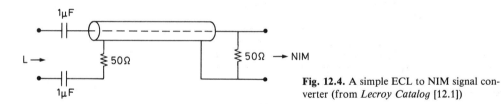

Fig. 12.4. A simple ECL to NIM signal converter (from *Lecroy Catalog* [12.1])

13. Signal Transmission

We now turn to the problem of transmitting pulse signals from one part of the electronics system to the other, or, more specifically, the interconnecting cables. This may seem somewhat trivial, at first, but this will be shown to be otherwise.

The goal of signal transmission, of course, is two-fold: (1) get the signal from point A to point B, and (2) preserve the information in the signal. Recalling that a pulse generally consists of a continuous spectrum of frequencies from 0 to infinity, this would mean that our interconnecting cable would have to be capable of transmitting an infinite range of frequencies uniformly and coherently over the required distance – in most systems, a few meters. Such an ideal cable, of course, does not exist. Stray capacitances, inductances and resistance, inherent in any configuration of conductors, will invariably attenuate some frequencies more than others, causing a distortion of the pulse at the receiving end. Indeed, sending a *fast* pulse signal through simple wire connections, for example, already results in intolerable distortion after only a few centimeters!

In practice, of course, it is not necessary to transmit an infinite range of frequencies. The Fourier spectrum of a rectangular pulse of width T is largely contained in the region $\Delta f \simeq 1/T$, and, as we have seen, most of the information will be reproduced if only this range is kept. This is not to imply that our problems are over, however. In deed, for a fast 2 or 3 ns pulse, this means uniformly transmitting all frequencies up to several hundred MHz. This is still a very large range and it is by no means obvious how transmission of even this finite range can be performed without attenuation of some frequencies. Fortunately, the theory of pulse signal transmission has been developed for some time now and has allowed the design of *transmission* lines which permit low-distortion transmission over long distances.

In nuclear electronics, the standard transmission line is the coaxial cable. These cables offer a number of advantages as opposed to other designs and our discussion will be focused primarily on this cable type, although much of what will be presented is generally applicable to other transmission lines as well. Since it is the physicist who makes these interconnections, (the design of the module circuits being left to the electronics engineer!), it is, of course, extremely important that he understand how these signals are transmitted and to recognize the problems which arise.

13.1 Coaxial Cables

The basic geometry of a coaxial transmission line is that of two concentric cylindrical conductors separated by a dielectric material. A cutaway section of a typical cable showing its construction is illustrated in Fig. 13.1. The outer cylinder, which carries the return current, is generally made in the form of wire braid, while the dielectric material is usually of polyethylene plastic or teflon, although other materials are sometimes

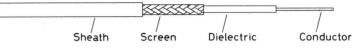

Fig. 13.1. Cut-away view of a coaxial cable

used. The entire cable is protected by a plastic outer covering. One advantage of this type of construction is that the outer cylindrical conductor, besides serving as the ground return, also shields the central wire from stray electromagnetic fields. Frequencies down to $\simeq 100\,\text{kHz}$ are effectively attenuated in most standard cables. A variety of cable sizes and designs are available commercially from several manufacturers and are briefly summarized in Table 13.1. The most commonly used cables are RG-58C/U (50 Ω) for fast signals and RG-58/U (93 Ω) for spectroscopy work. In recent years, however, the miniature RG-174/U cables have gained particular popularity in nuclear and high energy physics. Seventy-five ohm cable (RG59/U) is also used for high voltage transmission.

Table 13.1. Some common coaxial cable types and their characteristics (data from *LeCroy catalog* [13.1])

Type [RG]	Delay [ns/m]	Diameter [cm]	Capacitance [pF/m]	Max. operating voltage [kV]	Remarks
50 Ω, single braided cables:					
58U	5.14	0.307	93.5	1.9	Standard cable for fast NIM electronics
58A/U	5.14	0.305	96.8	1.9	
58C/U	5.06	0.295	93.5	1.9	
174/U	5.14	0.152	98.4	1.5	Miniature cable for fast NIM electronics
213/U	5.06	0.724	96.8	5.0	Formerly RG-8A/U
215/U	5.06	0.724	96.8	5.0	Same as 213/U but with armor; formerly RG-10A/U
218/U	5.06	1.73	96.8	11.0	Large, low attenuation cable formerly RG-17A/U
219/U	5.06	1.73	96.8	11.0	Same as 218/U but with armor; formerly RG-18A/U
220/U	5.06	2.31	96.8	14.0	Very large, low attenuation formerly RG-19A/U
221/U	5.06	2.31	96.8	14.0	Same as above but with armor; formerly RG-20A/U
50 μ, double braided cables:					
55B/U	5.06	0.295	93.5	1.9	Small size, flexible cable
221/U	5.06	0.470	93.5	3.0	Small size, microwave cable; formerly RG-5B/U
214/U	5.06	0.724	98.4	5.0	Formerly RG-9BU
217/U	5.06	0.940	96.8	7.0	Power transmission cable; formerly RG 14A/U
224/U	5.06	0.940	96.8	7.0	Same as 217/U but with armor; formerly RG-74A/U
223/U	5.06	0.295	93.5	1.0	formerly RG-55A/U
High voltage cables:					
59/U	5.14	0.381	68.9	2.3	$Z = 73\,\Omega$, standard HV cable for detectors
59B/U	5.14	0.381	67.3	2.3	$Z = 75\,\Omega$

Table 13.1 (continued)

Type [RG]	Delay [ns/m]	Diameter [cm]	Capacitance [pF/m]	Max. operating voltage [kV]	Remarks
93 Ω cables:					
62/U	4.0	0.635	44.3	0.75	Standard cable for slow NIM signals in spectroscopy work
62A/U	4.0	0.632	44.3	0.75	Standard cable for slow NIM signals
High temperature, 50 Ω single braided cables:					
178B/U	4.7	0.086	95.1	1.0	Miniature size cable
179B/U	4.7	0.160	65.6	1.2	
196A/U	4.7	0.086	–	1.0	Teflon dielectric
211A/U	4.7	1.575	95.1	7.0	Operation between −55° − +200 °C; formerly RG-117A/U
228A/U	4.7	1.575	95.1	7.0	Same as 211A/U but with armor; formerly RG-118A/U
303/U	4.7	0.295	93.5	1.9	
304/U	4.7	0.47	93.5	3.0	
316/U	4.7	0.15	–	1.2	Miniature cable
High temperature, 50 Ω double braided cables:					
115/U	4.7	0.635	96.8	5.0	Used where expansion and contraction are a problem
142B/U	4.7	0.295	93.5	1.9	Small-size flexible cable
225/U	4.7	0.724	96.8	5.0	Operation between −55° − +200 °C; formerly RG-87A/U
227/U	4.7	0.724	96.8	5.0	Same as 225/U but with armor; formerly RG-116/U

As will be seen in the following sections, pulses signals are transmitted through a coaxial line as a traveling wave. As such the coaxial line is nothing more than a wave guide.[1] In electronics, however, it is customary to view the coaxial cable as a circuit element and to consider the voltage and current in the cable rather than the E and B fields, which is perhaps more familiar to the physicist. The former approach is, of course, the more practical since voltage and current are directly accessible. We will therefore take this point of view in our discussion.

13.1.1 Line Constituents

By virtue of its geometrical configuration, (two conductors separated by a dielectric), coaxial cables necessarily contain a certain self-capacitance and inductance. From electromagnetic theory, it is easy to show that for two long concentric cylinders, these components are

---

[1] Signals are transmitted in degenerate TEM mode. The next mode begins at frequencies well above those of interest and does not interfere for our purposes.

$$L \simeq \frac{\mu}{2\pi} \ln\left(\frac{b}{a}\right) \ [\text{H/m}] \quad = 0.2 \, K_\text{m} \ln\left(\frac{b}{a}\right) \ [\mu\text{H/m}]$$

$$C \simeq \frac{2\pi\varepsilon}{\ln(b/a)} \ [\text{F/m}] \quad = \frac{55.6 \, K_\text{e}}{\ln(b/a)} \ [\text{pF/m}] \ , \tag{13.1}$$

where a and b are the radii of the inner and outer cylinders respectively, μ and ε the permeability and permittivity of the insulating dielectric, $K_\text{e} = \varepsilon/\varepsilon_0$ and $K_\text{m} = \mu/\mu_0$, the permittivity and permeability relative to the vacuum. For nonferromagnetic materials, of course, $K_\text{m} \simeq 1$. Typical values for L and C are on the order of $\simeq 100$ pF/m and a few tenths of μH/m.

In any real cable, however, there also exists a certain resistivity due to the fact that the conductors are not perfect, and a certain conductivity across the dielectric due to its *"imperfectness"* as an insulator. These components, like the capacitance and inductance, are distributed uniformly along the cable length and are generally small compared to the capacitive component. Nevertheless, for most applications, we can approximately represent a unit length of cable by the *lumped* circuit shown in Fig. 13.2. L and C are, respectively, the series inductance and capacitance per unit length, while R is the resistance per unit length and G is the conductance per unit length of the dielectric, represented here as a parallel resistance of $1/G \ \Omega$. These latter two quantities, as will be seen later, are responsible for signal losses in the cable. In the ideal case of a perfect lossless cable, R and G are zero.

13.2 The General Wave Equation for a Coaxial Line

With the aid of Fig. 13.2, we can now derive an equation for the voltage, V and the current, I, in the cable. Consider, therefore, a small unit length of cable, Δz, and let us calculate the difference ΔV and ΔI across this small distance,

$$\Delta V(z, t) = - R \, \Delta z \, I(z, t) - L \, \Delta z \, \frac{\partial I}{\partial t}(z, t)$$

$$\Delta I(z, t) = - G \, \Delta z \, V(z, t) - C \, \Delta z \, \frac{\partial V}{\partial t}(z, t) \ . \tag{13.2}$$

Dividing by Δz and taking the limit $\Delta z \to 0$, we find the differential equations

$$\frac{\partial V}{\partial z} = - R I - L \, \frac{\partial I}{\partial t}$$

$$\frac{\partial I}{\partial z} = - G V - C \, \frac{\partial V}{\partial t} \ . \tag{13.3}$$

By differentiating with respect to z and t, and substituting, the equations may be uncoupled to give

$$\frac{\partial^2 V}{\partial z^2} = L C \, \frac{\partial^2 V}{\partial t^2} + (L G + R C) \, \frac{\partial V}{\partial t} + R G V \tag{13.4}$$

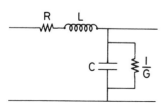

Fig. 13.2. Equivalent circuit for a unit length of transmission line

and an identical equation for I. This is the general wave equation for a coaxial cable (or, in fact, any other type of transmission line whose constituents are represented by Fig. 13.2).

13.3 The Ideal Lossless Cable

Before tackling the solutions to (13.4), let us first consider the simpler case of an ideal lossless cable where R and G are zero. For relatively short lengths of cable of a few meters or so, this, in fact, is a good approximation since the effects of R and G will be negligible for most purposes. The last two terms on the right-hand side of (13.4) vanish then leaving

$$\frac{\partial^2 V}{\partial z^2} = LC \frac{\partial^2 V}{\partial t^2}. \tag{13.5}$$

This can be recognized as the well-known wave equation.

Suppose now a simple sinusoidal voltage in time (i.e., one Fourier component), $V = V(z) \exp(i\omega t)$ is impressed on the cable. Substitution into (13.5) then yields

$$\frac{d^2 V}{dz^2} = -\omega^2 LC V = -k^2 V, \tag{13.6}$$

where we have set $k^2 = \omega^2 LC$. The space solutions are then of the form

$$V(z) = V_1 \exp(-kz) + V_2 \exp(kz)$$

which results in

$$V(z, t) = V_1 \exp[i(\omega t - kz)] + V_2 \exp[i(\omega t + kz)]. \tag{13.7}$$

This represents two waves, one traveling in the $+z$ direction, and the other in the opposite direction, $-z$. This second wave corresponds to a reflection and its presence or absence depends on the boundary conditions for the cable under question. As we will see later, reflections play an important role for signal transmission since they can distort the form of the original signal.

Examination of (13.7) will also show that the quantity, k, is the wave number and that the velocity of propagation is

$$v = \frac{\omega}{k} = \frac{1}{\sqrt{LC}}. \tag{13.8}$$

As long as the cable stays constant in in cross-section, the product LC is, in fact, independent of length and $LC = \mu\varepsilon$ where μ and ε are the permeability and permittivity of the dielectric. This is identical to the case for optical media. Thus, for a cable with free space as a dielectric, the velocity of propagation equals $1/\sqrt{\mu_0 \varepsilon_0} = c$, the speed of light in a vacuum.

The speed of signal propagation is more often expressed as its inverse, the time of propagation per unit length.

$$T = v^{-1} = \sqrt{LC}. \tag{13.9}$$

This quantity is known as the *delay* of the cable, and is typically on the order of $\simeq 5$ ns/m for standard 50 Ω cables currently found in the lab.

We might note here that the bandwidth of the ideal cable is infinite, that is, it is capable of uniformly transmitting all frequencies impressed upon it. In reality, of course, losses will limit this range as we will show later.

13.3.1 Characteristic Impedance

An important property of a transmission cable is its *characteristic impedance*. This is defined as the ratio of the voltage to the current in the cable, (including the phase relationship), i.e.,

$$Z_0 = \frac{V}{I} \tag{13.10}$$

which, with some manipulation of (13.3) and (13.7), can be shown to be

$$Z_0 = \sqrt{\frac{L}{C}} \tag{13.11}$$

for an ideal lossless cable. Examination will show that Z_0 has, in fact, the dimensions of impedance and is moreover purely resistive. It is, however, totally independent of the cable length and only dependent on the cross sectional geometry and the materials used. The characteristic impedance is, in fact, a rather special quantity in the sense that it cannot be measured with a normal resistive bridge, but behaves as a real impedance when connected to the output of a device. It is perhaps best interpreted as being the impedance offered to the propagation of the signal in the line.

For coaxial cables, an explicit calculation shows that

$$Z_0 = \sqrt{\frac{L}{C}} = 60 \sqrt{\frac{K_m}{K_e}} \ln \frac{b}{a} \quad [\Omega] \ , \tag{13.12}$$

where a and b are the inner and outer diameters of the conductors and K_m and K_e, the relative permeability and permittivity of the dielectric. At present, the standard cable used in fast nuclear electronics has a characteristic impedance of 50 Ω while cables of 93 Ω are generally used for slower spectroscopy pulses.

An interesting point to note is that all coaxial cables are necessarily limited to the range between $\simeq 50-200 \, \Omega$ characteristic impedance. This is because of the dependence of Z_0 on the logarithm of b/a in (13.12). To construct a coaxial cable of 1000 Ω, for example, would require a diameter ratio of 10^{11} – clearly an impractical if not impossible task! Moreover, there is an optimum value for the ratio b/a of $\simeq 3.6$ which minimizes losses (see Sect. 13.6). If this ratio is kept and K_e is on the order of 2.3, as for polyethylene, then $Z_0 \simeq 50 \, \Omega$.

13.4 Reflections

As we have seen from (13.7), the signal in a coaxial cable is, in general, the sum of the original signal and a reflected signal traveling in the opposite direction. For an arbitrary signal form f, we can write

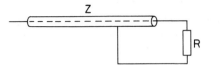

Fig. 13.3. Cable of characteristic impedance Z terminated by an impedance R

$$V = f(x - v t) + g(x + v t),\tag{13.13}$$

where g is the reflected wave form. The presence of reflections can have serious consequences. Obviously, if they overlap with the original signal, interference and distortion will result. Moreover, even if no overlap occurs, it is clearly undesirable to have *echos* of the original signal bouncing back and forth in the cables leading to spurious counts and confusion. Reflections, of course, occur when a traveling wave encounters a new medium in which the speed of propagation is different. In optical media this corresponds to a change in the index of refraction. By analogy, in transmission lines, reflections occur when the characteristic impedance of the line is abruptly changed.

These reflections can be calculated by considering the boundary conditions at the interface. Consider a cable of characteristic impedance Z terminated by an impedance R, the input impedance of some device, for example. Figure 13.3 illustrates such a configuration. As a signal travels down the line, the ratio of the voltage to current must always be equal to the characteristic impedance by definition. When the interface is encountered reflections are set up which adjust the ratio of V/I to the new characteristic impedance. These reflections, however, must also be compatible with the original characteristic impedance since they will also travel back down the original line in the opposite direction. We, thus, have the conditions

$$Z = \frac{V_0}{I_0}$$

$$R = \frac{V_0 + V_r}{I_0 + I_r}\tag{13.14}$$

$$Z = \frac{V_r}{-I_r},$$

where V_0 and I_0 are the voltage and current of the original signal and V_r and I_r those of the reflected signal. Note the negative sign for the reflected current!

From these equations, we find

$$\rho = \frac{V_r}{V_0} = \frac{-I_r}{I_0} = \frac{R - Z}{R + Z},\tag{13.15}$$

where ρ is known as the *reflection coefficient*. The polarity and amplitude of the reflected signal are thus dependent on the relative values of the two impedances. If R is greater than the cable impedance Z, then the reflection will always be of the same polarity but with an amplitude intermediate in value between 0 and the original pulse height. In the limiting case of infinite load impedance (an open circuit, for example), the reflected amplitude is equal to the incident amplitude. On the other hand, if R is smaller than the cable impedance, the reflection is opposite in polarity and intermediate in amplitude between 0 and the original. In the limit of zero load impedance (a short circuit), the reflection is equal and opposite to the incident pulse. In the special case of

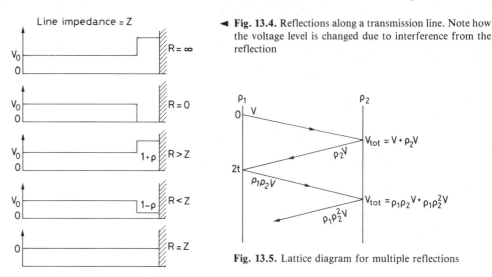

◀ Fig. 13.4. Reflections along a transmission line. Note how the voltage level is changed due to interference from the reflection

Fig. 13.5. Lattice diagram for multiple reflections

$R = Z$, ρ vanishes and there is no reflection whatsoever. Thus, only when the load and cable impedances are matched are interfering reflections avoided. The various cases are summarized in Fig. 13.4 for a simple step function pulse.

In the above example, it is easy to see what will also happen if the input end of the cable is also terminated by an impedance different from Z. Clearly, when the first reflected signal reaches the input end it will again be reflected but with some new coefficient ρ', and this again when it reaches R, etc. Multiple reflections are thus set up in the cable. An easy method of calculating the value of the signal at any time in the cable is with the simple *lattice* diagram shown in Fig. 13.5.

13.5 Cable Termination. Impedance Matching

As we have seen, signal distortion from reflections can only be avoided by matching the device impedances to the cable impedance. To a large extent, this problem has been removed by the fast NIM standard which requires that all input and output device impedances and cables be 50 Ω. However, occasions very often arise where an impedance mismatch cannot be avoided. One very common instance is when a fast signal must be viewed on an oscilloscope. Since oscilloscopes are high impedance devices ($\simeq 1\,\mathrm{M}\Omega$), direct entry of a fast NIM signal will result in an impedance mismatch and a false signal reading. (For a slow NIM signal, direct entry is compatible as given by the NIM standard, provided the cables are not too long.) In such cases, the cable can be *terminated* with an additional impedance of the appropriate value so as to adjust the total load impedance seen by the cable. For high impedance devices such as our oscilloscope, this is done by placing a resistance of 50 Ω in parallel with the device. The signal seen by the oscilloscope is then reflection free and appears as it would to a standard fast NIM module. Since the need to terminate cables arises fairly often, special 50 Ω terminators made so as to easily fit onto the cables are manufactured commercially. Figure 13.6 illustrates this method.

More generally, termination can be done in two ways: either by adding an impedance in series with the load or in parallel (shunt termination). As well, this can be

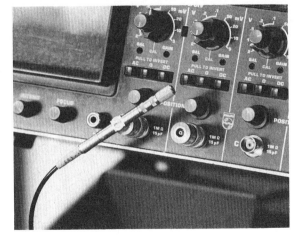

◄ **Fig. 13.6.** Cable termination at an oscilloscope

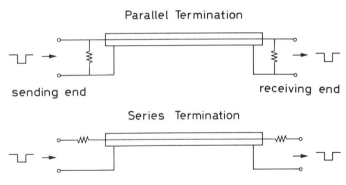

Fig. 13.7. Series and parallel termination

done at the sending end of the cable and/or at the receiving end. Figure 13.7 illustrates the various configurations. In most cases, a simple termination at the receiving end is usually sufficient. In the example of signal viewing at the oscilloscope, shunt termination at the receiving end is used. Nevertheless, certain situations may arise which require use of some other termination scheme with different values of the terminating resistance.

Example 13.1   A signal is to be sent from a coaxial cable of impedance Z_1 into another coaxial cable of impedance Z_2. What termination scheme should be used in order to avoid reflections?

Two cases arise:
a) $Z_1 < Z_2$
Here the impedance which cable 1 sees must be reduced. This implies adding a resistance R in parallel to cable 2, i.e.,

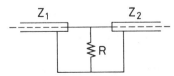

Since the combination must equal Z_1 we find

$$\frac{RZ_2}{R+Z_2} = Z_1$$

$$R = \frac{Z_1 Z_2}{Z_2 - Z_1}.$$

b) $Z_1 > Z_2$
Since the impedance seen by cable 1 must be increased, we add a resistance R in series.

Table 13.2. Cable termination schemes

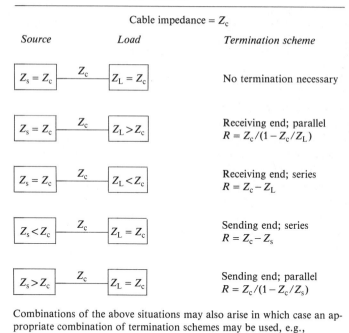

Cable impedance = Z_c

Source	*Load*	*Termination scheme*
$Z_s = Z_c$ — Z_c — $Z_L = Z_c$		No termination necessary
$Z_s = Z_c$ — Z_c — $Z_L > Z_c$		Receiving end; parallel $R = Z_c/(1 - Z_c/Z_L)$
$Z_s = Z_c$ — Z_c — $Z_L < Z_c$		Receiving end; series $R = Z_c - Z_L$
$Z_s < Z_c$ — Z_c — $Z_L = Z_c$		Sending end; series $R = Z_c - Z_s$
$Z_s > Z_c$ — Z_c — $Z_L = Z_c$		Sending end; parallel $R = Z_c/(1 - Z_c/Z_s)$

Combinations of the above situations may also arise in which case an appropriate combination of termination schemes may be used, e.g.,

$Z_s < Z_c$ — Z_c — $Z_L > Z_c$		Receiving end; parallel $R = Z_c/(1 - Z_c/Z_L)$ with sending end; series $R = Z_c - Z_s$

Then,

$$Z_2 + R = Z_1 \Rightarrow R = Z_1 - Z_2 .$$

Some other possible situations which often arise are summarized in Table 13.2 along with the termination scheme to be used.

13.6 Losses in Coaxial Cables. Pulse Distortion

Signal losses in a transmission line arise from resistance in the conductors and leakage through the dielectric. In addition, some loss may also result from electromagnetic radiation; however, this effect is small, especially in coaxial cables with their inherent shielding, and can be neglected for most purposes.

The effect of R and G on signal propagation may be seen by returning to (13.4) and once again impressing a sinusoidal signal, $V = V(z) \exp(i\omega t)$, on to the cable. Substitution then leads to the form,

$$d^2 V/dz^2 = (R + i\omega L)(G + i\omega C) V = \gamma^2 V , \tag{13.16}$$

where the complex number, γ,

$$\gamma = \alpha + ik = \sqrt{(R + i\omega L)(G + i\omega C)} \tag{13.17}$$

is known as the *propagation constant*. The general solution is then

$$V(z, t) = V_1 \exp(-\alpha z) \exp[i(\omega t - kz)] + V_2 \exp(\alpha z) \exp[i(\omega t + kz)] \tag{13.18}$$

which again represents two traveling waves, however, modified this time by the exponential factor $\exp(\pm \alpha z)$. The introduction of R and G, therefore, leads to an exponential attenuation of the signal with distance at a rate given by α. As can be seen from Fig. 13.8, however, α is generally small so that only in very long cables of many ten's of meters does signal loss begin to be a problem.

The above statement overlooks a more serious factor, however, and that is the dependence of α and $v = dk/d\omega$ on ω. This implies a differential attenuation of the frequency components which leads to a dispersion of the pulse packet.

The dependence on frequency enters in two ways: (1) explicitly through (13.17) and (2) through the fact that both R and G also vary with frequency: R via the *skin* effect and G via high frequency dielectric leakage.[2] The former is most significant at low frequencies (<10's kHz) where R and G are essentially constant. Evaluating α and k explicitly from (13.17), we find, in fact, the rather complicated expressions[3]

$$\begin{aligned}
\alpha^2 &= \tfrac{1}{2}[(RG - \omega^2 LC) + \sqrt{(R^2 + \omega^2 L^2)(G^2 + \omega^2 C^2)}] \\
k^2 &= \tfrac{1}{2}[-(RG - \omega^2 LC) + \sqrt{(R^2 + \omega^2 L^2)(G^2 + \omega^2 C^2)}].
\end{aligned} \tag{13.19}$$

At higher ω such that $R/\omega L \ll 1$ and $G/\omega C \ll 1$, however, α and k may be approximated by

$$\alpha \simeq \frac{1}{2}\left(R\sqrt{\frac{C}{L}} + G\sqrt{\frac{L}{C}}\right) \tag{13.20}$$

$$k \simeq \omega\sqrt{LC}.$$

The speed of signal propagation is thus $v \simeq 1/\sqrt{LC}$ as in the case of an ideal cable, and the attenuation constant is approximately independent of ω. For coaxial cables, this high frequency region begins at about $\omega \simeq 100$ kHz. As will be recalled, however, the frequency range of interest for fast pulses also begins at about this point. Thus the explicit ω dependence in (13.19) is not an important effect for fast nuclear pulses.

However, at and above these frequencies, R begins to vary with ω through the well-known *skin effect*. Indeed, as ω increases, the current in the conductors confines itself to thinner and thinner layers near the conductor surface. The effective cross-sectional area of the conductor is thus reduced and its resistance increased. For a coaxial cable, this results in a resistance per unit length which varies approximately as the square root of the frequency and inversely as the inner and outer radii, i.e.,

---

[2] The inductance also varies somewhat if the magnetic flux inside the conductors is taken into account. This contribution is small, however, and may generally be ignored. The capacitance, on the other hand, is strictly constant.

[3] The explicit ω dependence is removed in the special case of $R/L = G/C$. Such a cable would still attenuate but without distortion. Unfortunately, construction of such a cable is uneconomical and, moreover, does not remove the high-frequency ω dependence.

Attenuation of Cables

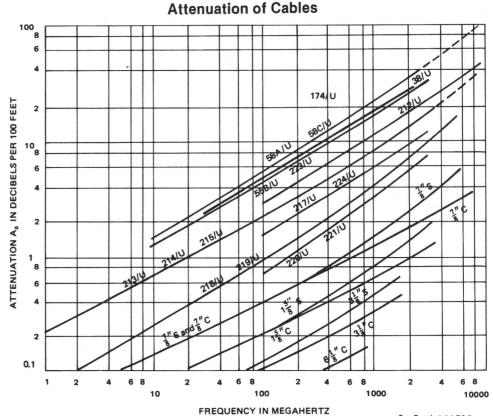

Fig. 13.8. Attentuation constants for various cables

S = Semi-rigid 50Ω.
C = Rigid copper coax.

$$R(\omega) = \frac{1}{2\pi} \sqrt{\frac{\omega\mu}{2\sigma}} \left(\frac{1}{a} + \frac{1}{b}\right) \quad \Omega \text{ per unit length,} \tag{13.21}$$

where σ is the conductivity, μ the permeability and a and b the inner and outer radii of the cable. For copper, this becomes

$$R(f) = 4.17 \times 10^{-8} \sqrt{f} \left(\frac{1}{a} + \frac{1}{b}\right) \quad \Omega \text{ per unit length,} \tag{13.22}$$

where $f = \omega/2\pi$. Taking the cable RG-58C/U with $b = 0.25$ cm as an example, we find R varies from $\simeq 2.4 \times 10^{-2}$ Ω/m at 100 kHz to $\simeq 2.4$ Ω/m at 1 GHz. Since R also depends on the cable dimensions, it is easy to show that attenuation through the skin effect is minimized when $b/a = 3.6$.

At high frequencies, there is also leakage across the dielectric. For materials such as polyethylene or teflon, this remains small up to several hundreds of MHz. But because of its linear dependence on ω, it rapidly overtakes the skin effect and becomes dominant at and above 1 GHz. Considering both effects, the high-frequency dependence of α can then be written as

$$\alpha(f) = a\sqrt{f} + bf, \tag{13.23}$$

where $f = \omega/2\pi$ and a and b are constants. Typical values of α for the various coaxial cable types at different frequencies may be found in Fig. 13.8.

For a cable with loss, it is also interesting to see what happens to the characteristic impedance. With nonzero R and G, the definition of Z_0 then gives,

$$Z_0 = V/I = \sqrt{\frac{R + i\omega L}{G + i\omega C}} \ . \tag{13.24}$$

Thus, Z_0 takes on an imaginary reactive part. If we ignore the effect of G, for the reasons given above, we can approximate (13.24) by

$$Z_0 = \sqrt{\frac{L}{C}} \left(1 + \frac{R}{i\omega L}\right)^{1/2} \simeq \sqrt{\frac{L}{C}} \left(1 + \frac{R}{2i\omega L}\right). \tag{13.25}$$

As a circuit element, therefore, the cable will behave as a resistance $R' \simeq \sqrt{L/C}$ in series with a capacitance $C' = 2\sqrt{LC}/R$. However, in the high-frequency approximation, the characteristic impedance reduces once again to (13.12).

13.6.1 Cable Response. Pulse Distortion

Equation (13.18) essentially gives the response of a lossy transmission line to a pure sinusoidal signal. For an arbitrary pulse waveform, however, the calculation becomes very complicated, generally requiring a numerical solution using computers. Nevertheless, in the relatively simple case of a unit step function, a closed analytic expression can be obtained which illustrates the essential effect of losses on pulse shape. For an initial step function described by

$$V(t) = \begin{cases} 0 & t < 0 \\ 1 & t \geq 0 \ , \end{cases} \tag{13.26}$$

therefore, we find the response

$$V(t) = \begin{cases} 0 & t \geq 0 \\ \mathrm{erfc}\left(\dfrac{1}{2\sqrt{\dfrac{t}{\tau_0}}}\right) & t \geq 0 \ , \end{cases} \quad \text{where} \tag{13.27}$$

$$\tau_0 = \frac{(x\alpha)^2}{\pi f}$$

with x: length of cable; α: attenuation constant in the region of validity; f: frequency at which α is evaluated.

The error function is defined as

$$\mathrm{erfc}(x) = 1 - \frac{2}{\sqrt{\pi}} \int_0^x \exp(-u^2) \, du \ .$$

For simplicity we have ignored the fixed delay caused by passing through the cable. Figure 13.9 plots this response as a function of t/τ_0. As can be seen, the primary feature

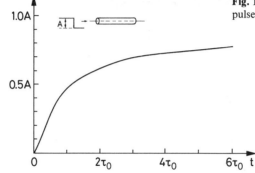

Fig. 13.9. Response of a lossy cable to a unit step function pulse (from *Fidecaro* [13.2])

is a loss in time response, the once sharp edge of the pulse now being softened to a slow rise. The parameter τ_0 can now also be seen as the time it takes to reach a height $\simeq \frac{1}{2}$ of maximum. If skin losses only are considered, then from (13.23)

$$\alpha_f \simeq a \sqrt{f} \tag{13.28}$$

so that

$$\tau_0 \simeq (xa)^2/\pi . \tag{13.29}$$

By use of the convolution theorem in Fourier analysis, the response of other waveforms may be calculated from the unit step response. Figure 13.10 shows the distortion of an ideal rectangular pulse of width Δ for various τ_0. In addition to the slower rise time, note how the centroid of the pulse is also displaced.

For the case of a photomultiplier pulse, *Brianti* [13.3] has made calculations considering both skin effects and dielectric leakage. His results, however, are based on the cable response to a $\delta(t)$-function rather than a step function. Some examples of these are shown in Fig. 13.11.

Fig. 13.10. Distortion of a rectangular pulse in a lossy cable (from *Fidecaro* [13.2])

Fig. 13.11. Distortion of a photomultiplier pulse in a lossy transmission cable (from *Brianti* [13.3])

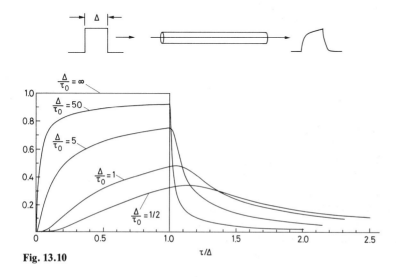

Fig. 13.10

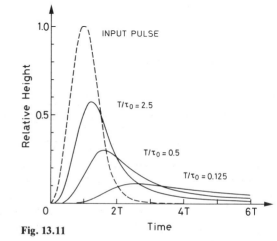

Fig. 13.11

14. Electronics for Pulse Signal Processing

The aim of this chapter is to survey some of the basic functions which can be applied to pulse signals for information processing. Since most of these functions have been largely standardized over the years by the nuclear electronics industry, our discussion will rest mainly on a descriptive level and only touch on the detailed circuitry where it is absolutely necessary. The assembling or combination of these functions into a system for a specific application is left to the following chapters. For pedagogic reasons also, we will concentrate on the simpler NIM modules and not enter into the more sophisticated digital processes which are found, for example, in CAMAC systems (see Chap. 18). Once a good understanding of the basic functions and their applications is acquired, however, the functioning of these latter modules will become evident. The references at the end of this chapter should also provide additional information for those desiring more detail.

All of the modules described here can be found commercially. Variations between modules by different manufacturers will, however, be encountered. Some models, for example, will combine several functions together into one module or offer additional features which may be useful in specific applications, while others will remain on a more rudimentary level. The difference is often seen in the price! For information concerning the detailed operation of any particular module, therefore, the reader should consult the manual provided by the manufacturer. However, the basic operations are usually so similar that after some experience many of the operating details may be guessed a good deal of the time.

In addition to a description of processing modules, we will also present a short section on some very simple circuits which can be constructed for signal manipulation, e.g., attenuation, splitting, etc. These come in very handy when setting up a nuclear electronics system and may help save some needless and time-consuming searching for an appropriate module.

14.1 Preamplifiers

The basic function of a preamplifier is to amplify weak signals from a detector and to drive it through the cable that connects the preamplifier with the rest of the equipment. At the same time, it must add the least amount of noise possible. Since the input signal at the preamplifier is generally weak, preamplifiers are normally mounted as close as possible to the detector so as to minimize cable length. In this way, pickup of stray electromagnetic fields is reduced and cable capacitance, which decreases the signal-to-noise ratio, minimized. In some detectors, notably the scintillation counter, considerable amplification already occurs before entry into the preamplifier. In such cases, the gain and low noise characteristics of the preamplifier are less critical and the preamplifier is used mainly to present the correct impedances to the detector and the electronics and to shape subsequent output pulses.

Three basic types of preamplifier exist:

1) voltage sensitive,
2) current sensitive,
3) charge-sensitive.

The current-sensitive preamplifier is generally used with very low impedance signal devices and is thus not very useful with radiation detectors, which are generally high impedance instruments. We will not discuss this type further, therefore.

Of the two remaining types, the voltage-sensitive preamp is the more conventional. This device amplifies any voltage which appears at its input. Since radiation detectors are essentially charge producing devices, this voltage appears through the intrinsic capacitance of the detector plus any other stray capacitances which may be in the input circuit, i.e.

$$V = \frac{Q}{C_{\text{tot}}}.$$

It is therefore important that the capacitance of the detector remain stable during operation. For PM's, proportional counters and Geiger-Muller tubes, this is the case. However, for semiconductor detectors variations in the intrinsic capacitance occur as temperature changes. This is caused by the leakage current in the semiconductor diode which is dependent on the temperature. It is therefore not advisable to use this type of preamplifier with semiconductor devices. Figure 14.1 shows a schematic diagram of the basic design of a voltage-sensitive preamp.

The shortcomings of the voltage-sensitive preamplifier can be avoided by using a charge-sensitive preamp. Figure 14.2 shows a schematic of the basic design for this type of amplifier. The basic idea is to integrate the charge carried by the incoming pulse on the capacitor C_f. By working out the voltages from the diagram, it can be seen that the output voltage is always proportional to

$$V_0 \simeq - \frac{Q}{C_f}$$

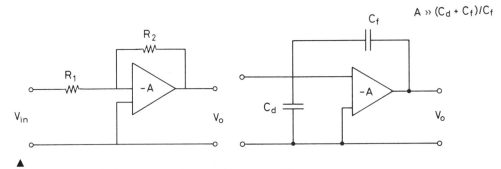

Fig. 14.1. Schematic diagram of a voltage-sensitive preamplifier

Fig. 14.2. Schematic diagram of a charge-sensitive preamplifier. To discharge the capacitor C_f, a resistor is also usually placed in parallel with C_f. This results in the exponential *tail* pulse

so that all dependence on the detector capacitance has been removed. Although originally designed for use with semiconductor detectors, the charge-sensitive amplifier has proven so superior that it is often used even when there is no need for its specific properties.

14.1.1 Resistive vs Optical Feedback

As we have seen, in the charge-sensitive preamplifier, the incoming charge is collected on a capacitor. This charge however, must eventually be removed. The simplest method is a slow discharge through a *resistance feedback* network. This produces the exponential *tail* pulse shown in Fig. 14.3a. The characteristic time constant varies from preamp to preamp but is generally quite long, on the order of $40 - 50\ \mu s$ or more.

For precision spectroscopy work, modern preamplifiers use *optical feedback* to enhance the bandwidth and reduce noise. In these designs, the feedback resistor is replaced by an optical coupling. The charging capacitor does not leak continually as in resistive feedback networks. Instead, the charge is kept until a certain fixed limit (usually a few Volts) is reached, at which point an internally generated current pulse of opposite sign is triggered and the capacitor discharged. During this process, a large negative pulse is generated in the amplification chain. To prevent analysis of this discharge pulse, an auxiliary *inhibit* signal is generated for use in blocking these pulses in the following electronics.

Figure 14.4 schematically diagrams the output of a preamp with optical feedback. Since there is no leakage, the pulses are flat-topped rather than exponential. When the fixed limit is reached, a large negative discharge pulse is seen after which the preamp capacitor begins recharging once again.

(a)   (b)

Fig. 14.3. (a) Exponential *tail* pulse from a preamplifier, (b) *pulse pileup*: a second pulse rides on the tail of the first

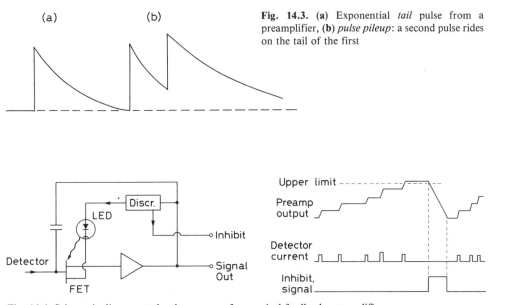

Fig. 14.4. Schematic diagram and pulse putput of an optical feedback preamplifier

14.2 Main Amplifiers

The amplifier serves two main purposes:

1) amplify the signal from the preamplifier, and
2) shape it to a convenient form for further processing.

In both cases, the amplifier must always preserve the information of interest. If timing information is required, a fast response is necessary. If pulse height information is desired, a strict proportionality between input and output amplitudes must be preserved (linear amplifier). In many of the latter, an adjustable gain over a wide range is provided so as to allow a scale adjustment in a spectrum analyzer.

For spectroscopy amplifiers one of the most important factors is the pulse shaping characteristic. In general, the pulse coming from the preamplifier can be characterized as an exponential with a long tail lasting anywhere from $\tau \simeq$ few $\mu s - 100 \, \mu s$. The amplitude of this pulse is proportional to energy. If a second signal should now arrive within the period, τ, it will ride on the tail of the first and its amplitude will be increased as shown in Fig. 14.3b. The energy information contained in this second pulse is thus distorted. This is known as *pile-up*. To avoid this effect, one must either restrict the counting rate to less than $1/\tau$ counts/s or shorten the tail by reshaping. This latter method is, of course, the more preferable option.

A second reason for pulse-shaping is the optimization of the signal to noise ratio. For a given noise spectrum, there usually exists an optimum pulse shape in which the signal is least disturbed by noise. [1] Tail pulses in the presence of typical noise spectra are not ideal, in fact and it would be more advantageous to have a Gaussian or triangular form. Thus, even at low counting rates where pile-up is not a problem, pulse-shaping remains important.

In contrast to the spectroscopy amplifier, the most important factor for fast amplifiers is preservation of the fast rise time of the signal which means maintaining a wide bandwidth. For this reason, these amplifiers generally perform very little or no pulse shaping and are limited to smaller gains ($\simeq \times 10$). Higher gains may be obtained by cascading several amplifiers, however, it is not generally recommended to go beyond gains of about $\times 1000$. Obviously, in applications where both good timing and pulse height information is desired, a conflict exists between the best *timing* shape and the best *signal-to-noise* shape. In such cases, a compromise must be made.

14.3 Pulse Shaping Networks in Amplifiers

Most pulse shaping networks in commercially available amplifiers today are based on two methods: *delay line* shaping and *RC differentiation-integration*. The basic elements making up these networks, i.e., the CR differentiator, the RC integrator and the

---

[1] The relation between pulse shape and noise can be best understood by looking at the signal and noise in terms of their Fourier components. Optimizing the signal to noise ratio, then, involves filtering out those frequencies where noise is at its greatest, i.e., a narrowing of the bandwidth. This narrowing, of course, also alters the distribution of frequencies in the signal resulting in a change of shape. The shape of the pulse, therefore, is an indication of the filtering scheme.

delay line, are described below in Sect. 14.23. The reader who is unfamiliar with these devices should first consult this section befores continuing.

In amplifiers, combinations of these simple circuits are generally used to limit the bandwidth and thereby improve the signal-to-noise ratio. This results, as we have noted, in a change of pulse shape. We briefly survey some of the most commonly found methods.

14.3.1 CR-RC Pulse Shaping

CR-RC pulse shaping is the most widespread technique and is performed by sending the signal through a cascaded CR differentiator and RC integrator. The pulse is thus filtered at low (differentiation) and high (integration) frequencies resulting in an improvement in signal-to-noise. A typical CR-RC circuit is shown in Fig. 14.5 along with the resultant shape for an incident step function pulse. Not that a unipolar pulse, i.e., one with only one polarity, is produced. In most cases, optimum signal-to-noise ratio is obtained with equal differentiation and integration time constants, however, the absolute value of this constant depends on the particular characteristics of the pulse. On many spectroscopy amplifiers, a choice of time constants is, in fact, offered by a front panel knob.

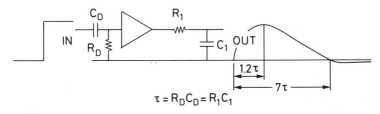

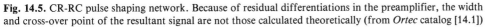

Fig. 14.5. CR-RC pulse shaping network. Because of residual differentiations in the preamplifier, the width and cross-over point of the resultant signal are not those calculated theoretically (from *Ortec* catalog [14.1])

14.3.2 Pole-Zero Cancellation and Baseline Restoration

One of the undesirable side-effects of RC pulse shaping is the presence of an undershoot in the shaped pulse. As shown in Fig. 14.6, this undershoot returns only very slowly to zero. This is obviously a problem for pulse height analysis for if a second pulse rides on this undershoot, an *amplitude defect* will arise with a subsequent distortion of the pulse height information in the second signal (Fig. 14.6).

The primary cause of CR-RC undershoot is the differentiation of finite length exponential or *tail* pulses from the preamplifier. Theoretically, no undershoot would occur if these tail pulses were infinitely long; however, in practice, real tail pulses are cutoff at some finite point. This undershoot, fortunately, may be corrected by a so-called *pole-zero* cancellation circuit. (This term arises from a Laplacian transform analysis of the circuit.) This is shown in Fig. 14.7 and involves adding a simple variable resistor in parallel with the capacitor in the CR stage. This circuit is adjusted by observing the signal on the oscilloscope and varying the resistance (usually by a front panel screw) until the undershoot disappears or is at a minimum. For spectroscopy work, a precise pole-zero adjustment is crucial. Note that with optical feedback preamplifiers, a pole-zero correction is *not* necessary since the output signals are not exponential in form.

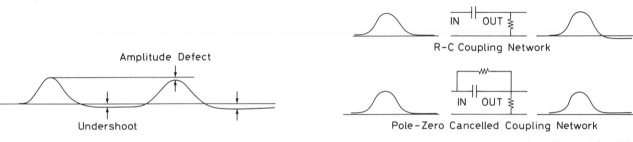

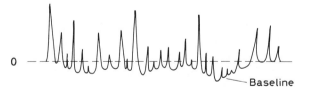

Fig. 14.6. Amplitude defect arising from undershoot in CR-RC pulse shaping

Fig. 14.7. Pole-zero cancellation circuit (from *Ortec* catalog [14.1])

Even with pole-zero cancellation, however, undershoot may still occur due to unwanted residual differentiations which can occur after the pole-zero correction. Indeed, one of the major culprits is the dc coupling capacitor present at the output of the amplifier. When coupled with the input impedance of the receiving module an effective CR differentiator is formed creating undershoot once again. To avoid this, the coupling capacitor is usually given a large value so that the time constant of the output capacitor is large compared to the signal. The differentiation effect is thus reduced along with the amplitude of the undershoot. In consequence, however, the duration of this smaller undershoot is lengthened. At low counting rates, where the pulses are widely separated in time, this procedure works well; however, as the count rate increases the pulses begin to fall on top of each other and the undershoot, though small, is multiplied several times. This is illustrated in Fig. 14.8. The net result of this is effective baseline fluctuations and amplitude distortion.

Fig. 14.8.
Baseline shift at high count rates

To remedy this situation, many amplifiers are equipped with *baseline restoring* circuits at the output stage. These circuits essentially shorten the long decay time of the undershoot by shorting the coupling capacitor to ground just after passage of the pulse. Separate baseline restoration units are also available for use with amplifiers not equipped with such corrective circuits.

14.3.3 Double Differentiation or CR-RC-CR Shaping

A solution to the problem of baseline shift at high count rates is to use bipolar pulses. These pulses can be formed by adding an additional CR stage to the CR-RC cascade to form a *double differentiating* network. This illustrated in Fig. 14.9.

Such a circuit forms a *bipolar* pulse with the each polarity lobe having approximately the same area. When passing through a coupling capacitor, such a pulse, then, leaves no residual charge, (as a unipolar pulse would), so that baseline shifts are avoided. At high count rates, in fact, bipolar pulses generally give better resolution than corrected unipolar pulses. However, at low rates, the unipolar pulse has the better signal-to-noise ratio.

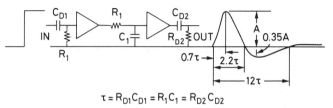

Fig. 14.9. Double differentiation pulse shaping network (from *Ortec* catalog [14.1])

$$\tau = R_{D1}C_{D1} = R_1C_1 = R_{D2}C_{D2}$$

Bipolar pulses are also useful in timing applications as the zero cross-over point provides a unique reference for triggering discriminators. This is explained in more detail in Chap. 17.

14.3.4 Semi-Gaussian Shaping

While a good signal-to-noise ratio is obtained with simple CR-RC shaping, a theoretical 18% improvement can be made if the pulse is given a Gaussian form. Shaping a pulse to an ideal Gaussian is not realizable electronically, however, an approximate semi-Gaussian shaping can be performed with a network consisting of a CR differentiation followed by many RC integrations, i.e., a CR-RC-RC-RC-... cascade. In general, four or five RC stages are sufficient to produce a form having close to the theoretical signal-to-noise improvement. The disadvantage of Gaussian pulses, infortunately, is that they are much wider than RC shaped pulses and thus run into overlap problems at high counting rates.

14.3.5 Delay Line Shaping

An alternative to RC shaping is the use of reflections from delay lines. This basic technique is outlined in Sect. 14.23.1. Delay line shaping has the advantage that the rise time of the pulse is left unaltered so that this method becomes ideal for fast amplifiers. The signal-to-noise ratio, however, is much poorer than that obtained in RC shaping so that delay-line shaping is used mainly to prevent signal overlap.

As in RC shaping, two delay-line stages may be cascaded to produce a bipolar pulse. This is illustrated in Fig. 14.10 which shows both single and double delay-line shaping.

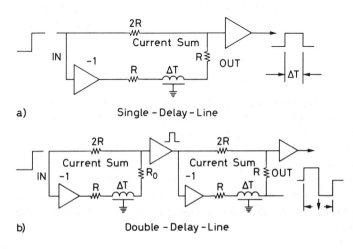

Fig. 14.10. (a) Single delay-line pulse shaping, (b) double delay line pulse shaping (from *Ortec* catalog [14.1])

14.4 Biased Amplifiers

The purpose of a *biased* amplifier is to expand a certain portion of a linear signal. This is useful in certain applications, particularly pulse height analysis with a multichannel analyzer, where it is sometimes desired to expand a certain portion of a spectrum for closer examination. To do this, the biased amplifier is equipped with a variable threshold which rejects all pulses below the set value and, in addition, subtracts this threshold value from all pulses greater than the threshold. Thus only the excess or remainder is accepted and this portion is then amplified to cover the full MCA input range.

Figure 14.11 illustrates how this works. Suppose the full range of acceptance of an MCA is 0 to 10 V and we would like to expand that portion of spectrum lying between 5 and 7.5 V. To do this, we set the threshold of the biased amplifier at 5 V and the amplifier gain to 4×. Incoming pulses below 5 V are thus blocked out while those above 5 V are displaced so that they lie between 0 and 5 V after acceptance. The range of interest now lies between 0 and 2.5 V. Amplifying by a factor 4, this range is thus expanded to cover 0 to 10 V which is the full range of our ADC. The pulses above this range are rejected or recorded as overflow by the MCA.

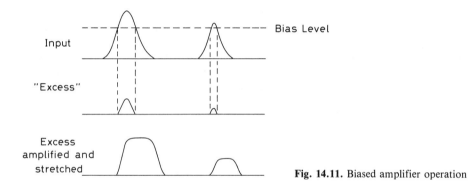

Fig. 14.11. Biased amplifier operation

For pulses close to the threshold, amplification of the excess sometimes results in a very narrowly peaked pulse. Since many MCA's have pulse rise time and width requirements, most biased amplifiers also have a pulse stretching stage included. In this way, consistently shaped pulses are produced at the output.

14.5 Pulse Stretchers

The pulse stretcher is a pulse shaping module which prolongs the duration of an analog signal at its peak value. This unit is used mainly to shape fast signals for acceptance by MCA's. In most units, the rise time and width are readjusted to a standard value with only the pulse height being preserved (Fig. 14.12).

14.6 Linear Transmission Gate

The *transmission* or *linear gate* is essentially a pulse signal switch which allows linear signals at its input to pass through only if there is a second coincident signal present at

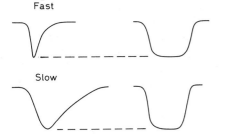

Fig. 14.12. Pulse stretcher function

the *gate* input. At all other times the input signal is blocked. The *gate* signal is usually a standard logic signal the width of which determines the time the gate stays open.

To illustrate some of the basic features of the linear gate, consider the simple circuit shown in Fig. 14.13. This is a unidirectional gate. The heart of this circuit is a diode which is kept in a non-conducting state by a bias which keeps point A at a lower potential than point B. Signals at the input with amplitudes lower than this bias are thus blocked from passing through. If now a gate signal of equal but opposite polarity to the bias is present along with the input signal, the signal is lifted up above the bias so that the diode becomes conducting and the signal is transmitted. If the gate signal is smaller than the bias only that portion above the zero level is transmitted so that an effective threshold exists. If the gate signal is larger than the bias, then the signal is transmitted along with the excess from the gate. This excess is called the gate *pedestal* as the signal appears to be transmitted on a pedestal (see Fig. 14.13). Depending on the application the pedestal may or may not be important. For this reason most linear gates provide a pedestal adjustment which allows the bias to be equalized to the amplitude of the gate signal.

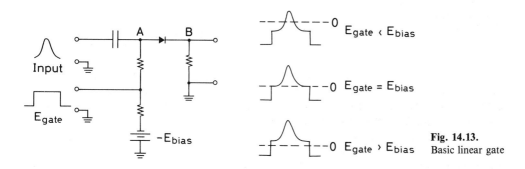

Fig. 14.13. Basic linear gate

Linear gates may be found as separate modules, but are very often provided on other modules, e.g. amplifiers, scalers, ADC's, etc., as an auxiliary feature. An amplifier or scaler, etc. in *gate* mode will therefore only function when there is a gate signal present. A selection of events, determined by whatever generates the gate signal, is thus achieved.

14.7 Fan-out and Fan-in

Fan-outs are active circuits which allow the distribution of one signal to several parts of an electronics system by dividing the input signal into several identical signals of the *same* height and shape. This should be distinguished from the passive pulse splitter discussed in Sect. 14.22.2 which divides both the signal and its amplitude.

The *fan-in*, on the other hand, accepts several input signals and delivers the algebraic sum at the output. These modules may be bipolar, i.e., accepting signals of both polarities, or of single polarity, i.e., accepting signals of one polarity only. Fan-ins are particularly useful for summing the outputs of several detectors or the signals from a large detector with many PM's.

Both fan-ins and fan-outs come in two varieties: linear and logic. The linear modules accept both analog and logic signals, whereas logic fan-outs and -ins are designed for logic signals only. In the case of a logic fan-in, the algebraic sum is replaced by a logical sum (i.e., OR).

14.8 Delay Lines

In making a coincidence measurement, it is important to assure that the electrical paths (and thus the times of propagation) along which two coincident signals travel to the coincidence unit are equal. Delay boxes provide adjustable delays which permit a lengthening or shortening of the electrical paths in a coincidence circuit. The boxes generally consist of variable lengths of cable which in a normal NIM module allow a $0 - 64$ ns delay. Several boxes may be cascaded to give delays of up to a few 100 ns. For longer delays of $\simeq 1\ \mu s$ or more, the length of cable becomes prohibitive and attenuation becomes a factor. In such cases, special low attenuation delay cables must be used or an active electronic circuit.

14.9 Discriminators

The *discriminator* is a device which responds only to input signals with a pulse height greater than a certain threshold value. If this criterion is satisfied, the discriminator responds by issuing a standard logic signal; if not, no response is made. The value of the threshold can usually be adjusted by a helipot or screw on the front panel. As well, an adjustment of the width of the logic signal is usually possible via similar controls.

The most common use of the discriminator is for blocking out low amplitude noise pulses from photomultipliers or other detectors. Good pulses, which should in principle be large enough to trigger the discriminator, are then transformed into logic pulses for further processing by the following electronics (see Fig. 14.14). In this role, the discriminator is essentially a simple analog-to-digital converter.

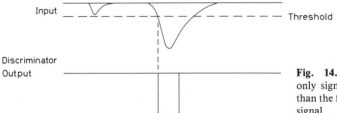

Fig. 14.14. Discriminator operation: only signals whose amplitude is greater than the fixed threshold trigger an output signal

An important aspect of the discriminator is the method of triggering. Because of its use in timing, it is important that the time relation between the arrival of the input pulse and the issuance of the output pulse be constant. In most discriminators, triggering occurs the moment the pulse crosses the threshold level. This is known as *leading edge* (LE) triggering. A more precise method is *constant fraction* (CF) triggering. These techniques are discussed in more detail in Chap. 17.

Two important parameters measuring the speed of the discriminator are the *double pulse* resolution and the *continuous pulse train* or *cw* rate. The double pulse resolution is the smallest time separation between two input pulses for which two separate output pulses will be produced. For fast discriminators, this is usually on the order of a few nanoseconds. The continuous pulse train rate is the highest frequency of equally spaced pulses which can be accepted by the discriminator. This rate can be as high as 200 MHz in fast discriminators.

14.9.1 Shapers

The shaper is a module which accepts pulses of varying width and height and reshapes these pulses into logic signals of standard levels and fixed width. To trigger the shaper, a minimum signal height is required and usually two or more inputs with differing threshold values are provided. Its function is thus identical to that of a discriminator.

The width of the shaper output signals is usually cable-timed, i.e., the width is determined by an external delay cable which the user chooses. In these models, the shaper is bistable so that it may also be operated as a flip-flop. The *set* signal is given at the normal input while the *reset* signal is taken from the *width-in* terminal.

14.10 Single-Channel Analyzer (Differential Discriminator)

The *single channel analyzer* (SCA) or *differential discriminator* (DD) is a device which sorts incoming analog signals according to their amplitudes. Like the discriminator, it contains a lower level threshold below which signals are blocked. In addition, however, there is also an upper level threshold above which signals are rejected. Thus only signals which fall between these two levels provoke a response from the SCA, i.e., a standard logic signal. This is illustrated in Fig. 14.15.

The opening between the upper and lower levels is usually called the *window*. With detectors where the output is proportional to energy, the SCA can be used to measure energy spectra by choosing a small, fixed window and systematically sweeping the window across the full pulse height range. The relative number of counts per unit time at each position can then be plotted to give a histogram of the spectrum.

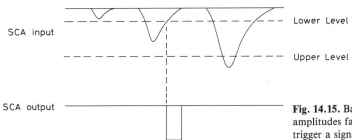

Fig. 14.15. Basic operation of a single channel analyzer (SCA): only signals whose amplitudes fall within the window defined by the upper and lower level threshold trigger a signal

The SCA generally has three working modes, although not all are always available on any given model.

1) Normal Mode or Differential Mode. In this mode, the upper (ULD) and lower (LLD) levels can be adjusted independently of each other, that is the settings of the LLD has no effect on the setting of the ULD and vice-versa. Thus if one wants to select signals with amplitudes between 1 and 2.5 V, for example, the lower level would be set at 1 V and the upper level at 2.5 V. Care should of course be taken to keep the upper level above the lower level.

2) Window Mode. Rather than setting the upper and lower levels separately, in this mode one sets the lower level and the window width, that is the distance between lower and upper levels. In the example above, one would again have the lower level at 1 V, but the *window* (often indicated as ΔE) at 1.5 V ($= 2.5 - 1.0$). This mode has a particular advantage in that the window width is preserved even if the lower level is moved. Thus with $\Delta E = 1.5$ V as above, we could change the LLD to, for example, 3 V, and the ULD would automatically be changed to 4.5 V ($= 3 + 1.5$). This mode is most suitable for spectrum analysis. One can define a certain resolution width, i.e., the window, and sweep this across the range of amplitudes, measuring the relative number of counts at each position without having to change two levels each time.

3) Integral Mode. Here, the upper level is completely removed from the SCA circuit altogether so that one simply has a discriminator with an adjustable lower level. The number of signals which pass is then just the integral of all the pulses from the threshold to the maximum limit of the SCA.

In its role as a pulse height analyzer, the stability and linearity of the SCA threshold becomes an important factor. The degree to which the threshold control and the actual threshold correspond to each other is referred to as *integral linearity*. A typical linearity curve and an ideal curve are shown in Fig. 14.16. In percent, it is defined as

$$L_i = 100 \frac{|\Delta V_{max}|}{V_{max}} \tag{14.1}$$

with L_i: integral linearity; ΔV_{max}: maximum deviation of real threshold from ideal threshold; V_{max}: maximum input voltage to the SCA.

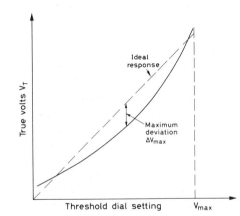

Fig. 14.16. Integral linearity of an SCA (after *Milam* [14.3])

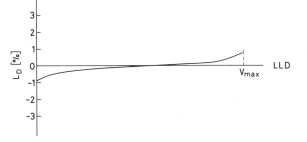

Fig. 14.17. Differential linearity of an SCA (after *Milam* [14.3])

Equally important is the *differential linearity* which gives a measure of the constancy of the window width as the LLD is changed. This is defined as

$$L_d = 100 \frac{\Delta V_w}{V_w} \tag{14.2}$$

with ΔV_w: max. change in window width as LLD is varied; V_w: average window width.

Figure 14.17 shows a plot of typical differential linearity. Clearly it is preferable to work in the middle of the SCA range than at the extremes.

An extension of the normal SCA is the *timing* SCA which also include circuits for correcting *walk* in the generation of the logic signal. The outputs from the SCA's are thus also suited for timing circuits. The various time-pickoff triggering techniques which can be used for this are described in Chap. 17.

14.11 Analog-to-Digital Converters (ADC or A/D)

The ADC is a device which converts the information contained in an analog signal to an equivalent digital form. This instrument is the fundamental link between analog and digital electronics. To give an example of its function, suppose an ADC accepts input pulses in the range of 0 and 10 V and is capable of outputting digital numbers from 0 to a maximum of 1000. (For simplicity, we will use decimal numbers in this example, although in most ADC's this will be expressed in binary form.) An input signal with an amplitude of 2.5 V will thus be converted to the digital number 250. Similarly, for a 150 mV pulse, we would find 15, and so on. The resolution of the ADC depends on the range of digitization. If numbers between 0 and 10000 were generated instead of 0 and 1000, a finer digitization and a higher resolution would be obtained.

ADC's may be of two types: *peak-sensing* or *charge sensitive*. In the former, the maximum of a voltage signal is digitized, as in our example above, while in the latter, it is the total integrated current. The latter is used with current-generating devices, e.g., a PM in current mode (usually fast detectors). Peak-sensing, on the other hand, is used with slower signals which have already been integrated, e.g., a PM in voltage mode. The time of integration or the time period over which the ADC seeks a maximum is usually determined by the width of a gate signal.

Electronically, many methods are in current use for analog to digital conversion. One of the simplest and oldest techniques, often used for spectroscopy ADC's, is the *ramp* or *Wilkinson* method. In this technique, the input signal is first used to charge a

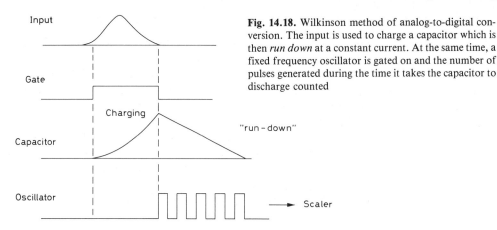

Input

Gate

Charging

Capacitor

"run – down"

Oscillator

Scaler

Fig. 14.18. Wilkinson method of analog-to-digital conversion. The input is used to charge a capacitor which is then *run down* at a constant current. At the same time, a fixed frequency oscillator is gated on and the number of pulses generated during the time it takes the capacitor to discharge counted

capacitor. This capacitor is then "*run down*", i.e. discharged at a constant rate. At the start of the discharge, a scaler counting the pulses from a constant frequency clock or oscillator is gated on. When the capacitor has completely discharged, the scaler is gated off. The contents of the scaler is then a number proportional to the charge on the capacitor. This method is illustrated by the block diagram in Fig. 14.18.

The most widely used technique is the *successive approximation* method. Here, the incoming pulse is compared to a series of reference voltages to determine the height of the pulse. For example, suppose the ADC accepts 0 to 10 V input pulses and a signal of 8 V arrives. The ADC first compares this pulse to a reference of 5 V. If the signal is greater than this value, which is true in our example, the first bit of the digital number is set to *1*. Then one-half the previous reference is added to make a new reference of 7.5 V and a comparison again made. Since the signal is greater, the second bit is also set to *1*. One-half is again added on to make 8.75 V. This time the signal is less than this value, so that the third bit is set to *0*. Now half is *subtracted* from the reference and a comparison is made. This continues until the required number of bits is obtained.

The time required for the digitization process is called the conversion time which, for most ADC's, is on the order of several tens of μseconds or more depending on the number of digits desired. Thus, they are slow devices when compared to other NIM modules. In the Wilkinson ADC, the conversion time is determined by the frequency of the clock. Obviously, the faster the clock, the faster the ramp can be run down (or the more digits can be converted). Currently, frequencies of 50 MHz to 200 MHz are used in this type of ADC. The successive approximation method is generally faster than the ramp type; however, the Wilkinson method is more linear and thus preferred for spectroscopy work.

In addition to these two types, there are *hybrid* ADC's which combine the Wilkinson and successive approximation types to obtain a compromise between speed and accuracy. The advent of modern integrated circuits has also made possible a number of very high-speed ADC's with conversion times less than 1 μs. Among these are the *flash* or *parallel* ADC which determines each bit of the digital number simultaneously rather than sequentially. This is made possible by sending the signal to a bank of voltage comparators connected in parallel with each comparator set at a different threshold (one for each bit). However, the larger the number of bits the larger the bank required. Other high-speed types are the *tracking* ADC, *parallel ripple* ADC and the *variable threshold flash* ADC. A review of these types is given in the article by *Henry* [14.2].

14.11.1 ADC Linearity

The linearity, or nonlinearity, of the ADC is defined in a manner similar to the SCA. *Integral* nonlinearity is the deviation from the ideal linear correspondence between pulse height and channel number. In most modern ADC's, this is less than 0.1% (i.e., 1 channel in 1000). The ADC *differential* nonlinearity, on the other hand, measures the nonconstancy in width of each channel. This parameter is of special importance for spectrum measurements where differing channel widths can distort the spectrum shape. Indeed, a perfectly flat input spectrum, for example, will take on a different shape since some channels will count more and others less. This distortion is not significant as long as the statistical uncertainty in each channel is greater than the differential nonlinearity. For example, if the differential linearity is 10%, the total counts per channel must exceed 100 before the effects of uneven channel width appear. For spectroscopy work, therefore, the ADC differential linearity is usually 1% or less which allows a maximum of 10^4 or more counts per channel.

Although they are given as two separate parameters, the integral and differential nonlinearities are related quantities since the integral nonlinearity is just the algebraic sum of the deviations of each channel from the ideal value. Because the deviations may be of opposite sign, however, it is possible for an ADC to have a good integral linearity, but a poor differential linearity. The differential linearity, therefore, is the more crucial of the two parameters.

14.12 Multichannel Analyzers

Multichannel analyzers (MCA) are sophisticated devices which sort out incoming pulses according to pulse height and keep count of the number at each height in a multichannel memory. The contents of each channel can then be displayed on a screen or printed out to give a pulse height spectrum.

The MCA works by digitizing the amplitude of the incoming pulse with an analog-to-digital converter (ADC). The MCA then takes this number and increments a memory channel whose address is proportional to the digitized value. In this way incoming pulses are sorted out according to pulses height and the number at each pulse height stored in memory locations corresponding to these amplitudes. The total number of channels into which the voltage range is digitized is known as the *conversion gain*. This determines the resolution of the MCA. Conversion gains of 128 up to 8 K or 16 K are often found on commercial MCA's. A block diagram is shown in Fig. 14.19.

The heart of the MCA is, however, the ADC, and in order to give it sufficient time to digitize the input signals, there are usually very strong requirements on the rise times and widths of the incoming analog signals. To facilitate this, there exist a number of modules used mainly for adapting signals for MCA's. These include the biased amplifier and the pulse stretcher which are also described in this chapter.

In addition to the ADC and memory, the MCA is also equipped with a discriminator or SCA, a linear gate and, in many cases, other useful provisions such as live time meters, etc. Some more sophisticated models are also equipped with microprocessors which allow a manipulation of the data stored in memory.

In some of the larger models, the MCA may also be operated in *multichannel scaling* mode (MCS). In this function, the MCA no longer acts a pulse height selector, but as a multi-channel scaler with each channel acting as an independent scaler. At the start

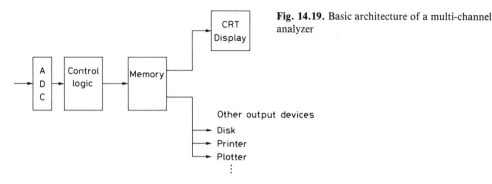

Fig. 14.19. Basic architecture of a multi-channel analyzer

of operation, the MCA counts the incident pulse signals (regardless of amplitude) for a certain *dwell time* and stores this number in the first channel. It then jumps to the next channel and counts for another dwell time period, after which it jumps to the next channel and so on. In MCS mode, therefore, the channels represent bins in time. This mode of operation is particularly useful for measuring the decay curves of radioactive isotopes, etc.

14.13 Digital-to-Analog Converters (DAC or D/A)

The opposite of the ADC is the *digital-to-analog converter* or DAC. Here, a digital signal is converted into an equivalent linear signal which can then be used to drive an analog device, e.g., a servo-mechanism. Like ADC's, there are several ways in which

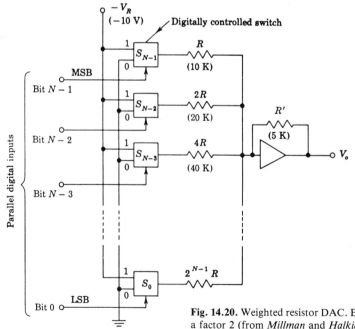

Fig. 14.20. Weighted resistor DAC. Each resistor value increases by a factor 2 (from *Millman* and *Halkias* [14.1])

this can be accomplished. One method is the binary weighted resistor technique shown in Fig. 14.20.

In this circuit, the binary word is passed through a network of parallel branches, one branch for each bit. At the beginning of the branches, there is an electronic switch which is connected to a constant reference voltage, V_r, if the state of the bit is 1, or grounded if the bit is 0. Each branch also contains a resistance weighted in proportion to the significance of the bit in the digital word, i.e., if the resistance of the most significant branch is R, then the next branch has a resistance $2R$, the following branch $4R$, etc. All branches are then summed by an amplifier to give an output voltage:

$$V_0 = (a_{n-1} 2^{-1} + a_{n-2} 2^{-2} + a_{n-3} 2^{-3} + \ldots + a_0 2^{-n}) V_r, \qquad (14.3)$$

where a_i is the state (1 or 0) of the ith bit and n is the number of bits. The *least significant bit* (LSB) is a_0 while the *most significant bit* (MSB) is a_{n-1}.

The disadvantage of the weighted resistor DAC is that its accuracy and stability depend on the precision of resistance values. For large numbers of digits, the absolute values of the later resistances can, in fact, become quite large. Moreover, the weighted resistor method is particularly susceptible to variations in temperature since it is difficult to find resistors of widely differing values with the same temperature characteristics.

An alternate method which overcomes many of the disadvantages of the weighted R technique is the $R - 2R$ ladder network. This network is essentially a current divider and is shown schematically in Fig. 14.21. Indeed, as can be verified by the reader, the resistance looking in any direction from any node is $2R$. If we take one binary bit, say a_{n-3} equal to 1 and all the rest 0, for example, then we see that it passes through 3 nodes and is divided in two each time. This results in the binary weighting given by (14.3). Unlike the weighted R method, the ladder method only requires resistances of R or $2R$. Moreover, the stability of the ladder depends only on the ratio of the resistances and not their absolute values. Variations are thus kept to a minimum and the resistance values remain reasonable.

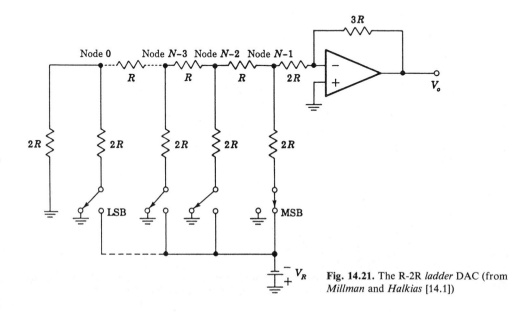

Fig. 14.21. The R-2R *ladder* DAC (from *Millman* and *Halkias* [14.1])

14.14 Time to Amplitude Converters (TAC or TPHC)

The TAC is a unit which converts a time period between two logic pulses into an output pulse whose height is proportional to this duration. This pulse may then be analyzed by a multichannel analyzer to give a spectrum as a function of the time interval. An ADC may also be placed after the TAC to digitize the output pulse. Units such as these are known as *time-to-digital converters* (TDC) and are also found commercially.

A time measurement by the TAC is triggered by a *START* pulse and halted by a *STOP* signal. One simple method used by TAC's is to begin a constant discharge of a capacitor at the arrival of a START signal and to cutoff this discharge when the STOP appears. The total charge collected is then proportional to the time difference between the START and STOP signals. This is illustrated in Fig. 14.22. More on TAC's can be found in Chap. 17.

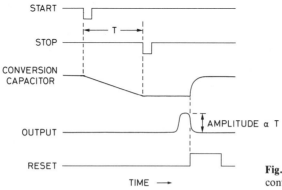

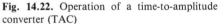

Fig. 14.22. Operation of a time-to-amplitude converter (TAC)

14.15 Scalers

The *scaler* is a unit which counts the number of pulses fed into its input and presents this information on a visual display. So-called *blind* scalers are those which do not have their own integrated display. Their contents may be read by a computer or fed into a separate display unit. In general, scalers require a properly shaped signal in order to function correctly; thus it is usually necessary to have a discriminator or a pulse shaper process signals from the detector before they can be counted by the scaler. Most commercial scalers are also available with a variety of auxiliary functions such as a gate, preset count, reset, etc.

14.16 Ratemeter

The ratemeter is a device which provides the instantaneous average number of events occurring at its input in a given unit of time (i.e. the frequency of events) and outputs this information in the form of a proportional dc voltage level. This output signal can then be used to drive a meter or a chart recorder or both. Like the scaler a standard logic pulse is usually required at the ratemeter input. A choice of integration times is usually provided as well as a selection of decay constants on the output signal. The latter quantity essentially adjusts the *reaction* time to instantaneous changes in the counting rate.

The ratemeter is a much used instrument appearing mainly in the form of radiation monitors!

14.17 Coincidence Units

The coincidence unit determines if two or more logic signals are coincident in time and generates a logic signal if *true* and no signal if *false.*

The electronic determination of a coincidence between two pulses may be made in a number of ways. One method is to use a transmission gate as we have seen (see Linear Gate). Another simple method often used is to sum the two input pulses and to pass the summed pulse through a discriminator set at a height just below the sum of two logic pulses. This method is shown in Fig. 14.23. Obviously, the sum pulse will only be great enough to trigger the discriminator when the input pulses are sufficiently close in time to overlap. The definition of *coincidence*, here, actually means coincident within a time such that the pulses overlap. This time period determines the *resolving time* of the coincidence and depends on the widths of the signals and the minimum overlap required by the electronics.

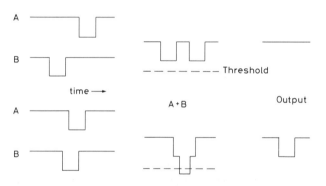

Fig. 14.23. The summing method for determining the coincidence of two signals. The pulses are first summed and then sent through a discriminator set at a level just below twice the logic signal amplitude

The coincidence unit is one example of a more general class of units known as the *logic gate.* These are units which perform the equivalent of Boolean logic operations on the input signals. The coincidence unit, for example, essentially, performs the logical "AND" operation. Other logic gates perform the "OR" operation, "NOT" and combinations of the above. These will be discussed in more detail in Chap. 16.

14.18 Majority Logic Units

A sophisticated and flexible version of the simple logic gates described above is the so-called *majority logic* unit. Such modules accept several input pulses and allow a selection of logical operations on the input. For example, in a majority logic unit which accepts four inputs: A, B, C, and D, one might expect to find the functions shown in Table 14.1.

Table 14.1. Majority logic functions for 4 inputs

Inputs connected	Function	
4	4-fold-AND	$A \cdot B \cdot C \cdot D$
	4-fold OR	$A + B + C + D$
	3-fold majority (any 3 of 4)	$A \cdot B \cdot C + B \cdot C \cdot D + A \cdot C \cdot D + B \cdot A \cdot D$
	2-fold majority (any 2 of 4)	$A \cdot B + A \cdot C + A \cdot D + B \cdot C + C \cdot D + B \cdot D$
3	3-fold AND	$A \cdot B \cdot C$
		$B \cdot C \cdot D$
		$C \cdot D \cdot A$
		$B \cdot A \cdot D$
	3-fold OR	$A + B + C$
		$B + C + D$
		$A + C + D$
		$A + B + D$
	2-fold majority (and 2 of 3)	$A \cdot B + A \cdot C + B \cdot C$
		$B \cdot C + B \cdot D + C \cdot D$
		$A \cdot C + A \cdot D + C \cdot D$
		$A \cdot B + A \cdot D + B \cdot D$
2	2-fold AND	$A \cdot B$
		$A \cdot C$
		$A \cdot D$
		$B \cdot C$
		$B \cdot D$
		$C \cdot D$
	2-fold OR	$A + B$
		$A + C$
		$A + D$
		$B + C$
		$B + D$
		$C + D$
1	Pulse shaper	A, B, C, D

14.19 Flip-Flops

The *flip-flop* is a two-input logic device which remains stable in either logic state until changed by an incoming pulse on the appropriate input. Such a device may be formed from two AND gates using feedback loops as shown in Fig. 14.24.

The two input signals are defined as *set* (S) and *reset* (R), while the output signals are Q and its inverse $\bar{Q}$. A *set* signal causes the Q output to go into logic state *1* (and $\bar{Q}$ into *0*), where it will stay until a *reset* signal arrives. The Q output then changes to logic state *0* where it will remain until another *set* arrives, etc. A shortcoming of this device is that the arrival of both a *set* and *reset* at the same time results in an undetermined state. This operation is undefined and it is impossible to predict what will happen. The most likely result is that no reaction whatsoever will occur. This type of flip-flop is known as an SR-flip-flop and is the simplest device of this genre. More sophisticated variants in-

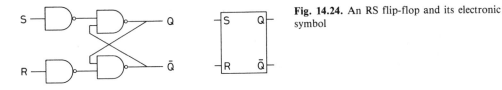

Fig. 14.24. An RS flip-flop and its electronic symbol

clude the JK-flip-flop, the clocked flip-flop, etc. A description of these may be found in any textbook on digital electronics.

The flip-flop is a basic element in digital electronics as it is the only device which essentially *memorizes* the input signal. It may thus be considered as a 1-bit *memory* or *latch*. By combining several flip-flops in parallel, a multi-bit memory for a digital word can then be formed. This is the basis for the register described below. Other uses of flip-flops are illustrated in the following chapters.

14.20 Registers (Latches)

Registers record the pattern of several input pulses (e.g., the signals from an array of counters or the bits of a binary word) and store this pattern in a buffer where it can be read by an external device such as a computer. Most registers are *coincidence* registers requiring the presence of a gate signal in order for the input signals to be recorded.

Registers may also be used to output logic signals. These are known as output registers and are most useful in computer-controlled systems. A bit-pattern written into the register buffer by the computer can then be output as logic signals to control the operation of other devices.

14.21 Gate and Delay Generators

Gate and delay generators or *timers* are triggerable devices which generate variable-width gate pulses ranging from a few nanoseconds to as long as a few seconds. The duration desired is usually chosen by a front panel helipot. Gate generators can be triggered by an input logic signal, and in some cases, manually via a front panel button. The gate pulse then may be used to activate a certain device, for example, a scaler for a chosen length of time. In this way, it serves as a *timer*. These modules are also equipped with an *end marker* signal which is a logic pulse issued at the end of the gate pulse. This feature can then be used as an active logic signal delay. Care, of course, must be taken to ensure the the accuracy of the signal widths and their stability.

14.22 Some Simple and Handy Circuits for Pulse Manipulation

While the NIM standard generally removes most problems of signal compatibility between modules, occasions still arise where some extra manipulation of the pulse signal must be made in order to satisfy some particular condition. Many times the problem can be solved by a simple circuit constructed from a few resistances and capacitors without need for a special module. We present here several simple circuits for pulse manipulations which can be easily constructed in this way.

14.22.1 Attenuators

Although relatively rare, occasions do arise where an attenuation of the signal is necessary. If signal amplitude is important, this should be done with a good linear attenuator. However, if only a simple reduction in amplitude is desired, an attenuator can be made from the circuits shown below.

For slow signals greater than say 100 ns rise time, the configuration in Fig. 14.25 may be used. The attenuation factor, α, is given by the ratio

$$\alpha = \frac{V_{\text{in}}}{V_{\text{out}}} = \frac{R_2}{R_1 + R_2} . \tag{14.4}$$

In order to meet the normal NIM impedance requirements for slow signals, the conditions

$$R_1 \gg Z_{\text{in}}$$

$$R_2 \ll Z_{\text{out}} ,$$

where Z_{in} and Z_{out} are the impedances of the input and output devices should also be imposed when choosing the resistances.

Because of high frequency attenuation, the circuit in Fig. 14.25 is not suitable for fast pulses. For these signals, the circuit in Fig. 14.26 is better suited. The values of the resistances are given by the equations:

$$R_1 = Z_L \frac{\alpha - 1}{\alpha + 1}$$

$$R_2 = Z_L \frac{2\alpha}{\alpha^2 - 1} ,$$

where α is the attenuation factor and Z_L is the load impedance as shown in Fig. 14.26.

14.22.2 Pulse Splitting

Pulse splitting involves a division of the input signal with a corresponding attenuation of the pulse amplitudes. This, as we have cautioned, should be distinguished from the

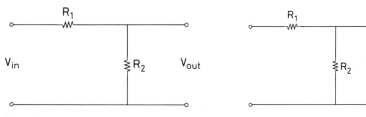

Fig. 14.25. Attenuator for slow signals (risetime > 100 ns)

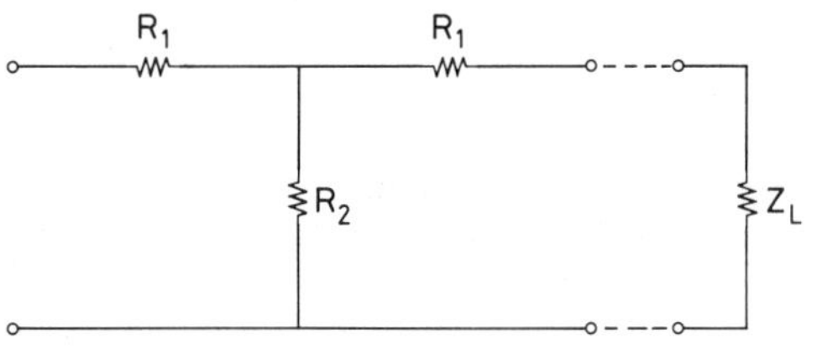

Fig. 14.26. Attenuator for fast pulses

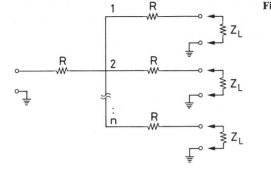

Fig. 14.27. Pulse splitter (or adder) for fast signals

fan-out which preserves the original pulse height. The pulse splitter may thus be used with analog signals (e.g., a PM signal), but *not* with logic signals which must maintain the correct levels.

For fast signals, which must satisfy impedance requirements, the circuit in Fig. 14.27 may be used. The circuit is symmetric so that the input signal is equally divided among the branches. Moreover, each branch sees the same impedance, as can be verified by the reader. If Z_L is the impedance to which the splitter must be matched, a simple calculation then shows that the resistance R should be

$$R = Z_L \frac{n-1}{n+1}, \tag{14.5}$$

where n is the number of branches.

In the case of slow pulses, splitting may be performed with a simple T-connector.

By inverting the input and outputs in Fig. 14.27, the opposite of the a splitter, a *pulse adder*, may be formed. Such a circuit sums the various inputs into one signal. In order for this to work, of course, the signals must be correctly timed to arrive at the same moment.

14.22.3 Pulse Inversion

Fast pulses may be transformed with the aid of simple transformer as shown in Fig. 14.28. Here the leads are simply reversed on the other end of the transformer.

For pulses slower than $\simeq 100$ ns, however, this method becomes nonlinear. The inversion of such slow pulses, then, requires an active circuit.

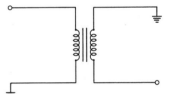

Fig. 14.28. A pulse inverter

14.23 Filtering and Shaping

14.23.1 Pulse Clipping

In high counting rate measurements, it is sometimes desired to shorten the duration of the pulse to avoid pile-up. A simple method for fast pulses is to use a well-timed cable reflection to destructively interfere with the pulse. Figure 14.29 illustrates this method. A well-chosen length of shorted cable is connected in parallel to the output. After splitting at the junction, one-half the pulse travels down this length of cable where it is reflected back in an inverted state. If T is the delay of this cable, this reflection rejoins

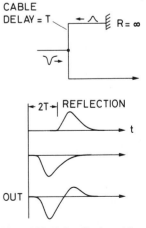

Fig. 14.29. Pulse clipping with reflection

the other half at a time $2T$ later, where it interferes with the pulse tail. This leaves a pulse of width $\simeq 2T$ as shown.

Clipping by reflections is a special case of a pulse shaping technique known as *delay line shaping* which is discussed in Sect. 14.3.5.

14.23.2 High-Pass Filter or CR Differentiating Circuit

A second method for pulse shortening is to use the CR high-pass filter shown in Fig. 14.30. This simple circuit, consisting of a single capacitor and resistor, acts as a low frequency filter, attenuating frequencies below

$$f \le \frac{1}{2\pi RC} .$$
(14.6)

Its effect on a step function pulse is illustrated in Fig. 14.30. As is seen, the flat part of the pulse (i.e., the top) is degraded and made to decay to the baseline, thereby shortening the pulse. In constrast, the fast rising part of the pulse, which depends on the higher frequencies, is not affected.

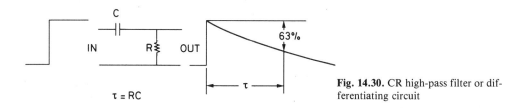

Fig. 14.30. CR high-pass filter or differentiating circuit

The CR high-pass filter is also commonly known as a CR *differentiator*. This latter name comes from the fact that it also performs the electrical analog of a mathematical differentiation, i.e., given a pulse form, the output will have the approximate form of its derivative. This can be verified mathematically. From Fig. 14.30, we have the equation

$$V_\text{in} = \frac{Q}{C} + V_\text{out} .$$
(14.7)

Differentiating both sides

$$\frac{dV_\text{in}}{dt} = \frac{1}{C} \frac{dQ}{dt} + \frac{dV_\text{out}}{dt} .$$
(14.8)

The term dQ/dt is just the current, i, so that

$$\frac{dQ}{dt} = i = \frac{V_\text{out}}{R} .$$
(14.9)

Equation (14.8) then becomes

$$\frac{dV_\text{in}}{dt} = \frac{1}{RC} V_\text{out} + \frac{dV_\text{out}}{dt} .$$
(14.10)

If RC is small such that the last term is negligible, then

$$\frac{dV_{in}}{dt} \simeq \frac{1}{\tau} V_{out}, \tag{14.11}$$

where $\tau = RC$ is known as the time constant of the circuit. If τ is large, however, then

$$\frac{dV_{in}}{dt} \simeq \frac{dV_{out}}{dt} \Rightarrow V_{in} \simeq V_{out}. \tag{14.12}$$

Thus, the circuit will only differentiate the input pulse if the time constant is small compared to the width of the pulse; otherwise, the pulse passes without a large change.

The CR differentiator is a basic element in many pulse shaping networks in amplifiers, although it is generally used in combination with other elements such as the RC integrator discussed below. This is elaborated upon in Sect. 14.3. For the nuclear physicist, the simple differentiator comes in handy for deriving short NIM pulses from long rectangular signals, for example, gates. This derived pulse may then be used to trigger some other part of the electronics. Figure 14.31 illustrates the effect of differentiation on a rectangular pulse. Two pulses are obtained, one at the leading edge and another inverted pulse, on the trailing edge. Under the NIM standard, this latter pulse will be ignored because of its polarity.

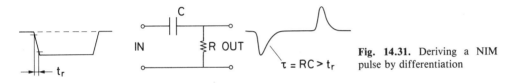

Fig. **14.31.** Deriving a NIM pulse by differentiation

When performing such a derivation, special care must be taken to ensure that the differentiated pulse attains the correct NIM level. This is done by choosing the time constant smaller than the width, as required, but larger than the rise time. In this way, the rising part of the pulse passes with no change and the differentiated pulse retains the original NIM pulse height. If the time constant is smaller than the rise time, however, the differentiated pulse begins to decay before reaching its full amplitude. This loss of amplitude is sometimes referred to as *ballistic deficit*.

14.23.3 RC Low-Pass Filter or Integrating Circuit

Complementing the CR low-frequency filter is the RC high-frequency filter shown in Fig. 14.32. Its effect on a step function is also shown. The lower cut-off frequency for this circuit is

$$f \geq \frac{1}{2\pi RC}. \tag{14.13}$$

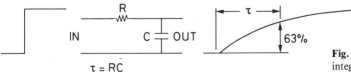

Fig. **14.32.** RC low-pass filter or integrating circuit

This circuit is also known as an RC *integrator* as it performs the electrical equivalent of an integration. This can be shown mathematically in the following manner. From Fig. 14.32, we have the equation,

$$V_{in} = iR + V_{out} .$$ (14.14)

Substituting,

$$i = \frac{dQ}{dt} = C\frac{dV_{out}}{dt}$$ (14.15)

we find

$$V_{in} = RC\frac{dV_{out}}{dt} + V_{out} .$$ (14.16)

If $\tau = RC$ is large, then

$$V_{in} \simeq \tau\frac{dV_{out}}{dt} \Rightarrow V_{out} \simeq \frac{1}{\tau}\int V_{in}\, dt .$$ (14.17)

If τ is small, however, we have

$$V_{in} \simeq V_{out} .$$ (14.18)

Thus in order for the integrator to work, the time constant must be large compared to the pulse. The principal application of the integrator is to smooth out fluctuations in a noisy signal as shown in Fig. 14.33. As a high-frequency filter, however, this integration also affects the fast rise time of the signal. Like the CR differentiator, the RC low-pass filter enjoys an equally important role in pulse shaping in amplifiers.

Fig. 14.33. Noise filtering by integration

15. Pulse Height Selection and Coincidence Technique

With this chapter, we begin our discussion of designing and setting up nuclear electronics systems for measurements in nuclear physics. This is a somewhat ambitious goal, as for a given application, there is no one particular system which must be used. Indeed, nowhere does the old adage of "more than one way of skinning a cat" apply more often than here. And if there are any hard and set rules, it is to use lots of imagination! There are nevertheless certain basic systems and methods which have proven their efficiency and efficacy over the years and which provide a foundation for more complex systems. Our method here will be to describe examples of these often used systems using them as illustrations of the how the various functions described in the previous chapter can be combined to give a particular result.

We will begin in this chapter with some simple analog systems and cover two particularly fundamental techniques: setting up and adjusting circuits for pulse height selection and coincidence determination. Once this is understood, the student can go on to the intricacies of electronic logic to which we devote a separate chapter. Before setting up these systems, however, we advise the student to familiarize himself with the use of the oscilloscope, if he has not already. A review of oscilloscope operations is given in Appendix A. A practice to be recommended, in fact, is to examine the input and output signals at each point in the system as it is being set-up. This not only leads to a better understanding of the system, but will also facilitate trouble-shooting later on.

15.1 A Simple Counting System

A basic measurement in nuclear or particle physics experiments is a simple counting of the number of signals from the detector. For example, measuring the activity of a source, or "*plateauing*" a counter. Let us examine a simple electronics set-up for this purpose. Figure 15.1 shows as schematic diagram of the processing units and connections necessary in this system.

In this set-up, the analog signal from the detector is shaped by a preamplifier and amplifier combination. The resulting signal is then sent through a discriminator which delivers a standard logic signal for every analog signal with an amplitude higher than the threshold. The logic signal is then sent to the scaler which counts each arriving

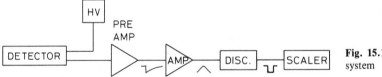

Fig. 15.1. A simple counting system

signal. One should also ensure here that the scaler is of the correct logic type, i.e., it accepts positive (TTL) or negative (NIM) signals, whichever may be the case. NIM to TTL converters may be used if there is an incompatibility.

For some scintillation counters, the direct output is sufficiently large such that amplification is not necessary. In such cases, the detector output may be connected directly to the discriminator.

The discriminator serves the dual purpose of excluding low level electronic noise and shaping the accepted signals to a form acceptable to the scalers. The threshold should be carefully adjusted so as to ensure that electronic noise is eliminated, but not so high as to also cut out good signals.

The scaler should be chosen according to the count rate desired. Evidently, the use of a scaler which is slower than the count rate of the experiment will result in lost counts and bad data! Thus, either take a faster scaler or reduce the count rate.

The system shown in Fig. 15.1 is operated manually, i.e., counting is started and stopped by hand. If two timers, one with a manual trigger, are available, and the scaler is provided with a gate and reset facility, the following circuit may be used for starting and stopping the scaler for fixed time intervals (Fig. 15.2).

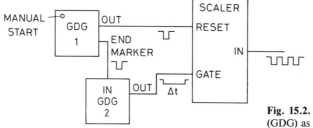

Fig. 15.2. Using two gate and delay generators (GDG) as a timer for a scaler

In this set-up, a short pulse from the first timer, triggered manually, is used to clear the scaler. The end marker then starts a gate pulse which opens the scaler for accumulation for a period equal to the gate width.

15.2 Pulse Height Selection

Often in may experiments, the reaction of interest or the products of the reaction are of a fixed energy or are limited to a small range of energies. For example, in the classic Rutherford scattering experiment, the energy of the probing α-particles is the same before and after scattering. Similarly, in experiments using positron annihilation, the energy of the two annihilation photons is fixed by kinematics. In such measurements, it would, of course, be advantageous to be able to pick out only those events with the correct energies and eliminate the rest. Interference from background reactions would then be reduced and a cleaner, more efficient measurement made. Assuming the use of an energy sensitive detector, such a *"filtering"* can be made by electronically selecting only those signals with the correct pulse height.

A simple system for such a purpose can be formed from Fig. 15.1, by replacing the discriminator with a single channel analyzer (SCA), as shown in Fig. 15.3. This module has adjustable upper (ULD) and lower (LLD) thresholds which define an acceptance *"window"*. Pulses with amplitudes falling into this window are accepted and converted

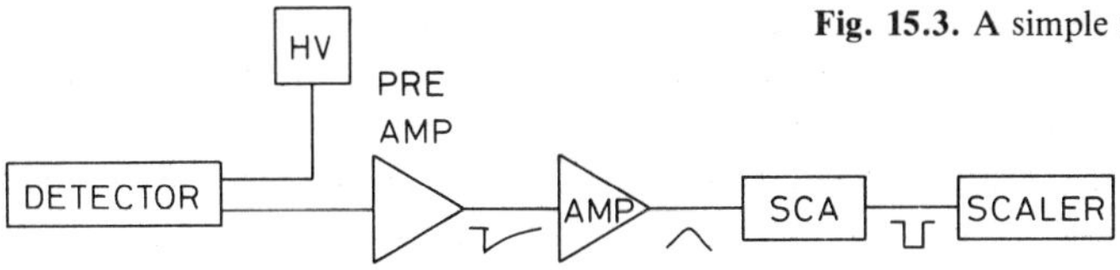

Fig. 15.3. A simple system for pulse height selection

to logic signals while everything else is rejected. The SCA thus acts as a pulse height filter.

In most commercial SCA's, the thresholds are calibrated in volts and are adjustable over the full input range of the module. In slow SCA's this range is usually 0 to 10 V, in accord with the NIM standard for analog signals. For fast SCA's, this is more commonly $0-5$ V or $0-1$ V. In either case, both thresholds should remain stable and linear with respect to the dial settings. While this is generally true in the central region, some fall-off may occur near the upper and lower extremes. For this reason, when setting a window, it is advisable to keep it near the center of the input range for the best results. This may require adjusting the gain of the amplifier or detector voltage so that the pulse heights of interest fall into this range.

15.2.1 SCA Calibration and Energy Spectrum Measurement

For energy selection, the thresholds must first be calibrated in terms of energy. In nuclear physics, this can be done by using the SCA to measure the energy spectra of sources with emissions of known energies.

To do this:

1) define a small energy window, $\Delta E = \text{ULD} - \text{LLD}$, for example, 0.1 or 0.05 V.

2) sweep this window across the full input voltage range in non-overlapping steps equal to the window width, and measure the number of counts per unit time at each position. For example, if $\Delta E = 0.10$ V, then starting at 0, measure at positions $0.0-0.10$, $0.10-0.20$, $0.20-0.30$ V, etc. This is a rather tedious operation and is more easily done if the SCA is equipped with a *window* mode of operation; otherwise, the ULD and LLD must be adjusted separately by hand.[1]

3) Plot the results versus LLD setting. A spectrum is thus obtained which should reveal peaks characteristic of the calibration source. A correspondence between LLD and energy can then be made. As an example, Fig. 15.4 shows the spectra from ^{22}Na and ^{60}Co sources taken with a NaI detector using the electronics set-up in Fig. 15.3. A window of 0.05 V was chosen and measurements were made for 10 s at each setting. (As such, the entire measurement took $\simeq 1$ h!!)

4) Identify the peaks with the aid of the isotope tables. A calibration curve can be drawn by taking the best fitting straight line through the points. This is shown in Fig. 15.4. The energy scale is indicated on the right-hand side of the graph.

One should note the linearity of the calibration. While a minimum of only two points is required to perform the calibration, it is generally a good idea to take more than this number in order to check for any nonlinearities or instabilities in time. Note

---

[1] When changing the threshold values, be careful of any play in the dial. Turning backwards to a previous position may not be the same as turning forwards to the same position.

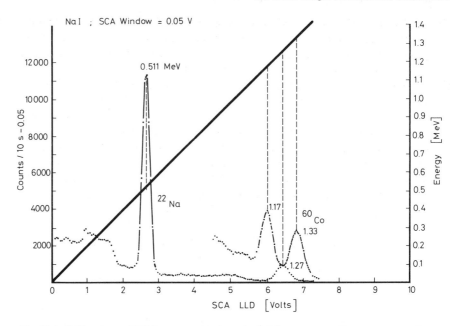

Fig. 15.4. Calibrating an SCA by measuring the pulse height spectrum of known sources. Spectra from ^{22}Na and ^{60}Co are shown along with a calibration line

also that the curve crosses through zero volts at zero energy. If this is not the case, there is most likely a dc offset being added to the signal somewhere in the electronics chain. This is generally not a serious problem as it simply shifts the spectrum to the right or left by a fixed amount.

This calibration is, of course, only valid for a fixed value of the amplifier gain and detector voltage. Changing any of these parameters changes the correspondence between energy and pulse height. In particular, if one follows the recommendation of centering the pulse heights of interest in the input range of the SCA by adjusting the amplifier gain, several calibrations may be necessary. In such a case, a preliminary calibration can first be made with a larger window, for example, and a final more detailed one made when the pulses are centered.

15.2.2 A Note on Calibration Sources

Before ending this section, a word about choosing calibration sources should be made. In general, sources of the same type of radiation as is to be measured in the experiment with energies as close as possible to the desired energy range should be chosen. This is to avoid any possible differences in detector response for different particle types and/or energies. For γ-rays, this is not usually a problem as many γ-ray or x-ray sources covering a relatively large energy range exist. For electrons, however, this becomes somewhat more difficult. The only natural sources of monoenergetic electrons are internal conversion sources and these are generally limited in number and in the range of energies they cover. Moreover, electron detectors are also usually very inefficient for γ-ray absorption by the photoelectric effect, so that the measurement of γ peaks may not be used either. To obtain a wider energy coverage, an alternate method is to measure the energy distribution of electrons coming from the Compton scattering of γ-rays in the detector. Since the maximum energy of this distribution (the Compton edge) is fixed kinematically and can be calculated from the known energy of the γ-ray (2.110), this point serves as a reference. In this way, the larger variety of γ-sources is also made available for calibration of electron detectors.

When using the Compton edge technique, a problem may arise in determining just where the Compton edge lies. Indeed, the finite resolution of the detector smears out the theoretically vertical edge into a sloping edge so that a sharp structure is no longer available. A common procedure is to choose the mid-point on this slope [15.1]. This appears to work for the Compton edges in Fig. 15.4 if we compare their mid-points to the calibration line. However, some investigators have found the "knee" or peak of the edge or a variable point dependent on energy [15.2] to be better. This can only be determined by trying the various schemes and seeing which gives the most consistent results.

15.3 Pulse Height Spectroscopy with Multichannel Analyzers

In the preceding section, we have described how a pulse height spectrum may be obtained with the use of an SCA. For those who have tried this technique, it is clear that while the method works, it is a tedious operation and not very convenient if many spectra are to be taken. For such work, the multichannel analyzer becomes indispensable. The basic functioning of this device is explained in Chap. 14. Here, we will only discuss some points about the use of the MCA.

Figure 15.5 shows a diagram of the basic set-up. Depending on the ADC and the detectors used, it may be necessary to use a pulse stretcher to shape the signals before entry into the MCA. This is usually the case with fast signals from an organic scintillator. In addition, a biased amplifier can be used if expanded portions of the spectrum are desired. These are optional however and are thus shown in outline in Fig. 15.5.

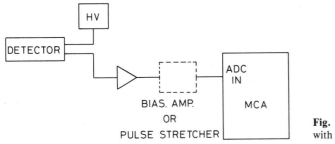

Fig. 15.5. Pulse height spectroscopy with a multichannel analyzer (MCA)

An important factor is the setting of the discriminator. This should not be too high, otherwise, valid portions of the spectrum will be cut out, nor so low as to flood the system with low energy noise. This latter case is a common error by first-time users, but shows up very quickly as an almost 100% dead time on the live-time meter (assuming the MCA is equipped with one!) or with a very high counting rate in the lowest channels.

Even with a correctly set discriminator, attention should be paid to the count rate. If the dead time is greater than $30-40\%$, then this rate is too high. This leads to pulse *pile-up* which results in broaden peaks and a deterioration in resolution. In such cases, the counting rate should be reduced by either increasing the distance between source

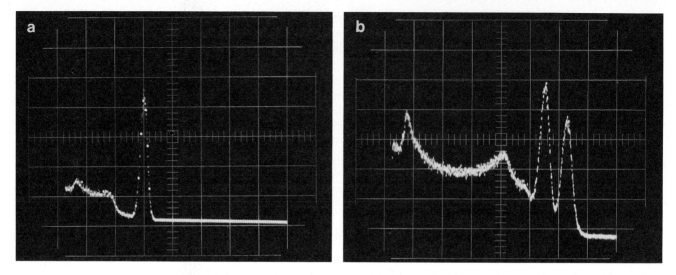

Fig. 15.6. Sample MCA pulse height spectra from a NaI detector: **(a)** ^{137}Cs, **(b)** ^{60}Co

and detector or using a weaker source. Scattering from surrounding materials should also be examined in the case of poor resolution.

A correct amplification of the input signal must also be found in order to have a correctly centered spectrum of sufficient resolution. Obviously, if the maximum signal amplitude is only 1 V and the MCA accepts from 0 to 10 V, the spectrum will be squeezed into the first 10% of the channels with a resolution ten times worse than can be obtained!

Many MCA's also allow a selection of the total number of channels (*conversion gain*) into which the spectrum is to be fitted. Choosing this number requires a little care. A spectrum with closely spaced peaks, for example, will never be resolved if too few channels are used. Using too many, on the other hand, results in a spectrum with large statistical fluctuations in each channel since the total number of counts is now divided among more channels. Small peaks or bumps may consequently be lost in these fluctuations. The remedy, here, it to count longer; however, time may not always be available.

For spectra showing sharp peaks, a general rule of thumb is to allocate 4 to 5 channels to the peaks in the region above FWHM (full width half maximum); see, for example, the spectrum in Fig. 15.6. Allowing more, of course, is always possible assuming time is not a factor. If the resolution of the detector is known beforehand and if the entire spectrum from 0 to maximum pulse height is desired, this criterion allows us to calculate a lower limit for the number of channels required. From the definition of resolution, we have

$$R \text{ (resolution)} = \frac{\text{FWHM}}{\text{energy of peak}} . \tag{15.1}$$

In terms of the MCA channels,

$$R = \frac{\text{FWHM}}{\text{position of peak}} = \frac{5}{\text{position of peak}} , \quad \text{position} = \frac{5}{R} . \tag{15.2}$$

A detector with a 10% resolution, therefore, requires at least

$$\text{position} \simeq \frac{5}{0.1} = 50 \text{ channels.} \tag{15.3}$$

For spectrum measurements in which only expanded portions are desired, this calculation is no longer valid.

Calibrating the MCA is similar to the SCA: the spectra from known sources can be measured and a calibration line drawn as in Fig. 15.4. The calibration line should be linear of the form

$$\text{pulse height} = a + b \cdot \text{channel}, \tag{15.4}$$

where a is the zero offset and b is related to the gain. If a line passing through zero is desired, this can be adjusted with the zero-offset control which shifts the spectra (and thus the calibration line) to the left or right without changing the spacing between the channels. If the slope of the line (i.e., the channel spacing) is to be changed, then the signal gain must be changed either by adjusting the amplifier gain or the detector voltage. For the beginning user, the difference between gain and offset and their relations to the channel spacing and positions should be made perfectly clear.

Although the ADC should be linear to much less than 0.1%, one must also include the nonlinearities of the detector, amplifier and any other instruments through which the analog signal passes before being analyzed. This can be calculated by adding the non-linearities quadratically, i.e.,

$$\delta^2(\text{tot}) = \delta^2(\text{MCA}) + \delta^2(\text{detector}) + \delta^2(\text{amp}) + \dots . \tag{15.5}$$

The resulting calibration line will thus be less linear than the MCA specification, but should still remain small. Nevertheless, if many channels are being used, say 1024, and the overall nonlinearity is 0.1%, there is still a 1 channel uncertainty. If a very precise calibration is desired in such cases, the calibration line may be fit with a quadratic form.

$$E = a \cdot (\text{channel})^2 + b \cdot \text{channel} + c . \tag{15.6}$$

Going beyond this, however, should not be necessary.

Example 15.1 *Mössbauer Spectrometry*

A good example of the use of the SCA and the MCA is Mössbauer spectrometry. Here, we are interested in measuring the transmission rate of a particular γ-ray (among

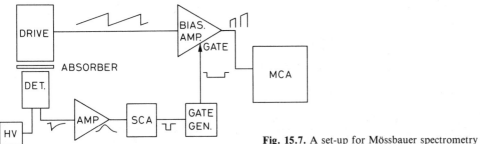

Fig. 15.7. A set-up for Mössbauer spectrometry

several emitted by the source) through an absorber as a function of the velocity of the source relative to the absorber. The source is mounted on a drive which delivers an electronic signal whose amplitude is proportional to velocity. For each transmitted γ-ray, then, we would like to measure this velocity and accumulate the total counts in a MCA. Figure 15.7 illustrates a typical set-up for this purpose.

The velocity signal is sent to the MCA via an amplifier in "*gate*" mode, i.e., a gate signal is required in order for the signal to pass through. At all other times, the signal is blocked. This gate is generated by a signal from an SCA which selects out the correct γ-ray. The spectrum obtained is then transmitted γ's versus source velocity. In place of the gated amplifier, a gated biased amplifier may also be used to allow the expansion of certain portions of the Mössbauer spectrum.

15.4 Basic Coincidence Technique

An extremely important technique in nuclear and particle physics is the electronic determination of coincidences. Like pulse height selection, coincidence in time between two or more events serves as a very powerful criterion for distinguishing reactions.

Figure 15.8 illustrates a simple coincidence measuring system. The basic technique is to convert the analog signal from the detectors into a logic signal and then to send these pulses to a coincidence module. The functioning of such a circuit is described in Sect. 14.17. If the two signals are, in fact, "*coincident*", then a logic signal is produced at the output.

The meaning of *coincident* here deserves some explanation. From the description of the circuit, it can be seen that a coincidence output signal is produced if any part of the

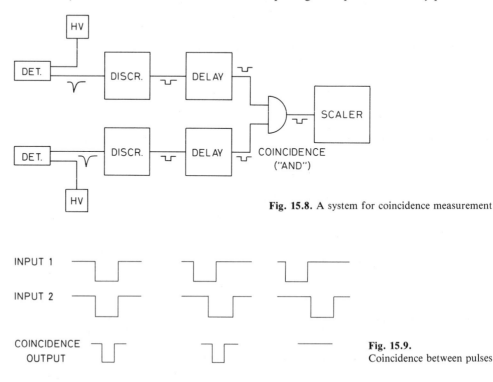

Fig. 15.8. A system for coincidence measurement

Fig. 15.9. Coincidence between pulses

two incoming signals overlap. (This is the ideal case, of course. In a real circuit, there is generally a minimum overlap necessary before it can be recognized as such. For the purposes of discussion, however, we will consider the coincidence circuit as ideal.) Thus all pulses arriving within a time equal to the sum of their widths are registered as coincident. Figure 15.9 shows some examples of coincident and non-coincident pulses.

In order for this set-up to work, however, it is necessary that the electrical path of each branch leading to the coincidence module be of equal length. This can be ensured by adding adjustable delays to each line, as shown in Fig. 15.8. In principle, only a delay is needed on the faster branch, however, one on each branch provides a bit more flexibility when adjusting the circuit.

15.4.1 Adjusting the Delays. The Coincidence Curve

To adjust the delays, a source of true coincident events is necessary. For γ-ray detectors a very common source is radiation from positron annihilation (e.g. ^{22}Na). Here, two photons of equal energy are emitted in opposite directions. If the detectors are placed face to face with the annihilation source between them, then these events may be used to adjust the delays.

If the detectors are inefficient for γ-ray detection but are thin, for example, plastic scintillators, coincidences may be obtained by placing the detectors close together and allowing a beam of charged particles (from a β source, for example) to pass through both counters. Be sure the particles have sufficient energy to penetrate through the counters, however! These two set-ups are illustrated in Fig. 15.10 and should be applicable most of the time. However, there will, of course, be situations in which neither of the above is suitable. In such cases, some imagination must be used!

With a source of coincident events, the relative values of the delays can be found by measuring the number of coincidences as a function of delay setting. Plotting the results on a graph gives what is then known as the *coincidence curve*. This is illustrated in Fig. 15.11.

Here we have plotted the delay setting of branch 1 on the positive x-axis and the delay of branch 2 on the negative x-axis since this represents a negative delay of branch 1 relative to branch 2. It should be noted that even with a zero delay setting on the box, a certain delay ($\simeq$ a fraction of a ns) is still present just from passing through the box. In our set-up with delay boxes on both branches, this is equalized. However, if only one

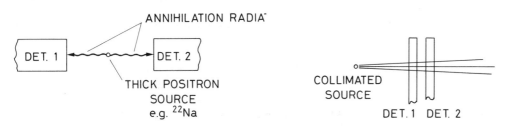

Fig. 15.10. Set-up for adjusting the coincidence between two detectors. For gamma ray detectors, the two photons from positron annihilation provide a useful source of coincident events. For charged particle detectors, a beam of electrons or other particles may be used provided the particle energy is high enough (and the counters thin enough) for the particles to pass through at least the first counter

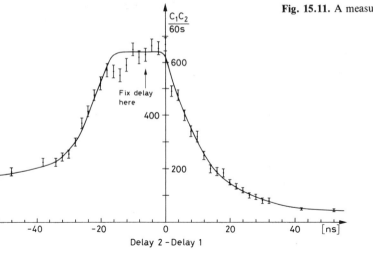

Fig. 15.11. A measured coincidence curve

box is used and the coincidence curve is made by placing the box on one branch and then the other, the *zero* points of each branch will *not* coincide because of this residual delay!

From this curve, it is clear that the correct setting for the delay is that in the middle of the plateau. Ideally, of course, this coincidence curve should be perfectly rectangular in shape as it, in principle, corresponds to the region of overlap between the two rectangular pulses. However, timing variations arising from the detectors and the electronics (as well as the less than perfect shape of the signals) cause a *jitter* in the time relation between the two signals which smears out the sides of the curve. If the width of the signals is smaller than these fluctuations true coincidences will, in fact, be lost. Measurements of the coincidence curve will therefore fluctuate giving the curve an aberrant shape. In such a case, the widths should be enlarged so as to encompass this *time jitter*. These effects are the limiting factors in a coincidence circuit and are discussed in more detail in Chap. 17.

The full width of this curve at half its maximum height (FWHM) is usually taken as the *resolving time* of the system. This time should generally be close to the sum of the two pulse widths. Obviously, the narrower this curve, the better is the capability of the circuit for distinguishing small time intervals and the less accidental coincidences. To increase this resolution, one might at first be tempted to decreases the widths of the pulses to the minimum. However, this is limited by the timing fluctuations which we have mentioned.

15.4.2 Adjusting Delays with the Oscilloscope

While not as precise as measuring a coincidence curve, a quick method of finding the approximate delay settings is to view the two pulses on a dual trace oscilloscope. By triggering on one pulse, the time relation of the second may be seen quite readily and appropriate delays can be added or subtracted to bring the pulses into an overlap condition. In doing this, of course, one should also ensure that the cable leads to the oscilloscope are both of the same length! Figure 15.12 illustrates this technique.

While this method is quite convenient, it requires a relatively intense source of coincidences, (especially when inefficient detectors are involved), in order for the second

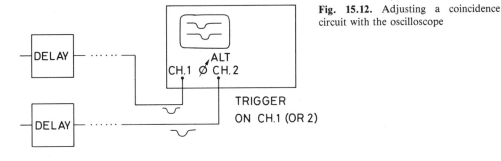

Fig. 15.12. Adjusting a coincidence circuit with the oscilloscope

signal to have sufficient intensity to be seen clearly on the oscilloscope screen. In the case of a weak signal, one must resort to measuring a coincidence curve.

15.4.3 Accidental Coincidences

In making a coincidence measurement, one must also consider the possibility of *accidental* coincidences occurring in the circuit. These may be due to uncorrelated background events in the detectors which happen to arrive within the resolving time of the circuit or to random noise which triggers the discriminators, etc. Clearly in any measurement, the number of such accidentals must be kept to a minimum.

The rate of accidentals in any coincidence setup may be estimated from the singles rate in each branch and the time resolution of the system. Suppose N_1 and N_2 are the singles counting rates for branches 1 and 2 respectively and σ is the resolution. Since any overlap in these pulses produces a coincidence, this means that the signals need only be within a period σ of each other in order to trigger the module. Assuming a constant singles rate, then, for each signal which arrives from branch 1, there will be $N_2\sigma$ pulses from branch 2 which fall into this allowable time period. Since there are N_1 pulses/unit time in branch 1, the total number of accidentals per unit time will be

$$\text{Accidentals} \simeq \sigma N_1 N_2 . \tag{15.7}$$

This rate may also be measured, and, in fact, corresponds to the baseline of the coincidence curve in Fig. 15.11. This automatically suggests a method for measuring the accidentals simultaneously with the true coincidences, i.e., form a second coincidence circuit with delays set so as to be completely off the coincidence curve. With current NIM modules which contain multiple outputs and multiple coincidence circuits within one physical module, such side measurements are facilitated.

15.5. Combining Pulse Height Selection and Coincidence Determination. The Fast-Slow Circuit

We have seen now how to set-up a system for pulse-height selection and a circuit for coincidence determination. The next obvious step is to consider combining the two. One way would be the set-up illustrated in Fig. 15.13.

The signals from the detectors are amplified and shaped, then sent to *timing SCA's* for pulse height testing. The logic pulses from these modules are corrected for *walk*

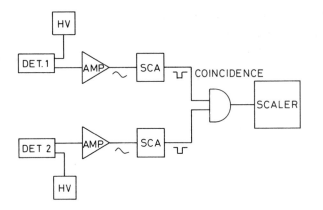

Fig. 15.13. A system with pulse height selection and slow coincidence

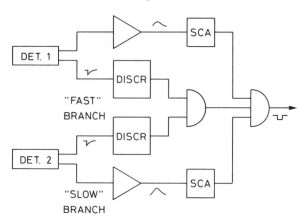

Fig. 15.14. A *fast-slow* coincidence system

effects (see Chap. 17) and may be tested for coincidence. Such a system generally gives good timing resolution and is adequate for most purposes.

It is clear, however, that in such a system the shaping of the pulse destroys some of the rise time information and does not represent the ideal situation for timing even with walk correction. Indeed, if the maximum in timing and pulse height resolution is desired, a so-called *fast-slow* system can be used. Such a system divides the signal into two branches, fast and slow, treats each separately and then combines the results. This system is schematically outlined in Fig. 15.14.

The *slow* branch is sent to a shaping amplifier and tested for pulse height in the usual manner. The *fast* branch, on the other hand, is passed directly (or through a fast amplifier) into a fast coincidence. This signal is then placed in a triple-fold coincidence with the slow branch. In this way, the cirteria of a fast coincidence is added to the slower coincidence of the SCA signals.

With scintillation counters, a method that is sometimes used is to take the slow signal from one of the later PM dynodes and the fast signal from the normal anode. The dynode signal is somewhat more linear since it is less saturated, while the anode signal has less time jitter because of its greater amplitude.

For fast scintillator signals, a recent development has been the construction of fast SCA's which allow both a fast timing signal and a good pulse height selection on unshaped signals. With such modules, the simpler set-up in Fig. 15.13 may then be used.

15.6 Pulse Shape Discrimination

In addition to pulse height information, the signals from some detectors also carry information in their shapes, (or more precisely in their rise and decay times). A prime example is the liquid scintillator NE213 which has the characteristic of emitting different pulse shapes in response to different types of particles. This is a result of the different ionizing powers of the particles which give rise to different excitation mechanisms in the scintillator and in consequence, different fluorescent decay times (see Sect. 7.2). A similar effect may be found in large ionization detectors. Particles of differing ionization power produce longer or shorter ionization trails in the detector volume which result in different charge collection times and hence different pulse shapes at the detector output. The technique of pulse shape discrimination (PSD) takes advantage of this property to allow a distinction between particle types when different incident radiations are present. One of the most common applications of PSD in in fast neutron detection where large gamma-ray backgrounds often accompany the neutrons. Using PSD, the gamma ray events may be selected out and suppressed leaving only the neutrons or vice versa.

Electronically, discriminating between different pulse shapes requires measuring the decay time of each pulse and this independently of the amplitude. In practice, this can be done efficiently only over a limited range of amplitudes (the *dynamic range*), which, for some applications such as spectroscopy, may imply certain restrictions. A second consideration is the speed of the PSD circuit. For high count rates, this becomes an important limiting factor. Not surprisingly therefore, a number of different circuits have been developed over the years. One of the most widely used methods is the *zero-crossing* system [15.3] shown in Fig. 15.15. This method offers a wide dynamic range and is relatively easy to implement; it is, however, limited to maximum count rates on the order of $20 - 30\,000$ counts/s.

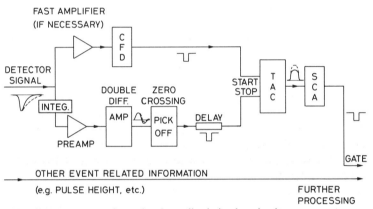

Fig. 15.15. A zero-crossing pulse shape discrimination circuit

In this system, the detector output signal is first split into two with one branch going to a fast time-pickoff discriminator (a constant fraction, for example), which triggers on the fast leading edge of the signal and starts a TAC or some other type of timing circuit. The second branch is integrated and sent through a preamplifier. This results in a pulse whose rise time depends on the decay time of the initial pulse (see, for example, Fig. 7.12). This signal is then doubly differentiated by a double delay amplifier which produces a bipolar pulse whose zero-crossing time depends on the input rise time. A zero cross-over pickoff module now triggers on this point and generates a STOP signal for the TAC. The time period measured by the TAC is thus proportional to the decay time of the detector pulse. Figure 15.16 shows a typical TAC spectrum from an NE213 detector in a field of neutrons and gammas. The two types of particles are clearly distinguished. To select a particular particle, an amplitude selection is performed on the TAC signal using an adjustable discriminator of SCA. The resulting logic signal may then be used to gate a recording instrument such as an MCA. Alternatively, the TAC pulse may be digitized and selected by computer, or stored on some sort of recording medium along with other event related information for analysis at a later time.

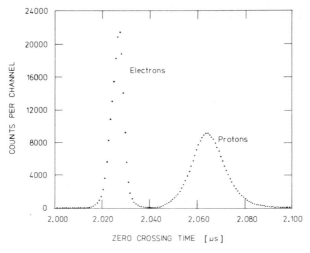

Fig. 15.16. A typical pulse shape timing spectrum from NE213 liquid scintillator in a field of neutrons and gamma rays (from *Perkins* and *Scott* [15.4])

When using pulse shape discrimination, there are a number of important points to consider. The first and most important is the efficacy of the discrimination. Quite obviously, one would like the PSD time spectrum (e.g., Fig. 15.6), to show narrow peaks associated with each particle type separated by as wide and deep a valley as possible. This allows a clean distinction between particle types to be made with a minimum of *breakthrough* i.e., events in the tails of the peaks which *"break"* through the separating cut to contaminate the events in the opposite region. A wide valley also makes the cut less sensitive to any drifts which might occur in the spectrum.

We can define the breakthrough more precisely as the fraction of unwanted events which are accidentally accepted as good ones. In general this should be less than 1%, although how much can be actually tolerated depends on the application. Even with a good 1% breakthrough fraction, however, one must also consider the relative amounts of each type of radiation. Indeed, if the undesired radiation type is 100 times more prevalent than the type being selected, for example, the total number of breakthrough events will be comparable to the total number of desired events − clearly an intolerable background!

Apart from the limits imposed by the intrinsic pulse shape discriminating properties of the detector itself, the quality of the timing spectrum depends very strongly on the time pick-off circuits used (see Chap. 17) and the dynamic range desired. Because of time *walk*, the timing spectrum will vary somewhat with the amplitude of the pulse. This effect is most marked for small pulses where time walk is the greatest. If a wide dynamic range is chosen, as in spectroscopy, it may be necessary to record events as a multi-dimensional spectrum, i.e., as a function of both amplitude and PSD time plus any other relevant parameters, and then make an amplitude dependent cut to select out the desired events. Another solution is to divide the amplitude range desired in two and make two separate measurements at two different gains. This latter solution, of course, requires normalizing the two measurements in some manner.

The total count rate on the detector is another factor which should be considered. At high rates, pile up effects, etc. can cause a worsening of time resolution and increase the breakthrough. With scintillation counters, PM gain shifts also occur.

16. Electronic Logic for Experiments

In the preceding chapter, we discussed a number of simple setups which treated the analog signals coming from the detectors. These signals were at some point converted into logic signals, either by a discriminator or an ADC, after which a simple analysis was performed, e.g. counting. In some of these systems, only events satisfying certain conditions were accepted. In the coincidence circuit, for example, this selection was made possible by the AND gate, which mathematically speaking, performed a Boolean logic operation upon the two detector signals. Then, according to the result, the event was accepted or rejected. This is a simple example of electronic logic and its use in selecting out certain events from many. Using combinations of logic gates, e.g., OR, AND, etc., it is evident that more complicated Boolean operations can be performed on the signals and more stringent selections made. In present day nuclear and particle physics experiments, this electronic logic can be extremely complex due to the large number of detectors and reactions which occur. And indeed many of these experiments cannot be performed otherwise!

Unfortunately, there are no simple prescriptions for constructing an electronic logic system and, in general, there is more than one (indeed many!) way of implementing it. To a certain extent, the configuration of a system depends on what electronics modules are available, cost, etc. Our purpose here is to present some simple examples and tricks which hopefully will guide the reader towards a better comprehension of such systems and help in his own formulation of other systems. Further examples are also given in the next chapter which treats the specific case of timing.

16.1 Basic Logic Gates: Symbols

Logic gates, as we have mentioned, electronically perform the equivalent of a Boolean operation on the input signals. The response of such units are defined, therefore, by truth tables. Figure 16.1 shows the truth tables for the common gates NOT, AND, OR and their negations along with their electronic symbols. For simplicity we have considered the case of just two input signals A and B, and the output of the gate C. Gates with more than two inputs are, of course, possible but it is an easy matter to extend the truth tables.

The NOT or negation is the simplest operation and is effectuated by simply inverting the state of the signal from 1 to 0 or vice-versa. Mathematically, the negation is symbolized by a *bar* over the signal, e.g., "NOT A" is written as $\bar{A}$. In an electronic logic diagram, the single negation operation is symbolized by an *op-amp* triangle with a small circle at its output. This circle signifies that the signal at that point is changed to the opposite logic state.

The NOT gate may be combined with any number of other gates to form new gates. Two often used combinations are the NAND (NOT AND) and NOR (NOT OR) which

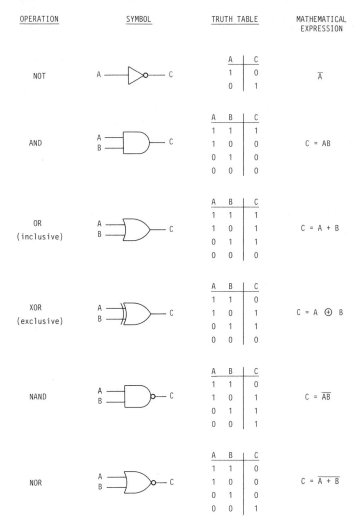

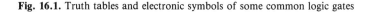

Fig. 16.1. Truth tables and electronic symbols of some common logic gates

are also shown. Note the negating circle at the output of the gates. Signals at the input of these gates may also be inverted. If one input of an AND is negated, for example, we then have an *anticoincidence* or *inhibit* gate as can be seen by working out the truth table.

Mathematically, the AND gate is similar to a multiplication, as can be seen from its truth table, so that it is often written as such. For example, A and B is written as $A \cdot B$ or simply AB. Similarly, the OR is equivalent to an addition and the operation A OR B is written as $A + B$. Note that two different OR gates are defined: the *inclusive* OR, which yields a positive response for A, B or both, and the *exclusive* OR in which a positive response is output only for A, B but not both. Their symbols are also indicated in Fig. 16.1. The usual OR is the inclusive case. The exclusive OR is expressed as an encircled plus as indicated in Fig. 16.1.

16.2 Boolean Laws and Identities

We shall review here some of the fundamental laws of Boolean logic and show how these can be directly translated into the more concrete terms of electronics. These laws are summarized in Table 16.1 for the three basic operations: AND, OR, and NOT. As we shall see, these three are sufficient, but not necessary, to build all other logic operations. In fact, only the NOT and *either* the AND or OR gate are required.

Let us see, for example, how we can construct an OR gate from the NOT and AND. The key is given by DeMorgan's Law:

$$\overline{AB} = \bar{A} + \bar{B} \ . \tag{16.1}$$

If we apply this to NOT-A and NOT-B, we find

$$\overline{\bar{A}\bar{B}} = A + B \ . \tag{16.2}$$

Electronically, this is performed by inverting the inputs of the AND gate to obtain $\bar{A}$ and $\bar{B}$ and the output to obtain $\overline{\bar{A}\bar{B}}$. This is shown in Fig. 16.2a.

Given the NOT and OR gates, we can also construct the AND in a similar fashion. For this, we use the other relation of DeMorgan,

$$\overline{A + B} = \bar{A}\bar{B} \ , \tag{16.3}$$

which, when applied to the inverses, yields the equivalence in Fig. 16.2b.

Table 16.1. Boolean identities and laws

Operations

OR	AND	NOT
$A + 0 = A$	$A0 = 0$	$A + \bar{A} = 1$
$A + 1 = 1$	$A1 = A$	$A\bar{A} = 0$
$A + A = A$	$AA = A$	$\bar{\bar{A}} = A$
$A + \bar{A} = 1$	$A\bar{A} = 0$	

Associative laws
$$(A + B) + C = A + (B + C)$$
$$(AB)C = A(BC)$$

Commutative laws
$$A + B = B + A$$
$$AB = BA$$

Distributive laws
$$A(B + C) = AB + AC$$

DeMorgan's laws
$$\overline{ABC\ldots} = \bar{A} + \bar{B} + \bar{C} + \ldots$$
$$\overline{A + B + C \ldots} = \bar{A}\bar{B}\bar{C}\ldots$$

Other identities
$$A + AB = A$$
$$A + \bar{A}B = A + B$$
$$(A + B)(A + C) = A + BC$$

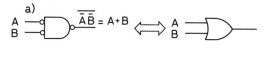

Fig. 16.2. (a) Constructing an OR from NOT and AND; (b) building an AND from NOT and OR

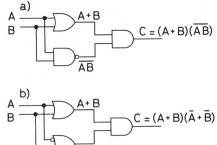

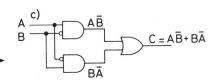

Fig. 16.3. Building an EXCLUSIVE OR from AND, ▶ OR and NOT gates

Let us continue in this vein and consider how an exclusive OR circuit may be constructed from the basic gates. For convenience we will henceforth consider both the AND and OR as given, even though one may be constructed from the other as we have seen above. Algebraically, an exclusive OR may be written as

$$C = (A+B)(\overline{AB}) \tag{16.4}$$

i.e., A or B but NOT (A and B). The reader can verify that this indeed satisfies the truth table in Fig. 16.1. We can implement this using two AND gates and one OR gate as shown in Fig. 16.3 a.

Using DeMorgan's Laws on $\overline{AB}$, we can also rewrite (16.4) as

$$C = (A+B)(\bar{A}+\bar{B}) \ . \tag{16.5}$$

This immediately suggests another implementation which is shown in Fig. 16.3 b.

Another way of formulating the exclusive OR is to notice that only the case $A \neq B$ is accepted, i.e., (A AND NOT B) or (B AND NOT A). Mathematically, this is written as

$$C = A\bar{B} + B\bar{A} \ , \tag{16.6}$$

which gives us yet another implementation (Fig. 16.3 c). Applying DeMorgan's Laws results in a fourth possibility which we leave to the reader to do as an exercise. The point here, however, is that there is generally more than one way to implement a logical function. While logically this makes no difference, on the practical level it allows one a choice in terms of hardware components.

We have seen now a few examples of how a logic operation on paper can be implemented in terms of electronic components. In the following sections, we will consider more specific examples relative to nuclear and particle physics experiments.

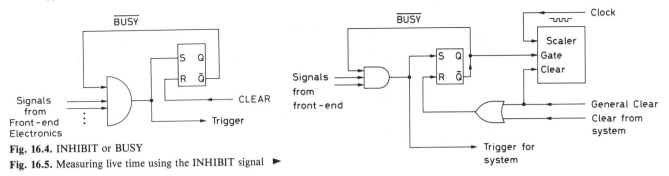

Fig. 16.4. INHIBIT or BUSY

Fig. 16.5. Measuring live time using the INHIBIT signal ▶

16.3 The Inhibit or Busy

In many systems, the arrival of an event triggers a series of processes, for example, a digitization by an ADC, followed by a writing of this and other information onto some recording medium, all of which requires a certain amount of time. During this period, it is necessary to block the system from any further events which might arrive and perturb the processing. A simple method is to use a BUSY or INHIBIT signal in conjunction with an AND gate. This is illustrated in Fig. 16.4. The arrival of the signal(s) passes through a coincidence (AND) gate which is kept open by the presence of the BUSY signal. After passing through, the signal resets a flip-flop which removes the BUSY signal. Any events arriving are then blocked by the coincidence requirement of the AND gate. When processing of the first event is finished, a CLEAR signal can be sent to the flip-flop which is set, thereby opening the gate once again.

Since the INHIBIT signal essentially controls the on/off state of the system, it also provides a means of measuring the live time (or dead time) of the system. As will be seen in Chap. 17, time can be measured by counting the total number of pulses from a fixed frequency generator (or *clock*) with a scaler. The total counts accumulated is then directly proportional to the time period elapsed. The live time, of course, is only that period during which the system is sensitive to arriving events. Therefore, we only want the scaler to count when the system is ready, i.e., when there is no INHIBIT. This suggests using the INHIBIT signal to gate the scaler (or conversely, the clock) on and off. This is illustrated in Fig. 16.5. Note the general CLEAR signal which resets the scaler as well as the flip-flop when starting data-taking. The CLEAR from the processing system, on the other hand, only resets the flip-flop leaving the contents of the scaler intact.

16.4 Triggers

Let us now consider the problem of triggers. In practically all physics experiments, one is generally faced with selecting out a particular reaction from background and/or other competing events which occur simultaneously. To do this, one must impose certain criteria which identify the reaction, e.g., coincidence among two or more detectors, number of outgoing particles, etc. The criteria, of course, depend on the detector set-up being used. Events satisfying these criteria, then activate other operations, e.g., recording instruments, etc., in the system. The electronic logic required for this selection is called the *trigger* in current nuclear and high-energy parlance. In many high-energy experiments, several different triggers are often present allowing an experiment to record more than one type of reaction at the same time. We will present a number of simple triggers for some simple detector setups to illustrate some of the general concepts.

16.4.1 One-Body Scattering

Consider the problem of 1-body scattering, i.e., reactions in which there is only one outgoing particle. A simple example is the elastic scattering of alpha particles from gold, the classic Rutherford experiment. A simple set-up for such an experiment is shown in Fig. 16.6. Here, a beam of alpha particles is incident upon a gold foil (of the appropriate thickness) and the scattered particle detected by a single detector located at an angle θ behind the foil. Since gold nuclei are much more massive compared to the alpha particle, the energy of the scattered particle should not be too different from its incident energy. This provides one criteria for elastic scattering. Using an SCA, we can set values for the upper and lower discriminators to select out this energy range and eliminate all other events. This then provides a rejection of background events.

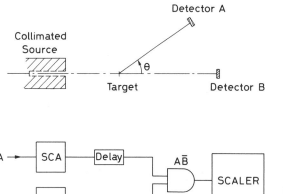

Fig. 16.6. Simple set-up for Rutherford scattering

Still, this criteria does not reject all the background. Indeed, particles scattered from other parts of the apparatus will also have close to the same energy and may also be accepted. A stricter requirement can be made by using a second detector located behind the target directly in the beam line. A true scattering therefore should give a signal in the first detector but not in the second. The logic for this trigger is now $A\bar{B}$, i.e., an anticoincidence. This circuit is shown in Fig. 16.6.

16.4.2 Two-Body Scattering

Consider now the case of two-body scattering in which we have two bodies which leave the target. Some examples are elastic electron scattering, elastic pp scattering, etc. The detection of such events immediately suggests the use of at least two detectors placed at appropriate angles with respect to the incident beam. And since they detect particles associated with the same event, we immediately think of placing the detectors in coincidence. The signature of a good event is thus AB. Figure 16.7 illustrates the beam and counter set-up.

We can make an even stricter selection by adding a third counter as we did in the case of Rutherford scattering. The trigger then would be $AB\bar{C}$. A more common proce-

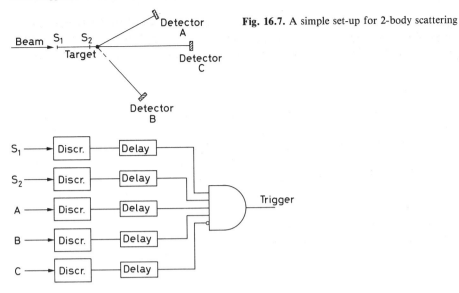

Fig. 16.7. A simple set-up for 2-body scattering

dure in scattering experiments is to add beam defining counters. This is done by placing two or more thin transmission counters of small surface area into the beam. Aligning them with respect to the target and requiring their coincidence in the trigger, the incoming particle trajectories can be better determined. If we call these counters S_1 and S_2, the trigger condition would then be $S_1 S_2 A B \bar{C}$. This circuit is illustrated in Fig. 16.7.

16.4.3 Measurement of the Muon Lifetime

Let us now turn to a more difficult experiment: the measurement of the muon lifetime. One easy source of muons are from cosmic rays. Muons are formed from the decay of pions which are created in the upper atmosphere by incident cosmic ray protons. Because the muons are more penetrating, they manage to reach the surface of the earth without being absorbed. They eventually stop in matter and decay via

$$\mu \rightarrow e + \nu + \bar{\nu}$$

with a lifetime of about 2 μs.

A set-up for measuring this decay time is shown in Fig. 16.8. The incoming muons are detected by the counters A and B which are thin enough to allow the muons to pass through. After B, the muons encounter a slab of lead in which some eventually stop

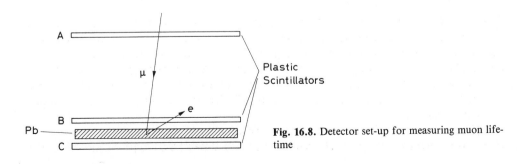

Fig. 16.8. Detector set-up for measuring muon lifetime

and decay. The emitted electron is then detected by the surrounding counters B or C. What we would like to measure therefore is the time interval between the signal for a stopped muon and the detection of the emitted electron.

What is the criteria for identifying a stopped muon? The arrival of a muon (or some other charged particle) is given by the coincidence AB, but this does not tell us if the muon stops. For this we must not have a signal from C. Thus, the trigger is $AB\bar{C}$. This can be used now to start a timing module such as a TAC (see Chap. 17 for a description of timing methods). To determine the STOP signal, we must consider the signature for an exiting electron. If the electron is emitted in the forward direction, then one signature is $C\bar{A}$. In the backward direction, however, we can have $B\bar{A}$ or BA, i.e., $(B\bar{A} + BA) = B$. The STOP signature is thus $C\bar{A}$ or B. The presence of B alone, however, is probably too loose a condition as events other than decay electrons, e.g., other incident muons, background events, noise in the B counter, etc., would also trigger a STOP signal. This would then lead to many false events which would increase the signal-to-noise ratio of the measurement. To be more selective, therefore, we will choose the STOP to be B or C but not A, i.e., $(C + B)\bar{A}$. This limits the solid angle of detection for the electron which decreases the efficiency but should give a better signal-to-noise ratio. Figure 16.9 shows how this logic is implemented. Although they are not shown in the diagram, delays in the circuit are most likely necessary in order to adjust the coincidences. Figure 16.9, of course, is not the only implementation of the STOP and we leave it to the reader to imagine others.

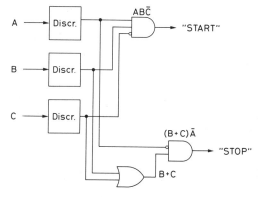

Fig. 16.9. Electronic logic for muon lifetime measurement

17. Timing Methods and Systems

Timing in nuclear and particle physics refers to the measurement of very small time intervals. Examples of its use include measurements of lifetimes of excited nuclear states or elementary particles, time-of-flight, etc. In this list also, we should include the determination of coincidences, which is essentially the determination of a zero time interval. The intervals we will be discussing, therefore, range from as little as a few pico-seconds to as large as a few micro-seconds.

Accurate measurements of very small time intervals require special techniques. In this chapter, we will review some of the basic problems encountered in timing, describe the current methods used for overcoming them, and illustrate some electronic techniques and systems for time-interval measurement.

17.1 Walk and Jitter

The most important factor in any timing system is its resolution, i.e., the smallest time interval that can be measured with accuracy. The resolution of a system can be measured in a number of ways. One method is to measure the time difference of two exactly coincident signals, i.e., the *coincidence curve* discussed in Chap. 15. As we saw, it is limited by fluctuations which occur in the time relation between the two signals. The major source of these variations occurs in the generation of the timing logic signal by the discriminator or SCA. Two principal effects arise: *walk* and *jitter*.

The *walk* effect is caused by variations in the amplitude and/or risetime of the incoming signals. Consider, for example, two signals of differing pulse height but exactly coincident in time as shown in Fig. 17.1. Suppose we introduce these signals into a discriminator with some fixed threshold. Because of the difference in amplitude, signal A will trigger the discriminator at time t_a and signal B at time t_b, although both are exactly coincident. This dependency on amplitude or rise-time causes the logic signal to *walk* about. Walk is a strong function of the triggering method used, in this case, *leading-edge* triggering. To minimize walk, a number of different triggering or *time-pickoff* methods have been developed. These are discussed in more detail below.

A second source of walk, although very much smaller in effect, is the finite amount of charge necessary to trigger the discriminator. In general, after reaching the discriminator threshold, a certain amount of charge must be integrated on a capacitor before a logic signal is emitted. This is represented by the shaded areas in Fig. 17.1. Because of the risetime and amplitude difference, this will also result in a walk effect.

Timing fluctuations are also caused by noise and statistical fluctuations in the original detector signal. Because of these random fluctuations, two identical signals will not always trigger at the same point, giving instead a time variation dependent on the amplitude of the fluctuations. This effect is usually referred to as time *jitter* and is illustrated in Fig. 17.2.

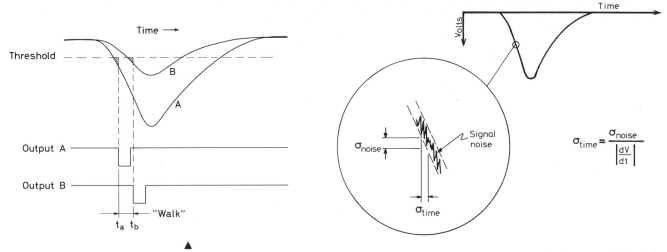

▲
Fig. 17.1. *Walk* in a discriminator or SCA. Coincident signals with different amplitudes cross the threshold at different times. An additional *walk* effect occurs because of the finite charge which must be integrated on a capacitor to trigger the discriminator or SCA

Fig. 17.2. Timing jitter. The timing error caused by jitter depends on the slope of the signal at the triggering point

In PM systems, for example, jitter is influenced by factors such as: (1) variations in the number of photons created in the scintillator, (2) the transit time of the photons and the electrons through the scintillator and PM, (3) gain variations in the electron multiplier, etc. Unlike walk, therefore, jitter arises from the intrinsic detection process itself. The effect of noise and statistical fluctuations on the timing resolution is easily calculated from Fig. 17.2. If σ_n is the variance in V due to noise and statistics, projecting this region onto the horizontal time scale yields the rms fluctuation in time,

$$\sigma_{time} = \frac{\sigma_n}{\left|\dfrac{dV}{dt}\right|} \tag{17.1}$$

Thus the rms timing resolution depends inversely on the slope of the signal − the faster the risetime, the better the timing jitter. In detectors with variable risetimes such as semiconductors or ionization instruments, this also suggests setting the triggering threshold at the point of greatest risetime to obtain the best results.

17.2 Time-Pickoff Methods

17.2.1 Leading Edge Triggering (LE)

The simplest method for deriving a timing logic signal is *leading-edge* (LE) triggering. This is the technique illustrated in Fig. 17.1 and, as we have seen, the logic signal is generated at the moment the analog pulse crosses the threshold. In SCA's this signal is generally delayed until the maximum of the pulse signal is first tested. An alternate method also used in some cases is to trigger on the falling edge.

This method is inherently subject to problems of walk, but can be used with good results if the amplitudes are restricted to a small range. With a 1 to 1.2 amplitude range, for example, resolutions as low as $\simeq 0.4$ ns can be obtained with a well-designed system of fast organic scintillators. At a 1 to 10 range, however, the walk effect balloons to as much as ± 10 ns.

Walk can also be minimized by using as low a threshold as permitted by the electronic noise. This is easily verified by lowering the threshold line in Fig. 17.1. In low noise systems, the leading edge may even be amplified, so as to get closer to the beginning of the signal. However, noise jitter is then increased as given by (17.1).

17.2.2 Fast Zero-Crossing Triggering

The *fast zero-crossing* technique was developed mainly to overcome the walk problem inherent in the LE method. Here, the pulse is first transformed into a bipolar pulse (through double delay line shaping, for example) and the trigger made on the zero-crossing point of the resulting bipolar pulse. This is illustrated in Fig. 17.3 which shows two pulses of differing amplitudes after being doubly differentiated. As can be seen, the two crossing points are precisely related in time and independent of the pulse amplitudes. As with LE timing, a maximum resolution of $\simeq 0.4$ ns is obtained if the amplitude range is restricted to $1:1.2$; however, at $1:10$, this maximum resolution only increases to $\simeq 0.6$ ns, an enormous improvement! Unfortunately, this method requires that the signals be of constant shape and rise-time. It is thus unsuited for signals from large volume semiconductor detectors or very large scintillators where such variations may occur.

17.2.3 Constant Fraction Triggering (CFT)

Probably the most efficient and versatile method available today is the constant fraction triggering technique. In this method, the logic signal is generated at a constant fraction of the peak height to produce an essentially walk-free timing signal. The basis for this idea arose from empirical tests which showed the existence of an optimum triggering level [17.1] for the best timing resolution. Depending on the type of signal, this level occurs at a certain fraction of the pulse height independent of the amplitude. Figure 17.4 shows how this works at a constant fraction of 50%.

The technique by which *CF* triggering is achieved is illustrated in Fig. 17.5. The incoming pulse V_a is first split into two with one part (V_d) delayed by a time τ_d equal to the time it takes for the pulse to rise from the constant fraction level to the pulse peak. The other part is inverted and attenuated by a factor k to give a pulse $V_c = -kV_a$. The two are then summed to produce a bipolar pulse, V_{out}. The point at which the signals cancel, i.e., the zero-crossing point, is then at a constant fraction k of the original signal height.

Unlike the zero cross-over technique, the CFT method does not require a bipolar pulse at the input, however, a constant rise-time is necessary. The efficiency of this technique is, nevertheless, very high yielding walk of as little as ± 20 ps over an amplitude range of 100 to 1.

17.2.4 Amplitude and Risetime Compensated Triggering (ARC)

As we have seen, the constant fraction method produces an essentially walk-free signal but requires that all pulses have the same rise time. This latter requirement can be re-

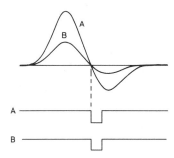

Fig. 17.3. Zero-crossing timing. Variations in the cross-over point are known as *zero-crossing walk*

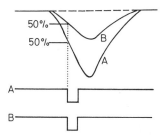

Fig. 17.4. Constant fraction discrimination

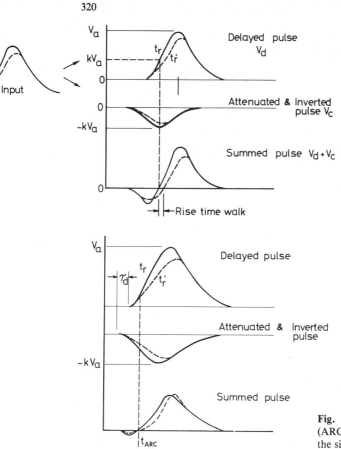

Fig. 17.5. Technique for constant fraction triggering. In order for this technique to work rise times of all signals must be the same. The dotted line shows the result with a different rise time signal

Fig. 17.6. Amplitude and risetime compensation (ARC) triggering. The zero-crossover occurs before the signal peak is reached

moved with a variant of CFT known as *amplitude and risetime compensation* (ARC) triggering. The difference is simply in the delay τ_d. In the true CFT method, τ_d must be long enough to allow the undelayed signal to reach its peak. In the ARC method, τ_d is made smaller than the rise time so that the summed signal crosses before the signal maximum is reached. The zero-crossing time thus depends only on the early portion of the signal where differences between pulse shapes are at a minimum. This is illustrated in Fig. 17.6. ARC triggering is the most precise method available today and is most useful with large volume semiconductor detectors where the pulses vary in shape as well as amplitude.

17.3 Analog Timing Methods

Assuming that the proper choice of time-pickoff method is chosen, let us now consider some electronic techniques for measuring the time difference between two signals. We divide these into analogic and digital techniques.

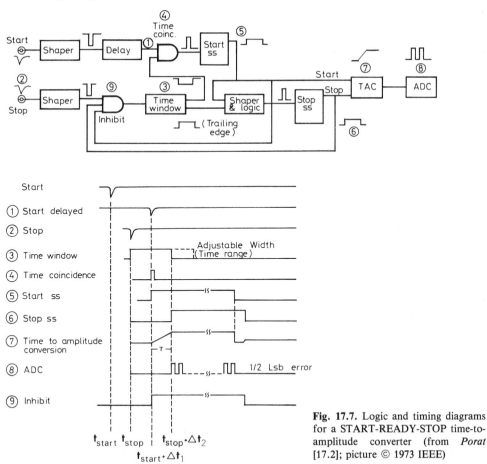

Fig. 17.7. Logic and timing diagrams for a START-READY-STOP time-to-amplitude converter (from *Porat* [17.2]; picture © 1973 IEEE)

17.3.1 The START-STOP Time-to-Amplitude Converter

The primary example of analogic time interval measurement is the START-STOP time-to-amplitude converter. The basic method is to relate the time interval between two events to the quantity of charge discharged by a capacitor during this period. As shown in Fig. 14.22, the arrival of the first signal (START) gates on the capacitor which discharges at a constant rate until the arrival of the STOP signal. The total charge collected thus forms an output signal whose height is proportional to the time difference between the START and STOP signals. The capacitor is then recharged and the next event awaited. If no STOP signal arrives, however, the output signal reaches its maximum amplitude which then causes an overflow condition. Such events, of course, cause a great deal of dead time. To remedy this, some START-STOP TAC's include additional logic circuits at the beginning to test for the presence of a STOP within the allowed time window before discharging of the capacitor begins. This technique, sometimes known as START-READY-STOP, is illustrated in Fig. 17.7.

17.3.2 Time Overlap TAC's

An alternate method reminiscent of coincidence circuits is the *time-overlap* technique. In this scheme, the overlap between two wide START and STOP pulses is measured

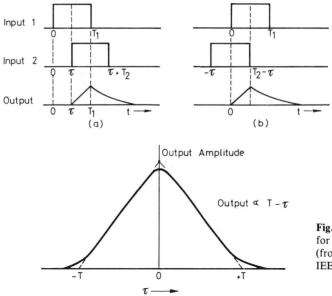

Fig. 17.8. The time overlap method for time-to-amplitude conversion (from *Porat* [17.2]; picture © 1973 IEEE)

and the difference taken. This is illustrated in Fig. 17.8. The capacitor is charged during the overlap period yielding a pulse whose height is proportional to

$$T - \tau \, , \tag{17.2}$$

where τ is the time interval to be measured and T is the full width of the pulse. Knowing T, the period τ then follows. This method, of course, is restricted only to periods smaller than the pulse with T. Intervals larger than this provoke an overflow signal or no signal at all. Note also that without further auxiliary logic, the method does not distinguish between which pulse arrives first.

17.4 Digital Timing Methods

17.4.1 The Time-to-Digital Converter (TDC)

To obtain a time interval measurement in digital form, one obvious method is to digitize the TAC using an ADC. However, more direct methods are available using counting techniques and stable oscillators. The basic principle here is to use the START signal to gate on a scaler which counts a constant frequency oscillator (or clock). At the arrival of a second STOP signal, this scaler is gated off to yield a number proportional to the time interval between the pulses.

Counting-type TDC's are easily constructed from simple logic modules found in the laboratory. Figure 17.9 illustrates one such system using AND and OR gates, simple SR flip-flops and timers. Here, the arrival of the START sets a flip-flop (Busy) which inhibits any further STARTs which may arrive while the TDC is in operation. At the same time, a gate signal is generated for the scaler, and a timer, set to a maximum time window, triggered. The scaler may be stopped in either of two ways: (1) by a *true* STOP which arrives within the defined time window, or, (2) by the end of this window. A *true*

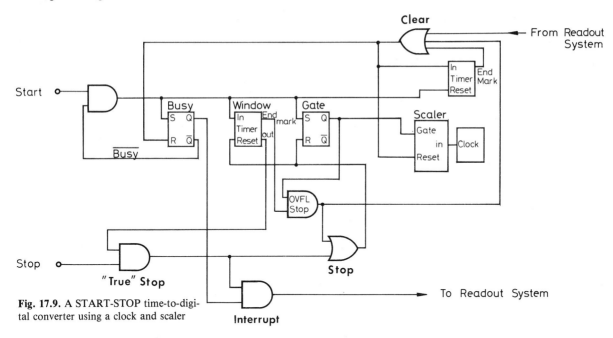

Fig. 17.9. A START-STOP time-to-digital converter using a clock and scaler

STOP is defined by the requirement that the timer still be running when the signal arrives. This is tested by the first AND gate in the STOP channel. If this condition is met, both the scaler and the window timer are gated off. An interrupt signal is then generated to trigger the readout system. After the value of the scaler is read and safely stored, the readout system generates a CLEAR signal which then resets the scaler and the BUSY flip-flop. The system then awaits the next event. If no STOP arrives within the time window, the end marker from the timer then automatically triggers a STOP and CLEAR without interrupting the readout system.

One must consider the possibility of events in which both a START and STOP signal arrive at the same time. Such an event is prohibited by the SR flip-flop logic and may cause the flip-flop to "*hang-up*" in an indeterminate state, blocking the system. This may be remedied by using a second "*watchdog*" timer which is triggered by each CLEAR, and reset by each START. The period of the timer should be set to a value large compared to the time between events. If no START arrives in such a long period, then the system is most likely blocked, and a CLEAR signal issued.

The resolution of the counting TDC depends on the frequency of the clock used: the higher the frequency, the smaller the time interval measured. For a given frequency however, this resolution can be doubled by using two clocks synchronized on opposite phases. This method is shown in Fig. 17.10.

17.4.2 The Vernier TDC

A second counting method is the *vernier* technique. The basic principle is illustrated in Fig. 17.11. Two oscillators of slightly different frequencies, f_1 and f_2, are used. The arrival of a START gates on the first clock while the second remains off. The moment the STOP arrives, the second clock is gated on and continues oscillating along with the first clock until the two are in phase. At this point, both are stopped. The contents of the two scalers are then related by

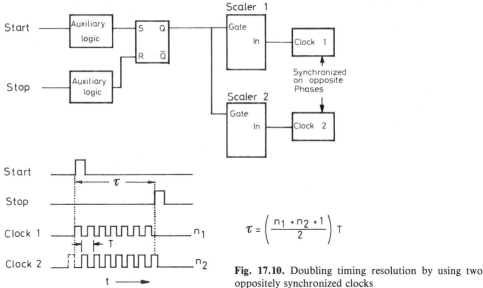

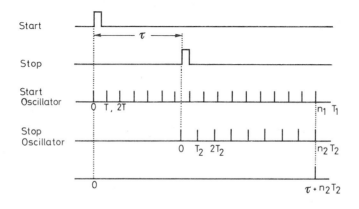

$$\tau = \left(\frac{n_1 + n_2 + 1}{2} \right) T$$

Fig. 17.10. Doubling timing resolution by using two oppositely synchronized clocks

$$n_1 T_1 = n_2 T_2 + \tau \; , \tag{17.3}$$

where n_1 and n_2 are the number of pulses counted, T_1 and T_2 are the periods of the two clocks and τ is the time interval to be measured. Solving for τ then is trivial, and

$$\tau = \frac{n_1}{f_1} - \frac{n_2}{f_2} \; . \tag{17.4}$$

For time intervals smaller than the period of the clocks, $n_1 = n_2 = n$, so that,

$$\tau = n \left(\frac{1}{f_1} - \frac{1}{f_2} \right) = n \frac{\Delta f}{f_1 f_2} \; . \tag{17.5}$$

From the above, we see immediately, then, that the resolution depends on the frequency difference Δf. If most vernier TDC's, this is usually 1% although this can be improved.

Fig. 17.11. Working principle of the vernier TDC

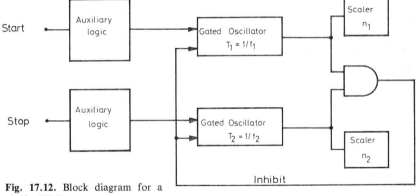

Fig. 17.12. Block diagram for a vernier TDC

Like the normal counting TDC discussed above, the vernier TDC may also be easily constructed from simple logic modules. One such set-up is shown in Fig. 17.12.

17.4.3 Calibrating the Timing System

Once a timing system has been chosen and constructed, it is necessary to calibrate the time scale and also to have a measure of system linearity and resolution. A simple method for calibrating the absolute channel width is to use a single source such as a fast pulse generator or a photomultiplier to drive both the START and STOP channels. The output signal is split in two with the STOP channel signal passing through a variable delay. Simple cable delays are the easiest and most accurate delay devices. The distance between peaks produced by the different delays then gives a calibration of the time scale. The width of each peak serves also as a measure of the time resolution. This is illustrated in Fig. 17.13.

In order to measure the linearity of the system, a source of random events uniformly distributed in time is necessary. While random pulse generators are commercially available, a simpler method is to use photomultiplier pulses produced by a radioactive source and a pulse generator with a frequency comparable to the counting rate. The clock pulses are applied to the START channel while the photomultiplier pulses are used for the STOP channel (see Fig. 17.14). The time intervals thus produced are uniformly distributed on the time scale and, within statistical errors, should produce the same number of counts in each channel. The degree of uniformity then provides a measure of the differential linearity of the system. From this, the integral linearity can be

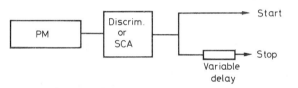

Fig. 17.13. Time scale calibration with a single source

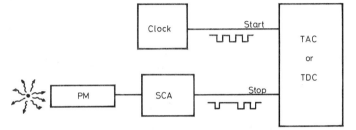

Fig. 17.14. Measuring differential and integral linearity using a clock and radioactive source

obtained by plotting the integral counts versus channel number. The accuracy of this method is generally very good and is only limited by the amplitude walk inherent in the triggering system used. An improvement can be made, however, by restricting accepted pulses to a small amplitude range. As described in Chap. 5, this method may also be used to determine the dead time of the system.

18. Computer Controlled Electronics: CAMAC

All (or almost all) experiments in nuclear and particle physics experiments today employ computer controlled data acquisition systems of some kind or another. Indeed, the quantity and rate as well as the complexity of the data which are generated in modern experiments make such systems mandatory. And in fact many experiments would be utterly impossible otherwise. The advantages of computer control are manifold. Besides simple data acquisition, systems can be made to monitor the apparatus, which may consist of hundreds or even thousands of detectors. Calibration procedures, for example, counter plateaux, timing curves, etc. may be performed automatically during system set-up. Online reconstruction and/or preliminary analyses of the raw data may also be made allowing the physicist to examine events as they arrive, etc.

Installing a computer controlled system, of course, requires interfacing the instruments of the computer. However, because of the variety of computer architectures, a different interface would have to be built for each instrument and each computer. This of course, brings us back to the problem of instrument compatibility first mentioned in Chap. 12. To alleviate this problem, a number of standardized systems have been developed. For nuclear and particle physics, the two standard systems are CAMAC and FASTBUS.

The CAMAC system was designed to complement the NIM system for nuclear electronics when it became clear that NIM was inconvenient for computer based systems. Introduced in 1969 by the European Standards on Nuclear Electronics (ESONE) Committee, CAMAC was designed simply as a standard system for the transmission of digital data, although its application to computer based systems was quite obviously in mind. It was quickly adopted by the U.S. NIM Committee and is now used worldwide in areas such as medical research, industry, in a addition to nuclear-particle physics.

Since the advent of CAMAC, however, the complexity and the volume of data generated in high-energy physics experiments have grown such that the capabilities of CAMAC are now also reaching their limits and it has become clear that future experiments will require an even faster system capable of handling even larger volumes of data. Consequently a new system, FASTBUS, first proposed in the late 1970's is now making its appearance on the commercial market. The system is much more complex and is still in the development stages, much as CAMAC was in the early 1970's.

On the opposite end of the spectrum, the explosion in microcomputer technology in recent years has also made possible the computerization of many smaller experiments for which systems such as CAMAC or FASTBUS are much too complicated and expensive to implement. For these experiments, a number of different data transfer systems are available. Among these are the GPIB bus (IEEE Std. 488), Multibus, VME, etc. The VME bus appears to be one of the more interesting offering speed and flexibility and has been used in conjunction with CAMAC and FASTBUS in some very large experiments.

As an introduction to computer control in nuclear and particle physics, we will consider the CAMAC system here in some detail. Much effort has been put into this standard over the years and the system is now well established both in terms of hardware and software. In contrast, FASTBUS, as we have already mentioned, is still in development and will not be treated here. The interested reader, however, may find several references in the bibliography.

18.1 CAMAC Systems

Like NIM, CAMAC is a modular system, the essential components of which are a crate and *plug-in* type modules. As shown in Fig. 18.1, the crate is composed of *slots* or *stations* into which the modules are inserted. At the rear of the module is a card-type connector which mates with a corresponding connector at the back of the slot. This connector contains 86 contact points (43 on each side) which couple the module to a series of parallel wires running along the backplane of the crate, linking each of the stations. This series of wires is known in CAMAC terminology as the *dataway* and is the essential feature of the CAMAC system. In more modern terms, the dataway would be known a *backplane bus*. This bus allows digital signals to be transmitted to or from plug-in modules which may be added on to the bus as desired. When the dataway is interfaced to a computer's data bus, the modules may then be accessed for control or readout by the computer processor. Also included in the dataway are power lines which supply the necessary operating voltages to the module. The function and characteristics of the dataway lines will be discussed in more detail in the following sections.

All communication within a CAMAC crate is overseen by a special module known as the *crate controller* (CC). This module acts essentially as a communication center managing the flow of information on the crate dataway. Commands or data issued by a computer to a module or vice versa must therefore pass through the crate controller. In general, within a given crate, the CC is the only module which can issue commands and is thus the *master* of the dataway. All other modules are *slaves* to the CC. Because of its

Fig. 18.1. A CAMAC crate and modules

special function, the crate controller always occupies the last two stations of the crate (numbers 24 and 25) which are specifically reserved for this purpose.

CAMAC systems may be configured in a variety of ways and at various levels of complexity. The basic system consists of one crate connected to a host computer. The interface to the computer is then usually included in the crate controller. The number of modules that can be used is then limited to those which can fit into one crate only. A larger and more complicated system might consist of several crates connected to the same computer. These crates can be linked to each other either in series or in parallel along a *branch highway*, which like the dataway, is a bus which carries signals to and from different crates. Analogous to the crate controller, a special unit known as a *branch driver* is needed to manage the flow of data along the branch highway. Still larger systems can be formed by configuring a multi-branch system. At this point it may also be necessary to couple more than one computer to the CAMAC system in order to handle the ever increasing quantity of data!

With the advent of microprocessors, compact systems may now also be formed by incorporating the microprocessor in a plug-in module or within the crate controller. The computer host then resides in the crate itself forming a *stand-alone* system. For small applications requiring a few applications modules, such systems offer an economical and more flexible alternative to more specialized apparatus performing the same function. Crate resident microprocessors may also be used advantageously in larger systems for a variety of purposes. They can be used, for example, to preprocess raw information directly from the crate modules without intervention from the main computer. Upon completion of the analysis, this reduced information can then be transferred to the main system computer for recording on magnetic tape or for further treatment. This minimizes the amount of data transfers and computing time on the system computer, thereby increasing speed. Microcomputers may also be used as auxiliary crate controllers for local control or processing. Here they may be incorporated into the crate controller itself or located in a normal CAMAC station. Figure 18.2 illustrates some of these configurations schematically.

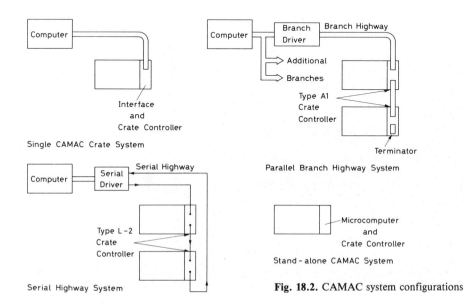

Fig. 18.2. CAMAC system configurations

18.2 The CAMAC Standard

The specifications for the CAMAC standard are described in a number of separate publications issued by the ESONE committee in Europe and the U.S. Government Printing Office in the United States. The principal documents are:

1) EUR 4100 (Europe) or TID-25875 (USA)
   This is the basic document describing mechanical and electrical standards for the CAMAC dataway and modules.
2) EUR 4600 (Europe) or TID-25876 (USA)
   This publication describes the electrical standard for multicrate systems using the parallel branch highway and crate controller Type A-1.
3) ESONE/SR/01 (Europe) or DOE/EV-0016 (USA)
   This report outlines a set of standard CAMAC software subroutines for use with high-level programming languages.

Most of these documents have been combined into a two-volume publication, EUR 8500en, available from the Commission of European Communities, Luxembourg and in a hardcover ANSI/IEEE publication SH-08482 (1982). In the following sections we will try to outline the basic features of CAMAC with a few illustrations so as to give the reader a more workable introduction to the system. For further details the reader is referred to the documents above.

18.2.1 Mechanical Standards

The standard single-width CAMAC module has dimensions of 221.5 mm in height and 305 mm in depth. Its width is exactly half that of a standard NIM module or 17 mm. CAMAC modules may also come in multiples of this single width.

The CAMAC crate is 19 inches in width and is equipped with 25 connector slots with slots 24 and 25 specifically reserved for the crate controller.

18.2.2 Electrical Standards: Digital Signals

The transmission of digital signals along the dataway in a CAMAC crate is performed through TTL logic signals. The required signals levels for this logic family are given in Table 18.1. A discussion of the actual signals available on the dataway and the protocol used is given in the next section.

Table 18.1. Dataway signal levels

	Logic 0	Logic 1
Input must accept	$+2.0$ to 5.5 V	0 to $+0.8$ V
Output must generate	$+3.5$ to 5.5 V	0 to $+0.5$ V

18.3 The CAMAC Dataway

The dataway is the nervous system of the CAMAC system. It consists of a series of parallel wires running along the backplane of the crate connecting each of the slots. Com-

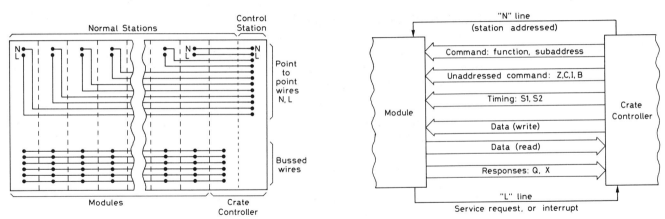

Fig. 18.3. Schematic diagram of the CAMAC dataway (Reprinted by permission from Kinetic Systems Corp., Lockport, Ill. [18.1])

munication between modules, crate controller and host computer is made via the dataway. Three types of wiring make up the dataway:

1) *Power Lines.* These lines are bussed to each station (i.e., each station is connected in parallel along these lines) and supply standard voltages of ± 6 V and ± 24 V. On some crates ± 12 V may also be found along with 117 VAC and 200 VAC.

2) *Bussed Signal Lines.* These include lines for data transfer, addressing, commands, and specific control signals. Most of the dataway lines are of this type. Their specific functions are discussed in more detail below.

3) *Point to Point Lines.* These are separate, dedicated lines connecting each crate station to station 25, the controller station. There are only two such lines: the crate address (N) and the *Look-at-Me* (L).

Figure 18.3 provides a schematic sketch of the dataway wiring. The dataway signals available may be classified into 6 categories: control, addressing, timing, data, status, and command. Some or all of these signals may be used by plug-in modules depending on their design and application. Table 18.2 briefly summarizes these signals and their functions.

Table 18.2. CAMAC dataway signals

Signal	Symbol	Function
Common control signals		
Initialize	Z	Sets module to a defined, initial state particularly when power is turned on (accompanied by the $S2$ and B signals)
Inhibit	I	Disables features for duration of signal
Clear	C	Clears registers, resets flip-flops, etc. (accompanied by the $S2$ and B signals)

Table 18.2 (continued)

Signal	Symbol	Function
Status signals		
Look-at-Me (LAM)	L	A signal from a module to the crate controller requesting service or attention. This is a dedicated point-to-point line. The presence of a LAM may be tested by using function $F(8)$
Response	Q	A one-bit reply signal issued by the module in response to certain commands from the crate controller
Command accepted	X	Indicates that the module is able to perform the action required by the command
Busy	B	Indicates dataway operation is in progress
Timing signals		
Strobe 1	$S1$	Signal used to control first phase of a dataway operation. At the issuance of $S1$ all signals must be settled on their respective lines
Strobe 2	$S2$	Governs phase two of dataway operation. At issuance of $S2$, dataway signals may change
Data signals		
Read	$R1-R24$	Signals for carrying data from a module. Twenty-four parallel uni-directional lines are allocated allowing a 24 bit-parallel word to be transferred in a single operation
Write	$W1-W24$	Signals used for carrying data to a module. Twenty-four parallel uni-directional lines are allocated
Address signals		
Station number	N	Selects module in crate. This is a dedicated point-to-point line
Subaddress	$A1,2,4,8$	Selects a specific section of the module. This depends on the structure of the module. Four lines are allocated to the subaddress thus allowing 16 possible subaddresses
Command signals		
Function	$F1,2,4,8,16$	Defines the function to be executed by the module. The value of F corresponds to specific functions defined in the following section. Five lines are allocated so that 32 code values are possible

Note: It is common practice to refer to the individual signal lines in a multiple-bit signal, for example, A or F, by the notation An or Fn where n is the corresponding bit number of the line, i.e., $n = 1,2,4,8, \ldots$. The value of A or F which is issued on the lines, on the other hand, is indicated by surrounding the value with parentheses, i.e., $A(m)$ or $F(m)$ where m is the decimal value of A or F.

18.3.1 Common Control Signals (Z, C, I)

The common control signals, *Initialize* (Z), *Clear* (C), and *Inhibit* (I) are generated by the crate controller, and are received by all units connected to the dataway (unaddressed operation). Their functions are as indicated in Table 18.2. The Z signal has priority over all other signals and is always accompanied by the Busy (B), Inhibit (I) and S2 signals. This is also the case for Clear. The Clear and Inhibit signals are free to be used as desired by the module designer for whatever features he wishes.

18.3.2 Status Signals

A *Look-at-Me* (LAM) signal is a request by a module to indicate that it needs servicing of some kind. An ADC may generate a LAM to indicate that it has finished converting a signal, for example, or a memory unit to indicate that it is full, etc. A LAM request may also have more than one origin inside a module. In some modules, a LAM register may be read containing information on the source or sources of the LAM request. Appropriate actions may then be taken upon analysis of this information. The LAM signal can be tested, cleared, disabled and enabled using the command operations.

The *Busy* signal is generated by the crate controller whenever a dataway operation is in progress to inhibit any competing operation during the duration of the current cycle.

The *Q-response* is a signal generated by a module in response to certain command operations requiring a yes/no type answer. Function F(8) to test the presence of a LAM, for example, results in $Q = 1$ if the LAM is set and $Q = 0$ if not. The Q signal is also used in block transfer operations.

The *X signal* is generated by a module in response to a command operation to indicate if it is capable of performing the requested operation. If yes, $X = 1$. If not, either because of a malfunction, nonexistence of the requested function, lack of power, etc., $X = 0$ is returned.

18.3.3 Timing Signals

To synchronize the sequence of events during a dataway operation, two strobe signals S1 and S2 are generated by the crate controller to indicate when data and command signals are settled on their respective lines. Modules may then trigger certain actions upon activation of S1 and S2.

18.3.4 Data Signals

Data are carried to a module via 24 unidirectional READ lines and from a module via 24 unidirectional WRITE lines. Data format is bit-parallel with a maximum word length of 24 bits. The READ and WRITE lines are activated using command operations.

18.3.5 Address Signals

These signals are used for addressing modules during a command operation. See Sect. 18.4 for further details.

Table 18.3 (from *LeCroy catalog* [13.1])

Pin Allocation at Normal Station

(Stations 1-24)

Bus-line	Free Bus-line	P1	B	Busy	Bus-line
Bus-line	Free Bus-line	P2	F16	Function	Bus-line
Individual patch contact		P3	F8	Function	Bus-line
Individual patch contact		P4	F4	Function	Bus-line
Individual patch contact		P5	F2	Function	Bus-line
Bus-line	Command Accepted	X	F1	Function	Bus-line
Bus-line	Inhibit	I	A8	Sub-address	Bus-line
Bus-line	Clear	C	A4	Sub-address	Bus-line
Individual line	Station Number	N	A2	Sub-address	Bus-line
Individual line	Look-at-Me	L	A1	Sub-address	Bus-line
Bus-line	Strobe 1	S1	Z	Initialize	Bus-line
Bus-line	Strobe 2	S2	Q	Response	Bus-line

	W24	W23	
	W22	W21	
	W20	W19	
	W18	W17	
24 Write Bus Lines	W16	W15	
	W14	W13	
W1 = least significant bit	W12	W11	
W24 = most significant bit	W10	W9	
	W8	W7	
	W6	W5	
	W4	W3	
	W2	W1	

	R24	R23	
	R22	R21	
	R20	R19	
	R18	R17	
24 Read Bus Lines	R16	R15	
	R14	R13	
R1 = least significant bit	R12	R11	
R24 = most significant bit	R10	R9	
	R8	R7	
	R6	R5	
	R4	R3	
	R2	R1	

Power Bus-lines					Power Bus-lines
	−12V d.c.	−12	−24	−24V d.c.	
	+200V d.c.	+200	−6	−6V d.c.	
	117V a.c. Live	ACL	ACN	117V a.c. Neutral	
	Reserved	Y1	E	Clean Earth	
	+12V d.c.	+12	+24	+24V d.c.	
	Reserved	Y2	+6	+6V d.c.	
	OV (Power Return)	0	0	OV (Power Return)	

(VIEWED FROM FRONT OF CRATE)

18.3.6 Command Signals

These signals are use to command addressed modules to perform certain functions. A five-bit code is used to indicate the desired function. The functions are discussed in more detail in Sect. 18.4.

18.3.7 Pin Allocations

Plug-in modules, as we have mentioned, obtain access to the dataway via an 86-pin PC card connector. These connectors are much more delicate than those on NIM modules,

Table 18.4 (from *LeCroy catalog* [13.1])

Pin Allocation at Control Station

(Station 25)

Individual patch contact		P1	B	Busy	Bus-line

Individual patch contact   P1   B   Busy   Bus-line

Left label	Left mid	Pin L	Pin R	Right mid	Right label
Individual patch contact		P1	B	Busy	Bus-line
Individual patch contact		P2	F16	Function	Bus-line
Individual patch contact		P3	F8	Function	Bus-line
Individual patch contact		P4	F4	Function	Bus-line
Individual patch contact		P5	F2	Function	Bus-line
Bus-line	Command Accepted	X	F1	Function	Bus-line
Bus-line	Inhibit	I	A8	Sub-address	Bus-line
Bus-line	Clear	C	A4	Sub-address	Bus-line
Individual patch contact		P6	A2	Sub-address	Bus-line
Individual patch contact		P7	A1	Sub-address	Bus-line
Bus-line	Strobe 1	S1	Z	Initialize	Bus-line
Bus-line	Strobe 2	S2	Q	Response	Bus-line

	L24	N24	
	L23	N23	
	L22	N22	
	L21	N21	
	L20	N20	
	L19	N19	
	L18	N18	
	L17	N17	
	L16	N16	
	L15	N15	
	L14	N14	
24 Individual Look-at-Me Lines	L13	N13	24 Individual Station Number Lines
L1 from Station 1, etc.	L12	N12	N1 to Station 1, etc.
	L11	N11	
	L10	N10	
	L9	N9	
	L8	N8	
	L7	N7	
	L6	N6	
	L5	N5	
	L4	N4	
	L3	N3	
	L2	N2	
	L1	N1	

Power Bus-lines		Pin	Pin		Power Bus-lines
	−12V dc.	−12	−24	−24V d.c.	
	+200V d.c.	+200	−6	−6V d.c.	
	117V a.c. Live	ACL	ACN	117V a.c. Neutral	
	Reserved	Y1	E	Clean Earth	
	+12V d.c.	+12	+24	+24V d.c.	
	Reserved	Y2	+6	+6V d.c.	
	OV (Power Return)	0	0	OV (Power Return)	

(VIEWED FROM FRONT OF CRATE)

so that CAMAC modules should be handled with more care. Table 18.3 lists the pin locations and definitions for a normal crate station slot. Station 25, which is reserved for the crate controller, has a different pin configuration and these are listed in Table 18.4.

18.4 Dataway Operations

A typical dataway operation generally involves a transfer of signals between a module or modules on one end and the crate controller on the other. The operations which can be made are of two types: command or unaddressed. A command operation involves a command signal which is to be sent to a specific module. An address for the module must therefore be specified. An unaddressed operation involves the issuance of common control signals, i.e., Initialize (Z), Inhibit (I) or Clear (C), which operate on all modules connected to these dataway lines. In both cases a Busy signal (B) is issued simultaneously by the crate controller to indicate that a dataway operation is in progress.

The basic form of a command operation is the set of signals NAF, where NA is the address and F is the command function code. The address is composed of N, the number of the crate station occupied by the module and a subaddress, A, which may take on values from 0 to 15. The subaddress refers to an internal part of the module and its significance depends on the specific construction of the module. In an eight-fold ADC, for example, (i.e., a module containing 8 ADC's with separate inputs), the subaddress may refer to the specific ADC section in the module, in which case, only $A(0)$ to $A(7)$ would be available. In another module, A may refer to a specific register containing certain information, etc. Note that since the address of the module corresponds to its location in the crate, changing the position of a module would require changing its address in the controlling computer program if the module is to be correctly accessed.

On the dataway of a crate, the N number is carried by dedicated point to point lines connecting each station to the crate controller station 25. For a given N, the crate controller will activate the N line corresponding to the station addressed. Simultaneously, the A subaddress is placed on the A lines which in contrast to the N lines are bussed to all stations as are the F lines which are also activated at same time.

The F code may take on any value from 0 to 31 and is the means by which commands may be transmitted to the module. The F values correspond to specific functions which can be performed by the plug-in module. Table 18.5 lists the values of F and the standard function definitions.

There are essentially three groups of commands. $F(0)$ to $F(8)$ are READ commands which use the R lines while $F(16)$ to $F(23)$ are WRITE commands using the W lines. $F(8)$ to $F(15)$ and $F(24)$ to $F(31)$, on the other hand, are control commands requiring either a yes/no response or none at all. In these cases, the R and W are not used and all necessary responses are given by the status of the Q line. To read the contents of a register in an ADC, for example, one would use the code $F(0)$. Similarly, to test if the Look-at-Me signal is set, $F(8)$ would be used. The response is then be given by the state of the Q signal. It should be noted that only a certain number of the code values have been standardized and that room has been provided for ad hoc functions to be defined by the module designer, if required. Moreover, for a given module, only those function codes relevant to the module's function will be implemented. The CAMAC functions available on any module, are generally given on the module's specification sheet.

Each addressed module during a command operation must generate an X signal in response. If the module recognizes the command and is able to execute it, an $X = 1$ response is generated. If the module is unable to perform the function because it is either not present, not equipped to execute the desired function, unpowered, unconnected, etc., $X = 0$ is returned.

18.4.1 Dataway Timing

Let us now look at what happens electronically during a dataway operation. Because each operation involves a series of signals, synchronization is very important and correct timing must be rigorously maintained in order to assure correct transfer of infor-

Table 18.5. CAMAC function codes

Code $F()$	Function
0	Read group 1 register[a]
1	Read group 2 register
2	Read and clear group 1 register
3	Read complement of group 1 register
4	Nonstandard
5	Reserved
6	Nonstandard
7	Reserved
8	Test look-at-me
9	Clear group 1 register
10	Clear look-at-me
11	Clear group 2 register
12	Nonstandard
13	Reserved
14	Nonstandard
15	Reserved
16	Overwrite group 1 register
17	Overwrite group 2 register
18	Selective set group 1 register
19	Selective set group 2 register
20	Nonstandard
21	Selective clear group 1 register
22	Nonstandard
23	Selective clear group 2 register
24	Disable
25	Execute
26	Enable
27	Test status
28	Nonstandard
29	Reserved
30	Nonstandard
31	Reserved

[a] Registers in CAMAC modules may be divided into two sets: groups 1 and 2, which are accessed by separate commands. Within each group a particular register is selected using the subaddress A, so that up to 32 registers may be accessed in a module. Information concerning system status, configuration etc. are generally be held in group 2.

mation. The timing diagram in Fig. 18.4. illustrates the sequence of events for a command operation.

At time t_0, the beginning of the operation, the NAF and B signals are simultaneously activated along with any R or W signals. Since it is difficult to have a perfect synchronization between the signals, timing margins are allowed. These are indicated by the shaded areas in the diagram. Thus, by time t_1, the NAF and B signals must all have reached the appropriate voltage levels. Between t_1 and t_2, the addressed module must react and initiate the Q and X status signals and at time t_3, these and the R or W signals must be at the required voltage levels. Here now, we see more clearly the significance of the strobe signals $S1$ and $S2$. $S1$ is initiated at t_3 and must be stable by t_4. At this time all command and data signals must be settled on their respective lines. Using $S1$, there-

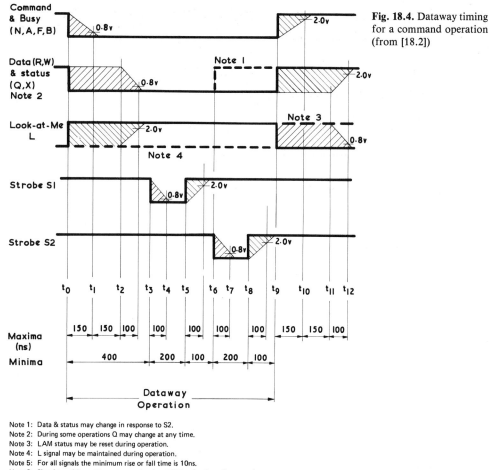

Fig. 18.4. Dataway timing for a command operation (from [18.2])

Note 1: Data & status may change in response to S2.
Note 2: During some operations Q may change at any time.
Note 3: LAM status may be reset during operation.
Note 4: L signal may be maintained during operation.
Note 5: For all signals the minimum rise or fall time is 10ns.
Note 6: Signal transitions at t_0 or t_9 may be absent if the signals on Command or Data lines are the same for the immediately preceding or following operations.

fore, gates may be opened on the module or at the computer interface to receive the data signals. At t_6, the strobe $S2$ is initiated and data or status signals may now change. Note that the LAM signal may or may not be present during this operation. From Fig. 18.4, we also see that the minimum cycle time for one complete CAMAC operation is 1 microsecond.

The sequence for an unaddressed operation is somewhat simpler and shorter. This is illustrated in Fig. 18.5. At time t_0, the B and either the C or Z and I signals are initiated and settled by time t_1. Between t_1 and t_6, the signals are processed in the modules if required. At t_6, the $S2$ line is activated. Note that in unaddressed operations, only the $S2$ strobe is required. However, S1 may be optionally generated if desired. The minimum time for an unaddressed cycle is 750 ns.

18.4.2 Block Transfers

In the preceding sections we saw that the minimum time for a single CAMAC data transfer was 1 μs. Under program control, the actual time is usually many times great-

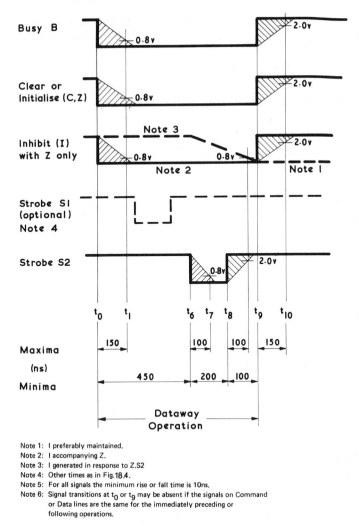

Fig. 18.5. Timing for unaddressed operations (from [18.2])

Note 1: I preferably maintained.
Note 2: I accompanying Z.
Note 3: I generated in response to Z.S2
Note 4: Other times as in Fig.18.4.
Note 5: For all signals the minimum rise or fall time is 10ns.
Note 6: Signal transitions at t_0 or t_9 may be absent if the signals on Command
or Data lines are the same for the immediately preceding or
following operations.

er, however, because the time needed by the computer to execute and issue the command must also be counted. Execution times by the central processor of the computer can be quite long and may vary from hundreds of microseconds to milliseconds. This proves to be highly disadvantageous when large blocks of data must be transmitted. Suppose, for example, we want to read the contents of a memory containing 4 K words of storage. Using simple command operations, we would have to execute the READ command 4 K times. Assuming a typical execution time by the processor to be on the order of 1 ms, the total time required to transfer the entire block would be on the order of several seconds!

To overcome this problem, most CAMAC systems today are equipped with *Direct Memory Access* (DMA) capability. The DMA unit is essentially a controller unit which makes a direct transfer of data between the CAMAC device and the computer memory or peripheral without intervention from the central processor. In a DMA transfer, a starting memory address corresponding to the beginning of the data block to be transferred is usually given along with the total number of data words to be transferred. The final address of the data block may also be given. These data are stored in memory registers. The function (READ or WRITE) is then defined and the DMA operation initiated. The DMA unit then takes control of the data bus, transfers the first word of the block, increments the address register and decrements the word counter. The cycle is then repeated until the word count is zero. In this manner an efficient transfer of a block of data is enabled with no commands from the computer processor.

Block transfers may be performed in one of three modes: *Address Scan, Repeat* or *Stop* mode. The *Address Scan* mode is used when a block of registers or modules are to be read (or written onto) sequentially. The modules involved need not be located at consecutive addresses, however, subaddresses within a given module must be so. The Q-response is used to determine if an address is occupied or not. For occupied N, $Q = 1$ is returned. With N held constant, the subaddress is then incremented and transfers made until $Q = 0$ is returned. A is then reset to $A(0)$ and N is incremented, etc.

Repeat mode is used to read a single register a fixed number of times. The readiness of the register for a read or write transfer is determined by the state of the Q-response. If $Q = 1$, the register is ready and an operation is made. If $Q = 0$, the register is not ready and repeated trials are made until $Q = 1$. The repeat mode is an inherently dangerous mode since it is possible for the DMA channel to "*hang up*" waiting for a positive Q response when none is forthcoming. Some additional criteria, such as a limit on the total number of tries waiting for a Q-stop, must therefore be imposed to protect against such occurrences.

The *Stop* mode is used for reading or writing a block of data whose size is determined by the CAMAC device. Sequential reads or writes are performed until an end-of-block condition is reached. This is signaled by a $Q = 0$ response.

Variations on these three modes are also possible which may be more useful for certain types of data acquisition.

18.5 Multi-Crate Systems – The Branch Highway

As we have already mentioned, CAMAC systems may contain more than one crate. In such cases, the crates are organized into branches with the crates on each branch being linked together along a data bus which in CAMAC terminology is known as a *branch highway*. And like the dataway, each branch is controlled by a branch driver which manages and coordinates the signal flow. Two types of branch highway are possible: *parallel* or *serial*.

The parallel highway is the more complicated of the two but offers the highest rate of data transfer. This is usually the choice for data acquisition systems. Crates gain access to the highway through ports in special crate controllers which contain the necessary circuitry for receiving and decoding the branch highway signals. These crate controllers, *Type A-1*, are standardized in document EUR 4600en. An enhanced version, *Type A-2*, may also be used and is discussed in document EUR 6500en. A maximum of 7 crates may be connected to the highway usually in the chain configuration shown in Fig. 18.2. Other configurations may be possible, however. The length of a parallel highway, is limited to relatively short distances of about 30 to 40 m. If longer distances are required, additional signal drivers must be inserted.

The parallel highway consists of a 132-wire bus (65 signal lines and their ground returns plus grounded shielding) and has a structure very similar to the dataway. Separate

READ and WRITE lines are available, along with address, control, timing, service demand and function lines.

In a multi-crate system, the address must be extended to include the crate number (C), so that the complete address is given by CNA. Multi-crate systems can have up to 7 crates so that C may take on values between 1 to 7. In a multi-branch system, the address must also include the branch number B, bringing the complete address to BCNA.

Along a parallel highway, the crate number C is carried by dedicated lines while N and A are bussed. N is decoded by the crate controller which then activates the appropriate station line in the crate, while A is simply retransmitted along the dataway lines.

Command operations along a branch are executed in a manner similar to a dataway operation although the timing gets somewhat more complicated. Among the unaddressed operations, however, only the *Branch Initialize* (*BZ*) is possible. Service demands on the branch highway, on the other hand, are more extensive and complicated. Here there are two types: the *Branch Demand* (BD) and the *Graded LAM* (GL). The Branch Demand is a general demand signal generated by the crate controllers indicating that one or more modules along the branch have set their LAMs and require service of some kind. In response to BD, the branch driver may then issue a request for more information by initiating a *Graded LAM* operation. A graded LAM is a 24-bit word whose bit-pattern allows the branch driver to identify the type of servicing being requested. During a GL operation, each crate controller identifies and tallies the LAMs in its respective crate and forms a GL word whose pattern corresponds to the types of LAMs set. The GL words from all crates are then combined and sent to the branch driver. Based on the GL pattern, appropriate services and actions may then be taken.

In contrast to the parallel highway, the serial highway offers a much lower rate of data transfer. However, it is a much simpler bus which may be extended to long distances without the need of signal boosters. As well, up to 62 crates may be connected to the serial highway.

All "*messages*" transmitted along the serial highway are organized as sequences of 8-bit bytes. Transmission of the bytes is performed in synchronism to a system clock which may have a maximum rate of 5.0 MHz. The information itself, however, may be transmitted bit by bit (*bit-serially*) along 1 signal line and an accompanying clock line or *byte-serially* along 8 parallel lines plus a clock line. In both cases, each sequence of 8-bits is preceded by a START bit and followed by a STOP bit to mark the beginning and end of each byte. Within each byte also, only bits 1 – 6 may be used for data. Bit 7 is used as a delimiter bit to indicate the beginning and end of the message while bit 8 may be used as a parity bit.

Like the parallel highway, the serial highway requires a serial driver and special *Type L-2* crate controllers. The latter have been standardized and further details may be found in the document EUR 6100 or ANSI/IEEE Std. 595 (1982).

18.6 CAMAC Software

While CAMAC has essentially resolved the hardware interface problem by introducing the dataway, the software to run CAMAC presents a more difficult problem. At the most basic level, software drivers are needed to control input-output operations at the CAMAC interface. This is highly machine dependent software, usually written in assembly language, as it must take into account the architecture and operating system of the computer as well as the CAMAC interface. Many manufacturers offering CAMAC interfaces also offer software driver packages along with higher-level languages for some of the more common computers and operating systems. For the computer not fitting into any of these categories, writing a CAMAC driver could mean a considerable investment of time and effort for the non-system programmer.

Once a driver is installed, applications programs may be written. This is best done in a high-level language (such as FORTRAN) where the logic is more easily seen and followed. To make the hardware and interface of the CAMAC system as transparent as possible to the user, a number of CAMAC routines, coded in assembly language for speed, and callable by the higher-level language are usually written to perform basic command and unaddressed operations. The user writing an acquisition program then need only make reference to these routines in order to perform any CAMAC operations.

At this level some standardization is possible, so that programs written in a higher level language can be more or less transportable from computer to computer. As a result, the ESONE/NIM committee in 1978, issued a recommended set of standard CAMAC subroutines for the non-specialist user programming in a high-level language. These routines were designed to buffer the programmer from unnecessary details of the interface and controller and to maximize transportability of the software. Only the required actions of these routines are specified so that they may be used with any high-level computer language. The actual implementation of the routines for a given computer and CAMAC interface is left to the CAMAC system programmer.

In the following paragraphs we will illustrate some of the routines offered by the standard. The entire set may be found in the document ESONE/SR/01 or DOE/EV-0016. Since FORTRAN is the most common programming language among physicists, we will use this language in the examples. Software implementations in other languages will, of course, show differences, however, the general ideas remain the same.

The subroutines are classified into three levels. The level A routines are the most fundamental performing simple CAMAC command operations:

Call CDREG(IREG,IB,IC,IN,IA)   Declare a register

This routine associates a variable IREG to a CAMAC address BCNA for later use in the program. Thereafter, all references to this address are made by simply using IREG rather than BCNA:

Call CFSA(IF,IREG,IDATA,IQ)   Perform a Command Action
Call CSSA(IF,IREG,IDATA,IQ)

These routines perform a command operation on the address IREG, where IF is the function code and IQ the returned Q-response. Data transferred during the operation (READ or WRITE) is contained in the array IDATA. The routine CFSA assumes a full 24-bit word while CSSA assumes a shorter 16-bit word.

The level B routine set consists of the level A routines plus routines performing unaddressed operations, such as:

Call CCCZ(IREG)   Initialize crate
Call CCCC(IREG)   Clear Crate
Call CCCI(IREG,L)   Set or Clear Inhibit

These routines perform the unaddressed operations on the crate addressed by IREG. The CCCI routine sets an Inhibit if L is *true* and clears it if L is *false*. Also included in level *B* are LAM declaration and handling routines for such functions as testing, clear, enable, etc. For example:

    Call CDLAM(LAM,IB,IC,IN,IA,INTA)   Declare LAM

This routine declares a LAM source in a manner similar to the register declaration routine. Through this routine the variable LAM is associated with the BCNA source. INTA is an array containing servicing information, etc. and is system dependent. Some of the handling routines are:

    Call CCLM(LAM,L)   Enable/Disable LAM

    Call CCLC(LAM)   Clear LAM

    Call CTLM(LAM,L)   Test LAM

where *L* is a logical variable.

The third level *C* consists of additional routines initiating block transfers, such as:

    Call CFMAD(IF,IADD,IARRAY,ICB)   DMA in Address Scan Mode

    Call CFUBR(IF,IREG,IARRAY,ICB)   Q-Repeat Mode

    Call CFUBC(IF,IREG,IARRAY,ICB)   Stop Mode

where IADD, IARRAY and ICB are all arrays. IADD contains the starting and stopping addresses of the scan mode, IARRAY the CAMAC data array and ICB control information, such as the word count, the number actually transferred, etc. A general action routine is also defined to execute a sequence of CAMAC commands:

    Call CFGA(IFA,IADR,INTC,IQA,ICB)   Perform sequence of commands

where FA is an array of CAMAC functions, IADR an array of CAMAC addresses at which the command will be executed, IQA an array of *Q*-responses and ICB an array containing various control information.

At level *C*, all the necessary routines are present for writing efficient data acquisition and control programs. In any implementation, also, there will be some system dependent routines which take account of special features, etc. This is particularly true for LAM handling and block transfer operations where efficiency is of paramount importance.

Let us now consider a few simple examples using the standard routines to illustrate CAMAC data acquisition. As above, we will assume a knowledge of FORTRAN. In the first program, we will read a scaler and print out the results:

```
      PROGRAM SCALER
C
C     First define scaler in branch 2, crate 1, slot 7
C
      CALL DREG(ISCAL,2,1,7,0)
C
C     Now reset scaler and clear INHIBIT
C
      CALL CCCC(ISCAL)
      CALL CCCI(ISCAL,.FALSE.)
C
C     Read 16-bit scaler register
C
```

```
            Call CSSA(0,ISCAL,IDATA,IQ)
C
C      Print IDATA and Q-response
C
            PRINT 10, IDATA,IQ
        10 FORMAT(' DATA = ',I10, ' IQ = ', I10)
C
            END
```

In the following program we illustrate simple LAM handling with an ADC. After setting up the ADC, a LAM, indicating the end of a conversion is awaited by entering into a loop. When the LAM test is positive, the contents of the ADC are read and printed.

```
            PROGRAM ADC
C
C
            INTEGER  INTA(2)
            LOGICAL L
C
C
            DATA INTA/0,0/0,0/
C
C      Define ADC address (B = 2, C = 1, N = 3)
C
            CALL CDREG(IADC,2,1,3,0)
C
C      Define LAM source
C
            CALL CDLAM(LAM,2,1,3,0,INTA)
C
C      Clear and Enable LAM
C
            CALL CCLC(LAM)
            CALL CCLM(LAM,.TRUE.)
C
C       Wait for LAM by looping
C
        1 CALL CTLM(LAM,L)
            IF(L.EQ..FALSE.) GO TO 1
C
C      LAM detected, read result and print out
C
            CALL CSSA(0,IADC,IDATA,IQ)
C
            PRINT 10, IDATA,IQ
        10 FORMAT(' DATA = ', I10, ' Q = ',I1)
C
            END
```

Because of their greater complexity, setting up and troubleshooting CAMAC systems are generally much more difficult than NIM systems. Indeed, not only is the hardware more complicated, one must also deal with software problems as well. And separating the two is not always easy! On the hardware side, more attention must be paid to contact problems than usual. Connectors should be clean and in good order. Because of its construction, some attention should be paid to alignment when modules are inserted into their slots or when connectors are mated. If a module meets resistance while being inserted into its slot, do not force it. It is easy to damage the card-connector; moreover, it is probably an indication of some mechanical misalignment. It is sufficient for the connector to be off by 2 or 3 mm to have intermittent problems. As a general safety rule also, CAMAC crates should first be turned off before inserting or removing modules. This is because of the proximity of the GND and voltage pins on the back connector.

Appendix

A. A Review of Oscilloscope Functions

A.1 Basic Structure

Despite its apparent complexity, the oscilloscope, in its most frequent application, may simply be considered as a voltmeter which visually displays the input voltage signal as curve in time. To understand its operation, a schematic diagram of its basic structure is presented in Fig. A.1.

Visual representation of the signal is provided by means of a cathode-ray tube (CRT) and two pairs of deflecting electrode plates which control the vertical and horizontal movement of the CRT electron beam. When a signal appears at the input, it is first amplified then split in two. One part is applied to the vertical-deflection plates, while the other is used to trigger a generator which applies a linear ramp voltage to the horizontal deflection plates. This latter voltage causes the beam to be swept horizontally across the screen at a constant speed. Since the beam is now also being deflected vertically in proportion to the input signal, a curve representing the waveform of the signal in time is produced. This is the simplest operation mode of the oscilloscope. Examples of further applications are given later in this appendix.

A.1.1 Bandwidth and Risetime

One of the most important operating characteristics of the oscilloscope is the bandwidth of the vertical amplifier. As explained in Sect. 11.4, this parameter essentially governs how fast a signal can be accepted and correctly displayed by the oscilloscope without distortion. This is especially important for signals from detectors such as organic scintillators which emit pulses of a few nanosecond rise time.

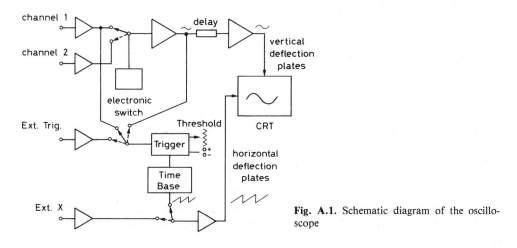

Fig. A.1. Schematic diagram of the oscilloscope

Because of the finite bandwidth of the amplifier, the oscilloscope has an intrinsic risetime of its own, that is, a signal with a perfectly vertical rising flank (0 risetime) will be displayed as a signal with a flank with a finite risetime t_{osc}. A rough relation between the bandwidth and this rise time is given by the formula

$$t_{osc}[ns] = \frac{350}{f_{3dB}[MHz]} \; , \tag{A.1}$$

where f_{3dB} is the bandwidth of the oscilloscope in Megahertz. The risetime displayed for a signal with a risetime t_{pulse} will thus be a combination of the risetime of the pulse and the oscilloscope risetime. An approximate formula for t_{disp} is given by

$$t_{disp}^2 = t_{pulse}^2 + t_{osc}^2 \; . \tag{A.2}$$

A 3 ns risetime pulse on a 100 MHz oscilloscope ($\Rightarrow$ 3.5 ns risetime) will thus be displayed as a pulse with a 4.6 ns risetime. Similarly, if a 10 ns risetime is measured on the same oscilloscope, the true risetime of the signal is

$$t_{pulse} = \sqrt{10^2 - 3.5^2} = 9.4 \text{ ns} \; . \tag{A.3}$$

A.2 Controls and Operating Modes

A.2.1 Input Coupling

As in any good voltmeter, the input impedance of the oscilloscope is very high, usually 1 MΩ, in parallel with a small capacitance on the order of $10-20$ pF. For fast signal viewing from 50 Ω cables, some oscilloscopes also provide a switch which allows selection of a 50 Ω input impedance so as to obviate cable termination. On most oscilloscopes also, two, sometimes more inputs, are available allowing the simultaneous display of several signals. Depending on how triggering is performed, this allows a visual comparison of two or more signals. At each input, the type of coupling may be selected:

AC – In this mode any constant dc level is suppressed and only nonzero ac frequencies are displayed.

DC – dc constant levels, as well as ac frequencies are displayed. Thus varying signals superimposed onto a constant dc level may be seen.

GND – The input signal is directly shunted to ground.

A.2.2 Vertical and Horizontal Sensitivity

VERTICAL SENSITIVITY – This controls the vertical scale. On most oscilloscopes a vernier control is available which allows a continuous adjustment between the indicated markings. Note that these markings are only valid when the vernier is in the *calibrated* position!

TIME – This determines the speed at which the beam is swept across the screen and thus the horizontal scale. Again, the indicated settings are only valid when in the *calibrated* position.

DELAYED SWEEP – In this mode, the time sweep is made after a certain delay time determined by the delay setting dial.

EXTERNAL X – When in this position, the time base generator is disconnected allowing the horizontal deflection of the beam to be controlled by an external voltage applied at the EXT-X or channel-2 input.

A.2.3 Triggering (Synchronization)

Triggering or synchronization is the most important adjustment to be made when using the oscilloscope. As we have seen, the horizontal sweep of the oscilloscope is activated only when there is a triggering signal satisfying certain conditions. This triggering signal may be the input signal itself or some other external signal, depending on the mode selected. The conditions for a trigger include the slope of the signal and its amplitude. In this manner a precise point on the signal may be selected for the beginning of the sweep. For a repetitive signal such as a sine wave, for example, this produces a steady trace on the display screen, always starting at the same point on the signal.

Signal Source

INTERNAL – Triggering is done by the input signal itself as outlined above. This is the normal mode of operation.

EXTERNAL – Triggering is made by an externally applied signal at the EXT. Trig. input.

LINE – Triggering is made on the ac line voltage.

Trigger Conditions

SLOPE – Selects the slope of the signal on which triggering should be made. Normally, positive signals are triggered on the positive slope and negative signals on the negative slope.

LEVEL or THRESHOLD – This defines the voltage level at which triggering of the sweep begins. Signals not reaching this minimum level are therefore not displayed. For most beginners, failure to obtain a trace is usually due to setting this level incorrectly.

Trigger Modes

NORMAL – In the *normal* mode, no sweep is made unless a triggering signal satisfying the desired conditions is present. At all other times the screen remains blank.

AUTO – A continuous sweep is performed even if no signal is present at the input. This is useful for finding the trace and adjusting its positioning on the screen. When a signal is present, synchronization is as in the *normal* mode.

SINGLE SWEEP – Only one single sweep is made.

EXTERNAL or X-input – In this position, the internal ramp generator is disconnected and the horizontal movement of the beam controlled by the voltage at the X-input.

A.2.4 Display Modes

CHANNEL 1 – Displays channel 1 only.

CHANNEL 2 – Displays channel 2 only, etc.

ALTERNATE – Displays channels 1 and 2 on alternate sweeps, i.e., one *full* sweep is made alternatively for each channel. This mode is useful for comparing the relation between two signals.

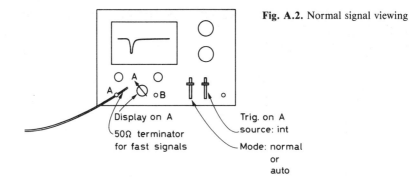

Fig. A.2. Normal signal viewing

CHOP − In this mode, the display is alternated between channels 1 and 2 *during* a sweep. This is usually done at a high frequency on the order of 500 kHz. The two signals are thus "*chopped*" in appearance.

ADD − The signals in channels 1 and 2 are added and displayed.

A.3 Applications and Examples

A.3.1 Signal Viewing

Figure A.2 shows the scheme for normal signal viewing. If no signal is observed, check the following points (not necessarily in the order listed!)

1) Check slope and trigger level.
2) Check that the trigger is on internal (or external, if that is what you are using).
3) Is the trace positioned on the screen? Put on *Auto* (if not already) and position trace with the horizontal and vertical position knobs. If there is still no trace, check the intensity.
4) Check the horizontal and vertical scales. Are they about the right order of magnitude for the signal you are observing?
5) Check that the input is not on OFF or GND.
6) Are you sure of your input signal? Try one that you know exists.

A.3.2 Comparison of Signals

Signals may be compared in time and amplitude using the setup in Fig. A.3. The oscilloscope sweep is triggered by either *A* or *B* according to the selection on the trigger

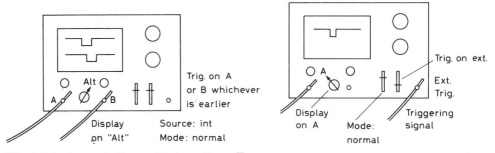

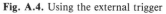

Fig. A.3. Setting a coincidence with the oscilloscope   **Fig. A.4.** Using the external trigger

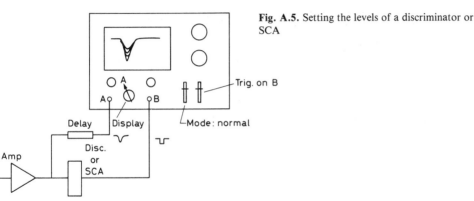

Fig. A.5. Setting the levels of a discriminator or SCA

source. With the display on *Alternate*, the relation in time of the second signal can be seen. This is the usual method for setting coincidences.

If the oscilloscope is not equipped with two channels an alternate scheme is to use the external trigger. This is shown in Fig. A.4.

A second example of signal comparison is checking the threshold of a discriminator or setting the levels of an SCA. This is shown in Fig. A.5. The output of the SCA or discriminator is used to trigger the scope for viewing the linear signal. Note that a delay must be added in order to take into account the processing time of the SCA or discriminator. Quite obviously only those linear signals which pass through the discriminator will be displayed since only these signals will have a trigger. In this way a certain amplitude may be selected or the level of the discriminator checked with respect to the noise.

B. Physical and Numerical Constants [a]

Avogadro's number	$N_A = 6.022045 \times 10^{23} \text{ mol}^{-1}$
Speed of light	$c = 2.997925 \times 10^{10} \text{ cm/s}$
Planck's constant	$h = 6.626176 \times 10^{-27} \text{ erg-s}$
	$\hbar = h/2\pi = 6.582173 \times 10^{-22} \text{ MeV s}$
	$\hbar c = 1.973285 \times 10^{-11} \text{ MeV cm}$
Electronic charge	$e = 1.602189 \times 10^{-19} \text{ Coul}$
Boltzmann's constant	$k = 8.61735 \times 10^{-11} \text{ MeV/K}$
	$= 1.380662 \times 10^{-16} \text{ erg/K}$
Classical electron radius	$r_e = e^2/m_e c^2 = 2.817938 \times 10^{-13} \text{ cm}$
Fine structure constant	$\alpha = e^2/\hbar c = 1/137.036$

Electron mass	$m_e = 0.511003 \text{ MeV} = 9.109534 \times 10^{-28} \text{ g}$
Proton mass	$m_p = 938.2796 \text{ MeV} = 1.672648 \times 10^{-24} \text{ g}$
Neutron mass	$m_n = 939.5731 \text{ MeV}$
Deuteron mass	$m_d = 1875.6280 \text{ MeV}$
Alpha particle mass	$m_\alpha = 3727.33 \text{ MeV}$
Muon mass	$m_\mu = 105.65916 \text{ MeV}$
Charged pion mass	$m_\pi = 139.5685 \text{ MeV}$

1 Joule $= 10^7$ erg	1 MeV $= 1.602189 \times 10^{-6}$ erg
1 eV/c^2 $= 1.782676 \times 10^{-33}$ g	1 amu $= 931.5016$ MeV
0°C $= 273.15$ K	1 inch $= 2.54$ cm
1 Tesla $= 10^4$ Gauss	1 barn $= 10^{-24}$ cm^2

[a] Source: Review of Particle Properties, Phys. Lett. **170B** (1986)

C. Resistor Color Code

Resistance values are coded by 4 colored bands around the resistor as shown below

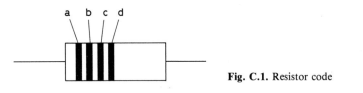

Fig. C.1. Resistor code

The value of the resistance is then

$$R = ab \times 10^c \pm d \; ,$$

where the colors have the following number values:

0	Black
1	Brown
2	Red
3	Orange
4	Yellow
5	Green
6	Blue
7	Violet
8	Gray
9	White
5%	Gold
10%	Silver
20%	No band

Example C.1   If the resistor has the band colors

a	green
b	orange
c	orange
d	gold

then the resistance is $R = 53 \times 10^3 \; \Omega$ with a tolerance of 5%.

References

Chapter 1

1.1   C. M. Lederer, V. S. Shirley (eds): *Table of Isotopes*, 7th Ed. (John Wiley & Sons, New York 1978)
1.2   E. A. Lorch: Int. J. Appl. Rad. and Isotop. **24**, 585 (1973)
1.3   K. W. Geiger, L. van der Zwan: Nucl. Instr. and Meth. **131**, 315 (1975)
1.4   A. O. Hanson: "Radioactive Neutron Sources" in *Fast Neutron Physics*, Vol. 1, ed. by J. B. Marion and J. L. Fowler (Interscience Publishers, New York 1963) pp. 3 – 48
1.5   E. Segre: *Nuclei and Particles* (W. A. Benjamin Co., New York 1977)

General References and Further Reading

Evans, R. D.: *The Atomic Nucleus* (McGraw-Hill Book Co., New York 1955)
Fermi, E.: Nuclear Physics (University of Chicago Press, Chicago 1950)
Hanson, A. O.: "Radioactive Neutron Sources" in *Fast Neutron Physics*, Vol. 1, ed. by J. B. Marion and J. L. Fowler (Interscience, New York 1963), p. 3 – 48
Segre, E.: Nuclei and Particles (W. A. Benjamin Co., New York 1977)

Chapter 2

2.1   J. D. Jackson: *Classical Electrodynamics* 2nd Ed. (John Wiley & Sons, New York 1975) Chap. 13
2.2   R. M. Sternheimer, S. M. Seltzer, M. J. Berger: Phys. Rev. **B26**, 6067 (1982); erratum in B27, 6971 (1983)
2.3   R. M. Sternheimer, M. J. Berger, S. M. Seltzer: At. Data and Nucl. Data Tables **30**, 262 (1984)
2.4   W. H. Barkas, M. J. Berger: "Tables of Energy Losses and Ranges of Heavy Charged Particles" in *Studies in the Penetration of Charged Particles in Matter,* National Academy of Sciences Publication 1133, Nuclear Science Series Report No. 39 (1964)
2.5   S. P. Ahlen: Rev. Mod. Phys. **52**, 121 (1980)
2.6   S. P. Ahlen: Phys. Rev. **A25**, 1856 (1982)
2.7   H. H. Andersen, J. F. Ziegler: *Stopping Powers and Ranges in All Elements,* 5 Vols. (Pergamon Press, New York 1977)
2.8   J. Linhard: Mat. Fys. Medd. Dan. Vid. Selsk. **28**, no. 8 (1954); J. Lindhard, M. Scharff: Phys. Rev. **124**, 128 (1961)
2.9   D. S. Gemmell: Rev. Mod. Phys. **46**, 129 (1974)
2.10 H. A. Bethe, J. Ashkin: "Passage of Radiations through Matter" in *Experimental Nuclear Physics*, Vol. 1, ed. by E. Segre (John Wiley & Sons, New York 1953)
2.11 T. Ypsilantis: Phys. Scripta **23**, 370 (1981); J. Litt. R. Meunier: Ann. Rev. Nucl. Sci. **23**, 1 (1973)
2.12 H. W. Koch, J. W. Motz: Rev. Mod. Phys. **4**, 920 (1959)
2.13 Y. S. Tsai: Rev. Mod. Phys. **46**, 815 (1974)
2.14 H. Davies, H. A. Bethe, L. C. Maximon: Phys. Rev. **93**, 788 (1954)
2.15 J. Marshall, A. G. Ward: Can. J. Research **A15**, 39 (1937)
2.16 L. Pages, E. Bertel, H. Joffre, L. Sklaventis: At. Data **4**, 1 (1972)
2.17 T. Baltakamens: Nucl. Instr. and Meth. **82**, 264 (1970)
2.18 E. Keil, E. Zeitler, W. Zinn: Z. Naturforsch. **15A**, 1031 (1960)
2.19 W. T. Scott: Rev. Mod. Phys. **35**, 231 (1963)
2.20 P. C. Hemmer, I. E. Farquahr: Phys. Rev. **168**, 294 (1968)
2.21 G. Knop, W. Paul: "Interaction of Electrons and Alpha Particles with Matter" in *Alpha-, Beta- and Gamma Ray Spectroscopy*, ed. by K. Siegbahn (North-Holland, Amsterdam 1968)

2.22   A. O. Hanson, L. H. Lanzl, E. M. Lyman, M. B. Scott: Phys. Rev. **84**, 634 (1951)

2.23   V. L. Highland: Nucl. Instr. and Meth. **129**, 497 (1975); errata in **161**, 171 (1979)

2.24   T. Tabata, R. Ito, S. Okabe: Nucl. Instr. and Meth. **94**, 509 (1971)

2.25   C. Tschalar: Nucl. Instr. and Meth. **64**, 237 (1968)

2.26   C. Tschalar: Nucl. Instr. and Meth. **61**, 141 (1968)

2.27   H. Bichsel: "Passage of Charged Particles through Matter" in *American Institute of Physics Handbook* 3rd Ed. (McGraw-Hill Book Co., New York 1972) Sect. 8-d

2.28   L. Landau: J. Phys. (USSR) **8**, 201 (1944)

2.29   S. M. Seltzer, M. J. Berger: "Energy Loss Straggling of Protons and Mesons: Tabulation of Vavilov Distribution" in *Studies in the Penetration of Charged Particles in Matter*, National Academy of Sciences Publication 1133, Nuclear Science Series Report. No. 39 (1964)

2.30   W. Borsh-Supan: J. Res. Nat'l Bur. Standards **65B**, 245 (1961)

2.31   B. Schorr: Computer Phys. Comm. **7**, 215 (1974)

2.32   B. Rossi: *High Energy Particles* (Prentice-Hall, New York 1952)

2.33   P. V. Vavilov: Sov. Phys. JETP **5**, 749 (1957)

2.34   O. Blunck, S. Leisegang: Z. Physik **128**, 500 (1950)

2.35   J. L. Matthews, D. J. S. Findlay, R. O. Owens: Nucl. Instr. and Meth. **180**, 573 (1980)

2.36   P. Shulek, B. M. Golovin, L. A. Kulyukina, S. V. Medved, P. Pavolich: Sov. J. Nucl. Phys. **4**, 400 (1967)

2.37   C. M. Davisson: "Interaction of Gamma Radiation with Matter" in *Alpha-, Beta- and Gamma Ray Spectroscopy,* ed. by K. Siegbahn (North-Holland Publ. Co., Amsterdam 1968)

2.38   Particle Data Group: "Review of Particle Properties" in Phys. Lett. **170B** (1986)

2.39   D. I. Garber, R. R. Kinsey: *Neutron Cross Sections* 3rd Ed., 2 Vols. BNL Report No. 325 (1976)

2.40   E. U. Condon, G. Breit: Phys. Rev. **49**, 229 (1936)

2.41   P. Reus: Nucl. Sci. and Eng. **93**, 105 (1986)

2.42   E. Richard-Cohen: Nucl. Sci. and Eng. **89**, 365 (1985)

General References and Further Reading

Ahlen, S. P.: "Theoretical and Experimental Aspects of the Energy Loss of Relativistic Heavily Ionizing Particles" in Rev. Mod. Phys. **52**, 1856 (1980)

Bethe, H. A., Ashkin, J.: "Passage of Radiations through Matter" in *Experimental Nuclear Physics*, Vol. 1, ed. by E. Segre (John Wiley & Sons, New York 1953)

Bichsel, H.: "Passage of Radiation" in *American Institute of Physics Handbook* 3rd Ed. (McGraw-Hill Book Co., New York 1972) Sect. 8-d

Davisson, C. M.: "Interaction of Gamma Radiation with Matter" in *Alpha-, Beta- and Gamma Ray Spectroscopy*, Vol. 1, ed. by K. Siegbahn (North-Holland, Amsterdam 1968)

Evans, R. D.: *The Atomic Nucleus* (McGraw-Hill Book Co., New York 1955)

Fano, U.: "Penetration of Protons, Alpha Particles and Mesons" in Ann. Rev Nucl. Sci. **13**, 1 (1963)

Feld, B. T.: "The Neutron" in *Experimental Nuclear Physics,* Vol. 2, ed. by E. Segre (John Wiley & Sons, New York 1953)

Knop, G., Paul, W.: "Interaction of Electrons and Alpha Particles with Matter" in *Alpha-, Beta- and Gamma Ray Spectroscopy*, ed. by K. Siegbahn (North-Holland, Amsterdam 1968)

Northcliffe, L. C.: "Passage of Heavy Ions Through Matter" in Ann. Rev. Nucl. Sci. **13**, 67 (1960)

Rossi, B.: *High Energy Particles* (Prentice-Hall, New York 1952)

Sternheimer, R. M.: "Interactions of Radiation in Matter" in *Methods of Experimental Physics*, Vol. 5, Part A, ed. by L. C. L. Yuan and C. S. Wu (Academic Press, New York 1961)

Studies in the Penetration of Charged Particles in Matter, National Academy of Sciences Publication 1133, Nuclear Science Series Rept. No. 39 (1964)

Chapter 3

3.1   *Radiological Health Handbook* (U.S. Dept. of Health, Education and Welfare 1970)

3.2   D. K. Meyers: "Low Level Radiation: A Review of Current Estimates of Hazards to Human Populations", Atomic Energy of Canada, Ltd. Report AECL 5715 (1982)

3.3   A. C. Upton: Scientific American **246**, 29 (Feb. 1982)

3.4   ICRP Publication 26, Annals of ICRP Vol. 1, No. 4 (Pergamon Press, Oxford 1977)

General References and Further Reading

Coggle, J. E.: *Biological Effects of Radiation* 2nd Ed. (Taylor & Francis, Ltd., London 1983)
Martin, A., Harbison, S.: *An Introduction to Radiation Protection* (Chapman and Hall, Ltd., London 1972)
Morgan, K. L., Turner, J. E.: "Health Physics" in *American Institute of Physics Handbook* 3rd Ed. (McGraw-Hill Book Co., New York 1972)
Pochin, E.: *Nuclear Radiation: Risks and Benefits* (Clarendon Press, Oxford 1983)
Shapiro, J.: *Radiation Protection* (Harvard University Press, Cambridge, Mass. 1981)
Upton, A. C.: "The Biological Effects of Low-Level Ionizing Radiation" in Scientific American **246**, 29 (Feb. 1982)

Chapter 4

4.1   F. E. James, M. Roos: Nucl. Phys. B**172**, 475 (1980)
4.2   E. Oliveri, O. Fiorella, M. Mangea: Nucl. Instr. and Meth. **204**, 171 (1982)
4.3   T. Mukoyama: Nucl. Instr. and Meth. **179**, 357 (1981)
4.4   T. Mukoyama: Nucl. Instr. and Meth. **197**, 397 (1982)
4.5   B. Potschwadek: Nucl. Instr. and Meth. **225**, 288 (1984)
4.6   A. Kalantar: Nucl. Instr. and Meth. **215**, 437 (1983)
4.7   M. Lybanon: Am. J. Phys. **52**, 22 (1984)
4.8   J. Orear: Am. J. Phys. **50**, 912 (1982); errata in **52**, 278 (1984); see also the remark by
       M. Lybanon: Am. J. Phys. **52**, 276 (1984)
4.9   W. T. Eadie, D. Drijard, F. E. James, M. Roos, B. Sadoulet: *Statistical Methods in Experimental Physics* (North-Holland, Amsterdam, London 1971)
4.10 P. R. Bevington: *Data Reduction and Error Analysis for the Physical Sciences* (McGraw-Hill Book Co., New York 1969)
4.11 F. E. James: "Function Minimization" in Proc. of 1972 CERN Computing and Data Processing School, Pertisau, Austria, 172, CERN Yellow Report 72-21
4.12 J. A. Nelder, R. Mead: Comput. J. **7**, 308 (1965)
4.13 Numerical Algorithms Group: *NAG Fortran Library*, Mayfield House, 256 Banbury Road, Oxford, OX2 7DE, United Kingdom; 1101 31st Street, Suite 100, Downers Grove, Ill. 60515, USA
4.14 F. E. James, M. Roos: "Minuit", program for function minimization, No. D506, CERN Program Library (CERN, Geneva, Switzerland)

General References and Further Reading

Bevington, P. R.: *Data Reduction and Error Analysis for the Physical Sciences* (McGraw-Hill Book Co., New York 1969)
Eadie, W. T., Drijard, D., James, F. E., Roos, M., Sadoulet, B.: *Statistical Methods in Experimental Physics* (North-Holland, Amsterdam, London 1971)
Frodesen, A. G., Skjeggstad, O., Tofte, H.: *Probability and Statistics in Particle Physics* (Universitetsforlaget, Oslo, Norway 1979)
Gibra, I. N.: *Probability and Statistical Inference for Scientists and Engineers* (Prentice-Hall, New York 1973)
Hahn, G. J., Shapiro, S.: *Statistical Models in Engineering* (John Wiley & Sons, New York 1967)
Meyers, S. L.: *Data Analysis for Scientists* (John Wiley & Sons, New York 1976)

Chapter 5

5.1   U. Fano: Phys. Rev. **72**, 26 (1947)
5.2   A. Foglio Para, M. Mandelli Bettoni: Nucl. Instr. and Meth. **70**, 52 (1969)
5.3   J. W. Muller: Nucl. Instr. and Meth. **112**, 47 (1973)
5.4   J. Libert: Nucl. Instr. and Meth. **146**, 431 (1977)
5.5   A. P. Baerg: Metrologia **1**, 131 (1965) and Ref. [5.3]
5.6   J. W. Muller: Internal Bureau of Weights and Measures Report BIPM-76/5 and BIPM 76/3 (1976) (Bureau International des Poids et Mesures, Pavillon de Breteuil, F-92310 Sèvres, France)

General References and Further Reading

Evans, R. D.: *The Atomic Nucleus* (McGraw-Hill Book Co., New York 1955)
Funck, E.: "Dead Time Effects from Linear Amplifiers and Discriminators in Single Detector Systems" in Nucl. Instr. and Meth. in Phys. Res. **A245**, 519 (1986)
Knoll, G. F.: *Radiation Detection and Measurement* (John Wiley & Sons, New York 1979)

Chapter 6

6.1   A. C. Melissinos: *Experiments in Modern Physics* (Academic Press, New York 1966)
6.2   P. Rice-Evans: *Spark, Streamer, Proportional and Drift Chambers* (The Richelieu Press, London 1974)
6.3   F. Sauli: *Principles of Operation of Multiwire Proportional and Drift Chambers*, CERN Yellow Report 77-09 (1977)
6.4   M. Kase, T. Akioka, H. Mamyoda, J. Kikuchi, T. Doke: Nucl. Instr. **227**, 311 (1984)
6.5   E. P. de Lima, M. Salete, S. C. P. Leite, M. A. F. Alves, A. J. P. L. Policarpo: Nucl. Instr. and Meth. **192**, 575 (1982)
6.6   M. M. F. Ribeirete, A. J. P. L. Policarpo, M. Salete, S. C. P. Leite, M. A. F. Alves, E. P. de Lima: Nucl. Instr. and Meth. **214**, 561 (1983)
6.7   B. Jean-Marie, V. Lepeltier, D. L'Hote: Nucl. Instr. and Meth. **159**, 213 (1979)
6.8   V. Palladino, B. Sadoulet: Nucl. Instr. and Meth. **128**, 323 (1975)
6.9   S. C. Brown: *Basic Data of Plasma Physics* (MIT Press, Cambridge, Mass. 1959)
6.10 M. E. Rose, S. A. Korff: Phys. Rev. **59**, 850 (1941)
6.11 T. Z. Kowalski: Nucl. Instr. and Meth. in Phys. Res. **A234**, 521 (1985)
6.12 G. F. Knoll: *Radiation Detection and Measurement* (John Wiley & Sons, New York, 1979) Chaps. 5 – 7
6.13 B. B. Rossi, H. H. Staub: *Ionization Chambers and Counters* (McGraw-Hill, New York 1949); D. H. Wilkinson: *Ionization Chambers and Counters and Counters* (Cambridge Univ. Press, Cambridge 1950)
6.14 J. Fischer, H. Okuno, A. H. Walenta: Nucl. Instr. and Meth. **151**, 451 (1978)
6.15 G. Charpak, R. Bouclier, T. Bressani, J. Favier, C. Zupancic: Nucl. Instr. and Meth. **62**, 235 (1968)
6.16 G. Charpak, D. Rahm, M. Steiner: Nucl. Instr. and Meth. **80**, 13 (1970)
6.17 Some more recent examples of MWPC construction are given in:
      J. C. Alder, B. Gabioud, C. Joseph, J. F. Loude, N. Morel, A. Perrenoud, J. P. Perroud, M. T. Tran, B. Vaucher, E. Winkelmann, D. Ranker, H. Schmitt, C. Zupancic, H. von Fellenberg, A. Frischknecht, F. Hoop, G. Strassner, P. Truol: Nucl. Instr. and Meth. **160**, 93 (1979);
      R. Crittenden, S. Ems, R. Heinz, J. Krider: Nucl. Instr. and Meth. **185**, 75 (1981);
      M. Kobayashi, S. Kurokawa, T. Fujitani, Y. Nagashima, T. Omori, S. Sugimoto, Y. Yamaguchi, J. Iwahori: Nucl. Inst. and Meth. in Phys. Res. **A245**, 51 (1986)
6.18 A. Breskin, C. Charpak, C. Demierre, S. Majewski, A. Policarpo, F. Sauli, J. C. Santiard: Nucl. Instr. and Meth. **143**, 29 (1977)
6.19 G. Charpak, G. Melchart, G. Petersen, F. Sauli: Nucl. Instr. and Meth. **167**, 455 (1979)
6.20 G. Charpak, F. Sauli: Ann. Rev. Nucl. Sci. **34**, 285 (1984)
6.21 G. Charpak, F. Sauli: Nucl. Instr. and Meth. **162**, 405 (1979)
6.22 A. Breskin, G. Charpak, F. Sauli, M. Atkinson, G. Schultz: Nucl. Instr. and Meth. **124**, 189 (1975)
6.23 A. Wagner: Phys. Scripta **23**, 446 (1981)
6.24 J. Fischer, A. Hrisho, V. Radeka, P. Rehak: Nucl. Instr. and Meth. in Phys. Res. **A238**, 249 (1985)
6.25 M. Basile, G. Bonvicini, G. Cara Romeo, L. Cifarelli, A. Contin, G. d'Ali, C. del Papa, G. Maccarone, T. Massam, F. Motta, R. Nania, F. Palmonari, G. Rinaldi, G. Sartorelli, M. Spinetti, G. Sussino, F. Villa, L. Vofano, A. Zichichi: Nucl. Instr. and Meth. in Phys. Res. **A239**, 497 (1985)
6.26 J. Va'vra: Nucl. Instr. and Meth. in Phys. Res. **A244**, 391 (1986); V. Commichau, K. H. Dederichs, M. Deutschmann, K. J. Draheim, P. Fritze, K. Hangarter, P. Hawelka, U. Herten, K. Hofman, D. Linnhofer, M. Tonutti, H. Weidkamp: Nucl. Instr. and Meth. in Phys. Res **A239**, 487 (1985)
6.27 R. J. Madaras, P. J. Oddone: Physics Today **37**, No. 8, 36 (Aug. 1984)
6.28 J. A. MacDonald (ed.): *The Time Projection Chamber,* AIP Conf. Proc. **108**, (American Inst. of Physics, New York 1984)

6.29   W. W. M. Allison, J. H. Cobb: Ann. Rev. Nucl. Sci **30**, 253 (1980)

6.30   J. N. Marx, D. R. Nygren: "The Time Projection Chamber" in Phys. Today **31**, No. 10 (Oct. 1978)

6.31   J. W. Lillberg: Nucl. Instr. and Meth. in Phys. Res. **A240**, 122 (1985);
       R. Richter: "Concepts for the Readout of Time Projection Chambers" in *The Time Projection Chamber*, ed. by J. A. MacDonald, AIP Conf. Proc. **108**, (American Institute of Physics, New York 1984);
       R. J. McKee: "Track Reconstruction of Normal Muon Decays in the LAMPF TPC" in *The Time Projection Chamber*, ed. by J. A. MacDonald, AIP Conf. Proc. **108**, (American Institute of Physics, New York 1984)

6.32   J. H. Marshall: Phys. Rev. **91**, 905 (1953); Rev. Sci. Instr. **25**, 232 (1952)

6.33   S. E. Derenzo, R. A. Muller, R. G. Smits, L. W. Alvarez: UCRL Report 19252 (1969)

6.34   R. A. Muller, S. E. Derenzo, G. Smadja, D. B. Smith, R. G. Smits, H. Zaklad, L. W. Alvarez: Phys. Rev. Lett **27**, 532 (1971)

6.35   S. E. Derenzo, A. R. Kirschbaum, P. H. Eberhard, R. R. Ross, F. T. Solmitz: Nucl. Instr. and Meth. **122**, 319 (1974)

6.36   K. Deiters, A. Donat, K. Lanius, R. Leiste, U. Röser, M. Sackiwitz, K. Trutzschler, E. Aprile, G. Motz, C. Rubbia, D. Koch, A. Stande: Nucl. Instr. and Meth. **180**, 45 (1981)

6.37   W. J. Willis, V. Radeka: Nucl. Instr. and Meth. **120**, 221 (1974)

6.38   C. W. Fabjan, T. Ludlam: "Calorimetry in High Energy Physics" in Ann. Rev. Nucl. Sci. **32**, 335 (1982)

6.39   A. Delfosse, O. Guisan, P. Muhlemann, R. Weill: Nucl. Instr. and Meth. **156**, 425 (1978)

6.40   C. Rubbia: CERN EP Internal Report 77-8 (1977);
       W. A. Huffmann, J. M. LoSecco, C. Rubbia: IEEE Trans. Nucl. Sci NS-**26**, 64 (1979)

6.41   P. J. Doe, H. J. Mahler, H. H. Chen: "The Liquid Argon Time Projection Chamber" in *The Time Projection Chamber*, ed. by J. A. MacDonald, AIP Conf. Proc. No. 108 (AIP, New York 1984) p. 84

6.42   E. Aprile, K. L. Giboni, C. Rubbia: Nucl. Instr. and Meth. in Phys. Res. **A241**, 62 (1985)

6.43   K. Masuda, T. Doke, T. Ikegami, J. Kikuchi, M. I. Lopes, R. Ferreira Marques, A. Policarpo: Nucl. Instr. and Meth. in Phys. Res **A241**, 607 (1985)

6.44   R. A. Holroyd, D. F. Anderson: Nucl. Instr. and Meth. in Phys. Res. **A236**, 607 (1985)

General References and Further Reading

Bartl, W., Neuhofer, G. (eds.): Proc. of the Vienna Wire Chamber Conference, Vienna, Austria, 1983, Nucl. Instr. and Meth. **217**, 1 (1983)

Brassard, C.: "Liquid Ionization Detectors" in Nucl. Instr. and Meth. **162**, 29 (1979)

Charpak, G.: "Evolution of Automatic Spark Chambers" in Ann. Rev. Nucl. Sci. **20**, 195 (1970)

Charpak, G.: "Multiwire and Drift Proportional Chambers" in Phys. Today **31**, No. 10 (Oct. 1978)

Charpak, G., Sauli, F.: "High Resolution Electronic Particle Detectors" in Ann. Rev. Nucl. Part. Sci. **34**, 285 (1984)

Fabjan, C. W., Fischer, H. G.: "Particle Detectors" in Rept. Prog. Phys. **43**, 1003 (1980)

Kleinknecht, K.: "Particle Detectors" in *Techniques and Concepts of High-Energy Physics*, ed. by T. Ferbel (Plenum Press, New York 1981)

Korff, S. A.: *Electron and Nuclear Counters* (Van Nostrand, New York 1955)

Marx, J. N., Nygren, D. R.: "The Time Projection Chamber" in Phys. Today **31**, No. 10 (Oct. 1978)

Rice-Evans, P.: *Spark, Streamer, Proportional and Drift Chambers* (Richelieu Press, London 1974)

Rossi, B. B., Staub, H. H.: *Ionization Chambers and Counters* (McGraw-Hill, New York 1949)

Sadoulet, B.: "Fundamental Processes in Drift Chambers" in Phys. Scripta **23**, 433 (1981)

Sauli, F.: "New Developments in Gaseous Detectors" in *Techniques and Concepts of High-Energy Physics II*, ed. by T. Ferbel (Plenum Press, New York 1983)

Sauli, F.: *Principles of Operation of Multiwire Proportional and Drift Chambers*, CERN Yellow Report 77-09 (1977)

Wilkinson, D. H.: *Ionization Chambers and Counters* (Cambridge Univ. Press, Cambridge 1950)

Chapter 7

7.1   Scintillators for the Physical Sciences, brochure No. 126P from Nuclear Enterprises, Inc., San Carlos, CA. 94070, USA

7.2   B. Bengston, M. Moszynski: Nucl. Instr. and Meth. **117**, 227 (1974)

7.3   Harshaw Scintillation Phosphor Catalog, The Harshaw Chemical Company, 6801 Cochran Road, Solon, Ohio 44149 USA

7.4   J. B. Birks: *The Theory and Practice of Scintillation Counting* (Pergamon Press, London 1964)
7.5   J. B. Birks: Proc. Phys. Soc A**64**, 874 (1951)
7.6   T. J. Gooding, H. G. Pugh: Nucl. Instr. and Meth. **7**, 189 (1960) and errata in **11**, 365 (1961)
7.7   R. L. Craun, D. L. Smith: Nucl. Instr. and Meth. **80**, 239 (1970)
7.8   G. Badhwar, C. L. Deney, B. R. Dennis, M. E. Kaplon: Nucl. Instr. and Meth. **57**, 116 (1967)
7.9   C. N. Chou: Phys. Rev. **87**, 904 (1952)
7.10   G. T. Wright: Phys. Rev. **91**, 1282 (1953)
7.11   T. A. King, R. Voltz: Proc. Roy. Soc. A, 289 (1966)
7.12   R. Voltz, G. Laustriat: J. Physique **29**, 159 (1968)
7.13   R. Voltz, H. du Pont, G. Laustriat: J. Physique **29**, 297 (1968)
7.14   D. Aitken, B. L. Leron, G. Yenicay, H. R. Zulliger: Trans. Nucl. Sci. NS-14, No. 2, 468 (1967)
7.15   R. S. Storey, W. Jack, A. Ward: Proc. Phys. Soc. **72**, 1 (1958)
7.16   R. B. Owen: IEEE Trans. Nucl. Sci. NS-**9**, 285 (1962)
7.17   F. J. Lynch: IEEE Trans, Nucl. Sci. NS-**22**, 58 (1975)
7.18   P. R. Bell: "The Scintillation Method" in *Beta and Gamma Ray Spectroscopy*, ed. by K. Siegbahn (North-Holland Publ. Co., Amsterdam 1955)

General References and Further Reading

Birks, J. B.: *The Theory and Practice of Scintillation Counting* (Pergamon Press, London 1964)
Meiling, W., Stary, F.: *Nanosecond Pulse Techniques* (Gordon and Breach, New York 1968)
Moszynski, M. Bengston, B.: "Status of Timing with Plastic Scintillation Counters" in Nucl. Instr. and Meth. **158**, 1 (1979)

Chapter 8

8.1   J. M. Schonkeren: "Photomultipliers" Philips Application Book Series ed. by H. Kater and L. J. Thompson (Philips Eindhoven, The Netherlands 1970)
8.2   EMI Photomultiplier Catalog 1979, EMI Industrial Electronics, Ltd., Bury Street, Ruslip, Middlesex, HA4 7TA, England
8.3   *Photomultiplicateurs,* RTC Photomultiplicateurs, 130 Av. Ledru-Rollin, 75540, Paris CEDEX 11, France
8.4   S. Dhawan: IEEE Trans. Nucl. Sci. NS – **28**, 672 (1981)
8.5   C. C. Lo, B. Leskovar: IEEE Trans. Nucl. Sci. NS – **28**, 698 (1981); E. Gatti, K. Oba, P. Rehak: IEEE Trans. Nucl. Sci NS-**30**, 461 (1983)
8.6   L. G. Hyman, R. M. Schwarz, R. A. Schluter: Rev. Sci. Instr. **35**, 393 (1964)
8.7   "Electron Tubes", Philips Data Handbook Part 9, Sept. 1982 (T9 09-82) (Philips, Eindhoven, The Netherlands 1982)
8.8   A. G. Wright: "Design of Photomultiplier Output Circuits for Optimum Amplitude or Time Response", Thorn-EMI Technical Note R/P 065, EMI Industrial Electronics, Ltd., Electron Tube Division, 243 Blyth Road, Hayes, Middlesex UB3 1HJ, England
8.9   M. D. Hull (ed.): "Fast Response Photomultipliers" Philips Application Book (Philips, Eindhoven, The Netherlands 1971)
8.10   R. Wardle: "Test Parameters and General Operating Rules for Photomultipliers", Thorn-EMI Technical Note R/P 067, EMI Industrial Electronics, Ltd. Electron Tube Division, 243, Blyth Road, Hayes Middlesex UB3 1HJ, England
8.11   B. H. Candy: Rev. Sci. Instr. **56**, 183 (1985)
8.12   S. Orito, T. Kobayashi, K. Suzuki, M. Ito, A. Sawaki: Nucl. Instr. and Meth. **216**, 439 (1983)
8.13   A. Sawaki, S. Ohsuka, T. Hayashi: IEEE Trans. Nucl. Sci. NS-**31**, 442 (1984)
8.14   F. Kinard: Nucleonics **15**, 92 (1957)
8.15   M. Yamashita: Rev. Sci. Instr. **51**, 768 (1980); Rev. Sci. Instr. **49**, 499 (1978)
8.16   J. P. Boutot, J. Mussli, D. Vallat: Adv. Electron. Electron Phys. **60**, 223 (1983)
8.17   C. C. Lo, B. Leskovar: IEEE Trans. Nucl. Sci. NS-**28**, 659 (1981)

General References and Further Reading

Candy, B.: "Photomultiplier Characteristics and Practice Relevant to Photon Counting" in Rev. Sci. Instr. **56**, 183 (1985)

Hull, M. D. (ed.): "Fast Response Photomultipliers", Philips Application Book (Philips, Eindhoven, The Netherlands 1971)

Schonkeren, J. M.: "Photomultipliers", Philips Application Book Series, ed. by H. Kater and L. J. Thompson (Philips, Eindhoven, The Netherlands 1970)

Chapter 9

9.1 W. E. Mott, R. B. Sutton: "Scintillation and Cerenkov Counters" in *Nuclear Instrumentation II*, Vol. XLV of *The Encyclopedia of Physics,* ed. by E. Creutz (Springer-Verlag, Berlin 1958)
9.2 R. L. Garwin: Rev. Sci. Instr. **23**, 755 (1952)
9.3 G. Keil: Nucl. Instr. and Meth. **89**, 111 (1970)
9.4 D. Brini, L. Pelli, O. Rimondi, P. Veronesi: Nuov. Cim. Suppl. **4**, 1048 (1955);
     P. A. Tove: Rev. Sci. Instr. **27**, 143 (1956);
     H. Hinterberger, R. Winston: Rev. Sci. Instr. **39**, 419 (1968);
     P. Gorenstein, D. Luckey: Rev. Sci. Instr. **34**, 196 (1963);
     W. Gibson: Rev. Sci. Instr. **35**, 1021 (1964);
     E. Verodini: Nucl. Instr. and Meth. **48**, 42 (1967)
9.5 A. I. Kilvington, C. A. Baker, P. Illinese: Nucl. Instr. and Meth. **80**, 177 (1970)
9.6 R. L. Garwin: Rev. Sci. Instr. **31**, 1010 (1960);
     G. Keil: Nucl. Instr. and Meth. **83**, 145 (1970);
     W. A. Selove, W. Kononenko, B. Wilsker: Nucl. Instr. and Meth. **161**, 233 (1979)
9.7 R. W. Engstrom, J. L. Weaver: Nucleonics **17**, 70 (1959)

Chapter 10

10.1   J. M. McKenzie: Nucl. Instr. and Meth. **162**, 49 (1979)
10.2   G. Charpak, F. Sauli: Ann. Rev. Nucl. Sci **34**, 285 (1984)
10.3   P. G. Rancoita, A. Seidman: La Riv. Nuov. Cim. **5**, Series 3, #7 (1982)
10.4   J. Millman, C. C. Halkias: *Integrated Electronics: Analog and Digital Circuits and Systems* (McGraw-Hill Book Co., New York 1972)
10.5   G. Ottaviani, C. Canali, A. Alberigi Quaranta: IEEE Trans. Nucl. Sci. NS-**22**, 1, 192 (1975)
10.6   G. Restelli: "Semiconductor Properties of Silicon and Germanium" in *Semiconductor Detectors*, ed. by G. Bertolini and A. Coche (North-Holland Publishing Co., Amsterdam 1968), p. 11 – 26
10.7   E. Haller: IEEE Trans. Nucl. Sci. NS-**29**, No. 3, 1109 (1982)
10.8   P. Siffert, A. Coche: "Behaviour of Lithium in Silicon and Germanium" in *Semiconductor Detectors*, ed. by G. Bertolini and A. Coche (North-Holland Publishing Co., Amsterdam 1968) pp. 27 – 52
10.9   A. Coche, P. Siffert: "N-P Detectors" in *Semiconductor Detectors*, ed. by G. Bertolini and A. Coche (North-Holland Publishing Co., Amsterdam 1968) pp. 103 – 129
10.10 D. J. Skyrme: Nucl. Instr. and Meth. **57**, 61 (1967)
10.11 G. Cavalleri, G. Fabri, E. Gatti, V. Svelto: Nucl. Instr. and Meth. **21**, 177 (1963)
10.12 G. F. Knoll: *Radiation Detection and Measurement* (John Wiley, New York 1979)
10.13 E. Laegsgaard: Nucl. Instr. and Meth. **162**, 93 (1979)
10.14 B. Hyams, U. Koetz, E. Belau, R. Klanner, G. Lutz, E. Neugebauer, A. Wylie, J. Kemmer: Nucl. Instr. and Meth. **205**, 99 (1983)
10.15 Particle Data Group: "Review of Particle Properties", Phys. Lett. **170B** (1986)
10.16 G. Anzivino, R. Horisberger, L. Hubbeling, B. Hyams, S. Parker, A. Breakstone, A. M. Litke, J. T. Walker, N. Bingefors: Nucl. Instr. and Meth. in Phys. Res **A243**, 153 (1986)
10.17 E. Gatti, P. Rehak: Nucl. Instr. and Meth. **225**, 608 (1984)
10.18 P. Rehak, E. Gatti, A. Longoni, J. Kemmer, P. Holl, R. Klanner, G. Lutz, A. Wylie: Nucl. Instr. and Meth. in Phys. Res. **A235**, 224 (1985)
10.19 R. Bailey, C. J. S. Damerill, R. L. English, A. R. Gillman, A. L. Lintern, S. J. Watts, F. J. Wickens: Nucl. Instr. and Meth. **213**, 201 (1983)
10.20 R. Klanner: Nucl. Instr. and Meth. in Phys. Res. **A235**, 209 (1985)
10.21 PGT Germanium detector manual, PGT corporation, Princeton, N.J.
10.22 R. Singh: Nucl. Instr. and Meth. **136**, 543 (1976)
10.23 R. C. Whited, M. M. Schieber: Nucl. Instr. and Meth. **162**, 113 (1979)
10.24 C. Scharanger, P. Siffert, A. Holzer, M. Schieber: IEEE Trans. Nucl. Sci. NS-**27**, 276 (1980)

10.25   A. Holzer: IEEE Trans. Nucl. Sci. NS-**29**, 1119 (1982)

10.26   Proc. of 5th Int'l Workshop on Mercuric Iodide Nuclear Radiation Detectors, Jerusalem 1982, Nucl. Instr. and Meth. in Phys. Res. **A213** (1983)

10.27   F. S. Goulding, D. A. Landis: IEEE Trans. Nucl. Sci. NS-**29**, No. 3, 1125 (1982)

10.28   EG&G – Ortec Catalog 1983 – 84 (Ortec, 100 Midland Rd. Oak Ridge, Tenn. 37830 USA)

10.29   H. Spieler: IEEE Trans. Nucl. Sci. NS – **29**, No. 3, 1142 (1982)

10.30   H. W. Kraner: Nucl. Instr. and Meth. in Phys. Res. **A225**, 615 (1984);
         H. W. Kraner: IEEE Trans. Nucl. Sci. NS – **29**, No. 3, 1088 (1982)

10.31   H. W. Schmitt, W. M. Gibson, J. H. Neiler, F. J. Walter, T. D. Thomas: "Absolute Energy Calibration of Solid State Detectors for Fission Fragments and Heavy Ions" in Proc. IAEA Conf. on The Physics and Chemistry of Fission, Salzburg (1965) p. 531

10.32   H. W. Schmitt, W. E. Kiker, C. W. Williams: Phys. Rev. **B137**, 837 (1965)

10.33   M. Ogihara, Y. Nagashima, W. Galster, T. Mikumo: Nucl. Instr. and Meth. in Phys. Res **A251**, 313 (1986)

General References and Further Reading

Bertolini, G., Coche, A. (eds.): *Semiconductor Detectors* (North-Holland Publishing Company, Amsterdam 1968)

Dearnaly, G., Northrop, D. C.: *Semiconductor Counters for Nuclear Radiations*, 2nd Ed. (E. & F. N. Spon, Ltd., London 1966)

Equer, B., Primont, M.: "Les détecteurs semi-conducteurs en physique des particules" in Gif 85, Ecole d'été de physique des particules, Université Clermont-Ferrand, 9 – 13 Sept. 1985 (in French)

Ewan, G. T.: "The Solid Ionization Chamber" in Nucl. Instr. and Meth. **162**, 75 (1979)

Ferbel, T. (ed.): *Silicon Detectors for High Energy Physics*, Proc. of a Workshop held at Fermilab, Oct. 15 – 16 1981

Goulding, F. S., Landis, D. A.: "Semiconductor Detector Spectrometer Electronics" in *Nuclear Spectroscopy and Reactions*, Part A, ed. by J. Cherny (Academic Press, New York 1974) p. 289 – 343

Goulding, F. S., Pehl, R. H.: "Semiconductor Radiation Detectors" in *Nuclear Spectroscopy and Reactions*, Part A, ed. by J. Cherny (Academic Press, New York 1974) p. 413 – 481

Kittel, C.: *Solid State Physics,* 5th Ed. (John Wiley & Sons, New York 1976)

Knoll, G. F.: *Radiation Detection and Measurement* (John Wiley, New York 1979)

Pehl, R. H.: "Germanium Gamma Ray Detectors" in Physics Today **30**, No. 11, 50 (1977)

Proc. of the Fourth Symposium on Semiconductor Detectors, Munich, FRG, March 3 – 5 1986, in Nucl. Instr. and Meth. in Phys. Res **A253** (1987)

Proc. of Third European Symposium on Silicon Detectors, Nucl. Instr. and Meth. in Phys. Res. **A226** (1984)

Rancoita, P. G., Seidman, A.: "Silicon Detectors in High Energy Physics: Physics and Applications" in La Riv. Nuov. Cim. **5**, Series 3, #7 (1982)

Sze, S. M.: *Physics of Semiconductor Devices* (John Wiley & Sons, New York 1969)

Chapter 11

General References and Further Reading

Millman, J., Taub, H.: *Pulse, Digital and Switching Waveforms* (McGraw-Hill Book Co., New York 1965)

Nicholson, P. W.: *Nuclear Electronics* (John Wiley & Sons, London 1974)

Chapter 12

12.1   L. Costrell: "NIM Standard" in *Instrumentation in Applied Nuclear Chemistry* ed. by J. Krugers (Plenum Press, New York 1973)

General References and Further Reading

Costrell, L.: "NIM Standard" in *Instrumentation in Applied Nuclear Chemistry*, ed. by J. Krugers (Plenum Press, New York 1973)

Standard Nuclear Instrument Modules, U.S. AEC Report TID-20893 (Rev. 3) Dec. 1969, U.S. Government Printing Office, Washington D.C. 20402

Chapter 13

13.1   LeCroy 1985 Catalog, LeCroy Research Systems Corp., 700 South Main St., Spring Valley, New York 10977, USA
13.2   G. Fidecaro: Nuov. Cim. suppl. to Vol. 15, Series X, 254 (1960)
13.3   G. Brianti: CERN Yellow Report 65-10 (1965)

General References and Further Reading

Coekin, J. A.: *High Speed Pulse Techniques* (Pergamon Press, Oxford 1975)
Chipman, R. A.: *Transmission Lines,* Schaum's Outline Series (McGraw-Hill Book Co., New York 1968)
Lewis, I. A. D., Wells, F. H.: *Millimicrosecond Pulse Techniques* (Pergamon Press, London 1954)

Chapter 14

14.1   EG & G-Ortec Catalog, EG & G-Ortec, PSD Division, 100 Midland Road, Oak Ridge, Tenn. 37380 USA
14.2   T. W. Henry: IEEE Trans. Nucl. Sci. NS-**20**, No. 5, 56 (1973)
14.3   J. K. Milam: "Single Channel Analyzers" in *Instrumentation in Applied Nuclear Chemistry,* ed. by J. Krugers (Plenum Press, New York, London 1973)
14.4   J. Millman, C. C. Halkias: *Integrated Electronics: Analog and Digital Circuits and Systems* (McGraw-Hill Book Co., New York 1972)

General References and Further Reading

Ayers, J. B.: "Preamplifiers" in *Instrumentation in Applied Nuclear Chemistry*, ed. by J. Krugers (Plenum Press, New York 1973)
Hatch, K. F.: "Amplifiers" in *Instrumentation in Applied Nuclear Chemistry,* ed. by J. Krugers (Plenum Press, New York 1973)
Knoll, G. F.: *Radiation Detection and Measurement* (John Wiley & Sons, New York 1979)
Milam, J. K.: "Single-Channel Analyzers" in *Instrumentation in Applied Nuclear Chemistry,* ed. by J. Krugers (Plenum Press, New York 1973)
Nicholson, P. W.: *Nuclear Electronics* (John Wiley & Sons, London 1974)
Ross, W. A.: "Multichannel Analyzers" in *Instrumentation in Applied Nuclear Chemistry,* ed. by J. Krugers (Plenum Press, New York 1973)

A good deal of information may also be found in the catalogs and technical notes from the various nuclear electronics manufacturers.

Chapter 15

15.1   K. F. Flynn, L. E. Glendenin, E. P. Steinberg, P. M. Wright: Nucl. Instr. and Meth. **27**, 13 (1964)
15.2   H. H. Knox, I. G. Miller: Nucl. Instr. and Meth. **101**, 519 (1972)
15.3   T. K. Alexander, F. S. Goulding: Nucl. Instr. and Meth. **13**, 244 (1961)
15.4   L. J. Perkins, M. C. Scott: Nucl. Instr. and Meth. **166**, 451 (1979)

General References and Further Reading

Delaney, C.: *Electronics for the Physicist* (Penguin Books, Harmondsworth 1969) Chap. 10
Melissinos, A. C.: *Experiments in Modern Physics* (Academic Press, New York 1966)
Wapstra, A. P.: "The Coincidence Method" in *Alpha-, Beta- and Gamma Ray Spectroscopy,* Vol. 1, ed. by K. Siegbahn (North-Holland, Amsterdam 1968)

A thorough analysis of pulse shape discrimination with liquid scintillators for fast neutron detection is given in:

Perkins, L. J., Scott, M. C.: "The Application of Pulse Shape Discrimination in NE213 Neutron Spectrometry" in Nucl. Instr. and Meth. **166**, 451 (1979)

Chapter 16

General References and Further Reading

Horowitz, P., Hill, W.: *The Art of Electronics* (Cambridge University Press, Cambridge 1980)
Millman, J., Halkias, C. C.: *Integrated Electronics: Analog and Digital Circuits and Systems* (McGraw-Hill Book Co., New York 1972)

Chapter 17

17.1   D. A. Gedecke, W. J. McDonald: Nucl. Instr. and Meth. **58**, 253 (1968)
17.2   D. Porat: IEEE Trans. Nucl. Sci. NS-**20** No. 5, 36 (1973)

General References and Further Reading

Hoogenboom, J. A. M.: "Timing Circuits" in *Instrumentation in Applied Nuclear Chemistry,* ed. by J. Krugers (Plenum Press, New York 1973)
Paulus, T. J.: "Timing Electronics and Fast Timing Methods with Scintillation Detectors" in IEEE Trans. Nucl. Sci NS-**32**, No. 3, 1242 (1985)
Porat, D. I.: "Review of Sub-nanosecond Time-interval Measurements" in IEEE Trans. Nucl. Sci. NS-**20**, No. 5, 36 (1973)
Spieler, H.: "Fast Timing Methods for Semiconductor Detectors" in IEEE Trans. Nucl. Sci NS-**29**, No. 3. 1142 (1982)
Williams, C. W.: "Time Measurement" in *Methods of Experimental Physics* Vol. 2, Part B, ed. by E. Bleuler and R. O. Haxby (Academic Press, New York 1974) p. 170 – 219

Chapter 18

18.1   Camac Catalog 1982 – 83, Kinetic Systems Corp., 11 Maryknoll Drive, Lockport, Ill. 60441, USA
18.2   *CAMAC Updated Specifications*, Report No. EUR 8500en (Office for Official Publications of the European Communities, Luxembourg 1983)

General References and Further Reading

Burckhardt, D.: "An Introduction to Fastbus", CERN Rept. DD/84/8 (1984) *FASTBUS: a Modular High Speed Data Acquisition System for High Energy Physics and Other Applications,* ESONE Committee FB/01 and US-NIM Committee DOE/ER-1089
CAMAC updated specifications, Report No. EUR 8500en in two volumes (1983), available from Office for Official Publications of the European Communities, Luxembourg; or ANSI/IEEE publication SH – 08482 (1982)
Hartill, D. L.: "Electronic Control Devices" in *Techniques and Concepts of High Energy Physics,* ed. by T. Ferbel (Plenum Press, New York 1981)
Pointing, J.: "Old and New Standards for the Data Acquisition; Camac and Fastbus" in Proc. Int'l School of Physics "Enrico Fermi", Varenna 1981 (North-Holland, Amsterdam 1983) p. 105
Scharff-Hansen, P.: "Real-Time Data Acquisition with Mini Computers" in Proc. Int'l School of Physics "Enrico Fermi" Varenna 1981 (North-Holland, Amsterdam 1983) p. 230

Appendix A

General References and Further Reading

Erk, R. van: *Oscilloscopes, Functional Operation and Measuring Examples* (McGraw-Hill Book Co., New York 1978)
Maeder, D.: "Equipment Testing" in *Methods of Experimental Physics,* Vol. 2, Part B, 2nd Ed., ed. By E. Bleuler and R. O. Haxby (Academic Press, New York 1974)

Subject Index